STERLING
Education

Florida
Biology EOC

Questions & Explanations

End-of-Course Assessment

3rd edition

Customer Satisfaction Guarantee

Your feedback is important because we strive to provide the highest quality prep materials. Email us questions, comments or suggestions.

info@sterling–prep.com

We reply to emails – check your spam folder

3 2 1

ISBN-13: 979-8-8855723-9-2

Sterling Education materials are available at quantity discounts.

Contact info@sterling–prep.com

Sterling Education
6 Liberty Square #11
Boston, MA 02109

Published by Sterling Education

Printed in the U.S.A.

STERLING
Education

Thousands of students use our study aids to achieve academic success!

Passing the Biology 1 EOC assessment tests is essential for earning a Scholar Designation Diploma in Florida. This book helps you develop and apply knowledge to quickly choose the correct answer on EOC Biology 1 exam. Solving targeted practice questions builds your understanding of fundamental biological concepts and is a more effective strategy than merely memorizing terms.

The high-yield practice questions in this book cover topics tested on biology end-of-course assessment test. Instructors with years of teaching experience prepared these questions by analyzing the exam content and developing practice material that builds your knowledge and skills crucial for success on the test. Our editorial team of standardized test experts reviewed and systematized the content to match the current Florida state requirements.

The detailed explanations describe why an answer is correct and – more important for your learning – why another attractive choice is wrong. They teach the scientific foundations and details of essential topics needed to answer test questions. Read the explanations carefully to understand how they apply to the question and learn important biology principles and the relationships between them. With the practice material contained in this book, you will significantly improve your score.

We wish you great success and look forward to being an important part of your biology preparation!

250927akp

Florida Biology EOC Essential Review provides a comprehensive review of topics tested on biology end-of-course assessment exams. The content covers foundational principles and concepts necessary to answer test questions.

Visit our Amazon store

For online practice resources visit
https://www.sterling-prep.com

If you benefited from this book, please leave a review on Amazon so others can learn from your input. Reviews help us understand our customers' needs and experiences while keeping our commitment to quality.

Table of Contents

Florida Biology EOC

Practice Questions

Ecosystems, Biomes, Energy Flow and Nutrient Cycles

1. The lowest level of environmental complexity that includes living and nonliving factors is:

A. ecosystem
B. biosphere
C. biome
D. community

2. How does an area's weather differ from the area's climate?

A. Weather does not change much, and an area's climate may change many times
B. Weather is the area's daily conditions, while climate is the area's average conditions
C. Weather involves temperature and precipitation, while climate involves only temperature
D. Weather depends on where it is located on Earth, while the area's climate does not

3. One type of symbiosis is:

A. parasitism
B. predation
C. competition
D. succession

4. Climate zones are the result of differences in:

A. thickness of the ozone layer
B. greenhouse gases
C. angle of the sun's rays
D. heat transport

5. The greenhouse effect is:

A. a natural phenomenon that maintains Earth's temperature range
B. the result of the differences in the angle of the sun's rays
C. primarily related to the levels of ozone in the atmosphere
D. a phenomenon that has only occurred in the last 50 years

6. The tendency for warm air to rise and cool air to sink results in:

A. regional precipitation
B. the seasons
C. ocean upwelling
D. global wind patterns

7. An ecosystem where water covers the soil or near the soil surface for at least part of the year is:

A. estuary
B. salt marsh
C. mangrove swamp
D. wetland

8. Which of the following is a biological aspect of an organism's niche?

A. composition of soil
B. amount of sunlight
C. predators
D. the water in the area

9. An organism's niche is:

A. the range of temperatures that the organism needs to survive
B. a complete description of the place an organism lives
C. the range of physical and biological conditions in which an organism lives and the way it obtains what it needs to survive and reproduce
D. all the physical factors in the organism's environment

10. No two species can occupy the same niche in the same habitat at the same time:

A. unless the species require different biotic factors
B. because of the competitive exclusion principle
C. unless the species require different abiotic factors
D. because of the interactions that shape the ecosystem

11. Plants are:

A. omnivores
B. herbivores
C. primary consumers
D. primary producers

12. How do most primary producers make their food?

A. By breaking down remains into carbon dioxide
B. By converting water into carbon dioxide
C. By using chemical energy to make carbohydrates
D. By using light energy to make carbohydrates

13. Compared to land, the open oceans:

A. are nutrient-poor environments
B. are rich in silica, and iron
C. have less zooplankton
D. contain abundant oxygen

14. Several species of warblers can live in the same spruce tree ONLY because they:

A. can find different temperatures within the tree
B. occupy different niches within the tree
C. have different habitats within the tree
D. do not eat food from the tree

15. All the interconnected feeding relationships in an ecosystem make up a food:

A. web
B. network
C. chain
D. framework

16. A symbiotic relationship in which both species benefit is:

A. mutualism
B. parasitism
C. commensalism
D. omnivorism

17. A wolf pack hunts, kills and feeds on a moose. In this interaction, the wolves are:

A. predators
B. mutualists
C. prey
D. hosts

18. The total amount of living tissue within a given trophic level is:

A. energy mass
B. biomass
C. organic mass
D. trophic mass

19. An interaction in which an animal feeds on plants is:

A. symbiosis
B. predation
C. herbivory
D. carnivory

20. A symbiotic relationship is when one organism is harmed, and another benefit is:

A. synnecrosis
B. predation
C. mutualism
D. parasitism

21. What animals eat producers and consumers?

A. autotrophs
B. chemotrophs
C. omnivores
D. herbivores

22. Ecosystem services include:

A. food production
B. production of oxygen
C. solar energy
D. all the above

23. Organisms that can capture the energy and produce food are:

A. omnivores
B. heterotrophs
C. autotrophs
D. consumers

24. What is one difference between primary and secondary succession?

A. Secondary succession begins with lichens, and primary succession begins with trees
B. Primary succession modifies the environment while secondary succession does not
C. Secondary succession begins on the soil, while primary succession begins on newly exposed surfaces
D. Primary succession is rapid and secondary succession is slow

25. A term that means the same thing as a *consumer* is:

A. carbohydrate
B. heterotroph
C. autotroph
D. producer

26. Primary succession would likely occur after:

A. severe storm
B. lava flow
C. earthquake
D. forest fire

27. Which of the following organisms is a detritivore?

A. fungus
B. snail
C. crow
D. caterpillar

28. Matter can recycle through the biosphere because:

A. biological systems do not deplete matter but transform it
B. biological systems use only carbon, oxygen, hydrogen, and nitrogen
C. matter does not change into new compounds
D. matter is assembled into chemical compounds

29. A tropical rain forest may not return to its original climax community after which type of disturbances?

A. volcanic eruption
B. flooding after a hurricane
C. burning of a forest fire
D. clearing and farming

30. A collection of the organisms living in a place, together with their nonliving environment, is a(n):

A. ecosystem
B. biome
C. population
D. community

31. Which biome is characterized by low temperatures, little precipitation, and permafrost?

A. tropical dry forest
B. tundra
C. temperate forest
D. desert

32. A bird stalks, kills and then eats an insect. Based on behavior, which ecological terms describe the bird?

A. herbivore ↔ decomposer
B. autotroph ↔ herbivore
C. carnivore ↔ consumer
D. producer ↔ heterotroph

33. Which two biomes have the least precipitation?

A. boreal forest and temperate woodland
B. tundra and desert
C. tropical savanna and tropical dry forest
D. tundra and temperate shrubland

34. Which represents box 5 of the food web in the figure?

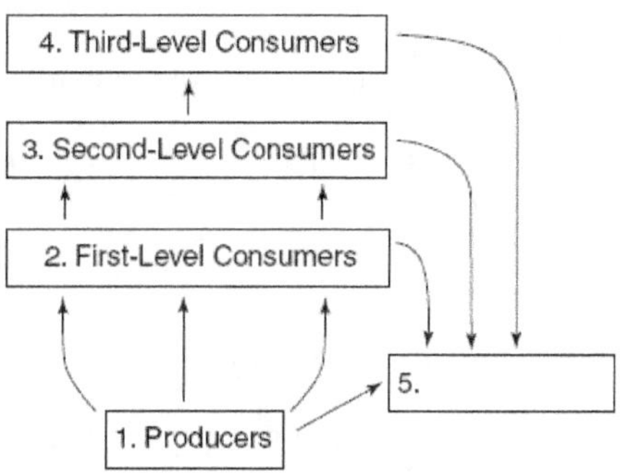

A. decomposers
B. carnivores
C. scavengers
D. herbivores

35. The average, year-after-year conditions of temperature and precipitation in a region are:

A. zonation
B. microclimate
C. weather
D. climate

36. Nitrogen fixation is carried out primarily by:

A. producers
B. consumers
C. bacteria
D. plants

37. The rate at which producers create organic matter is:

A. primary succession
B. nitrogen fixation
C. primary productivity
D. nutrient limit

38. Which landforms are NOT classified into a major biome?

A. coastlines
B. islands
C. prairies
D. mountain ranges

39. The North Pole and the South Poles are:

A. not classified into major biomes
B. part of aquatic ecosystems
C. classified as tundra biomes
D. not home to any animals

40. The type of interaction in which one organism captures and feeds on another is:

A. mutualism
B. predation
C. competition
D. parasitism

41. Is it likely to find zooplankton in the aphotic, benthic zone of an ocean?

A. No, zooplankton cannot undergo chemosynthesis in the dark without oxygen in the water
B. No, zooplankton feed on phytoplankton that cannot photosynthesize without light
C. No, zooplankton cannot photosynthesize in the dark without oxygen in the water
D. Yes, zooplankton is chemosynthetic autotrophs

42. Carbon cycles through the biosphere in all the following processes EXCEPT:

A. decomposition of plants
B. burning of fossil fuels
C. transpiration
D. photosynthesis

43. The nutrient availability of aquatic ecosystems is:

A. number of different animal species living in the water
B. amount of rainfall the water receives
C. number of other organisms in the water
D. amount of nitrogen, oxygen, and other elements dissolved in the water

44. What type of organism forms the base of many aquatic food webs?

A. phytoplankton
B. mangrove trees
C. secondary consumers
D. plants

45. The movements of energy and nutrients through living systems are different because:

A. nutrients flow in two directions while energy recycles
B. energy forms chemical compounds while nutrients are lost as heat
C. energy flows in one direction while nutrients recycle
D. energy is limited in the biosphere while nutrients are always available

46. Boreal Forest biomes are:

A. found near the equator
B. also known as taiga
C. home to more species than all other biomes combined
D. hot and wet year-round

47. Freshwater ecosystems that often originate from underground sources in mountains or hills are:

A. lakes
B. wetlands
C. rivers and streams
D. ponds

48. Which is one way a wetland differs from a lake or pond?

A. Water does not always cover a wetland as it does a lake or pond
B. Wetlands are salty, while lakes and ponds are freshwater
C. Water flows in a lake or pond but never flows in a wetland
D. Wetlands are nesting areas for birds, while lakes and ponds are not nesting areas

49. A wetland with a mixture of fresh water and saltwater is:

A. pond
B. river
C. stream
D. estuary

50. The permanently dark zone of the ocean is:

A. aphotic zone
B. intertidal zone
C. photic zone
D. coastal zone

51. Each of the following is an abiotic factor in the environment EXCEPT:

A. temperature
B. rainfall
C. plant life
D. soil type

52. Estuaries are commercially important because:

A. fossil fuels are found in estuaries
B. lumber trees grow in estuaries
C. hotels are often built-in estuaries
D. abundant fish species live in estuaries

53. Animals that get energy by eating the carcasses of animals killed by predators or died by natural causes are:

A. detritivores
B. heterotrophs
C. omnivores
D. scavengers

54. Which of the following statements is NOT true about the open ocean?

A. The open ocean begins at the low-tide mark and extends to the end of the continental shelf
B. Most of the photosynthetic activity on Earth occurs in the open ocean within the photic zone
C. The open ocean has low levels of nutrients
D. Organisms in the deep ocean are exposed to frigid temperatures

55. The branch of biology focused on interactions among and between organisms and their environment is:

A. paleontology
B. ecology
C. microbiology
D. entomology

56. Which of the following descriptions about the organization of an ecosystem is correct?

A. Species make up communities that comprise populations
B. Species make up populations that comprise communities
C. Communities make up species that comprise populations
D. Populations make up species that comprise communities

57. The photic zone:

A. is deep, cold, and permanently dark
B. extends to where the light intensity is reduced to 50% compared to the surface
C. extends to the bottom of the open ocean
D. extends to a depth of about 600 feet

58. Which of the following is an example of mutualism?

A. nematodes
B. bread mold
C. tapeworms
D. lichens

59. Organisms that must obtain nutrients and energy by eating other organisms are:

A. heterotrophic
B. eukaryotic
C. herbivores
D. autotrophic

60. Complex animals break down food through the process of:

A. cephalization
B. intracellular digestion
C. complete metamorphosis
D. extracellular digestion

Notes for active learning

Notes for active learning

Populations, Communities and Conservation Biology

1. A developer wants to build a new housing development in or around a large city. Which of the following plans would be LEAST harmful to the environment?

A. Building a neighborhood in a meadow at the edge of the city
B. Filling a wetland area and building oceanfront condominiums
C. Clearing a forested area outside of the city to build houses
D. Building apartments at the site of an abandoned factory in the city

2. Assemblages of different populations that live together in a defined area are:

A. communities
B. species
C. ecosystems
D. habitats

3. There are more than 165 cactus plants per square kilometer in an area of the Arizona desert. To which population characteristic does this information refer?

A. age structure
B. population density
C. growth rate
D. geographic range

4. Using resources in a way that does not cause long-term environmental harm is:

A. subsistence hunting
B. biological magnification
C. monoculture
D. sustainable development

5. What does the range of a population teach the observer that density does not?

A. The deaths per unit area
B. The births per unit area
C. The areas inhabited by a population
D. The number that lives in an area

6. Which ecological inquiry method is an ecologist using when she enters an area periodically to count the population numbers of a particular species?

A. modeling
B. experimenting
C. observing
D. questioning

7. Which of the following is NOT a factor in the population growth rate?

A. demography
B. emigration
C. death rate
D. Immigration

8. Which of the following is an example of population density?

A. number of deaths per year
B. number of bacteria per square millimeter
C. number of births per year
D. number of frogs in a pond

9. An example of a biotic factor is:

A. sunlight
B. soil type
C. competing species
D. average temperature

10. An example of a non-renewable resource is:

A. coal
B. fish
C. sunlight
D. wind

11. A mathematical formula designed to predict population fluctuations in a community is:

A. ecological model
B. ecological observation
C. biological experiment
D. biological system

12. The 1930s Dust Bowl in the Great Plains was caused by:

A. using renewable resources
B. poor farming practices
C. deforestation
D. contour plowing

13. The movement of organisms into a range is:

A. population shift
B. carrying capacity
C. immigration
D. emigration

14. Which of the following is NOT a basic method ecologists use to study the living world?

A. modeling
B. observing
C. experimenting
D. animal training

15. If immigration and emigration numbers remain equal, which is the most important contributing factor to a slowed growth rate?

A. decreased birth rate
B. constant birth rate
C. increased birth rate
D. constant death rate

16. When farming, overgrazing, climate change, or seasonal drought change farmland into land that cannot support plant life, it is:

A. deforestation
B. monoculture
C. desertification
D. depletion

17. Which factor might NOT contribute to an exponential growth rate in a given population?

A. reduced resources
B. less competition
C. higher birth rates
D. lower death rates

18. Which of the following is NOT a sustainable development strategy for managing Earth's resources?

A. selective harvesting of trees
B. crop rotation
C. desertification
D. contour plowing

19. Which are two ways a population can decrease in size?

A. Emigration and increased birth rate
B. Decreased birth rate and emigration
C. Increased death rate and immigration
D. Immigration and emigration

20. Farmers can reduce soil erosion by:

A. plowing up roots
B. grazing cattle on the land
C. contour plowing
D. increasing irrigation

21. Which age structure is likely for a population that has not completed the demographic transition?

A. 10 percent of people aged 50–54
B. 5 percent of people aged 10–14
C. 15 percent of people under age 15
D. 50 percent of people under age 15

22. Which two factors increase population size?

A. births and immigration
B. deaths and emigration
C. births and emigration
D. deaths and immigration

23. In a logistic growth curve, exponential growth is the phase when the population:

A. growth begins to slow down
B. growth stops
C. reaches carrying capacity
D. grows quickly

24. An example of sustainable resource use is the use of predators and parasites to:

A. control pest insects
B. eat unwanted plants
C. harm natural resources
D. pollinate plants

25. The graph shows the growth of a bacterial population. Which of the following describes the growth curve?

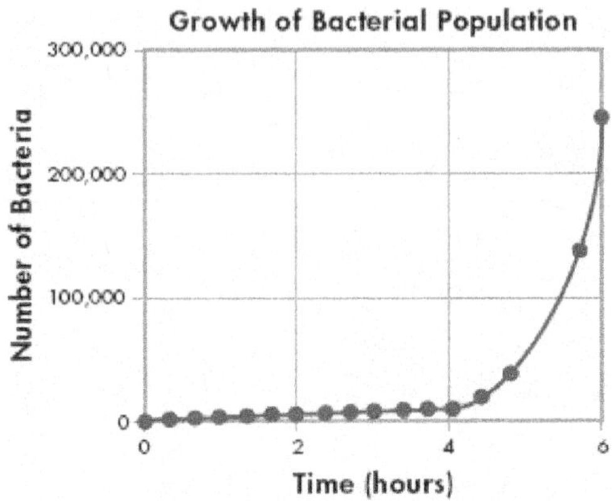

A. demographic
B. exponential
C. logistic
D. limiting

26. DDT was used to:

A. form ozone
B. feed animals
C. kill insects
D. fertilize the soil

27. All renewable resources:

A. are living
B. can be recycled or reused
C. are unlimited in supply
D. can regenerate or be replenished

28. Density-dependent limiting factors include:

A. blizzards
B. damming of rivers
C. disease
D. earthquakes

29. If a population grows larger than the carrying capacity of the environment:

A. birth rate must fall
B. death rate must fall
C. birth rate may rise
D. death rate may rise

30. One property that makes DDT hazardous is that it is:

A. deadly to herbivores
B. subject to biological magnification
C. volatile pesticide
D. insecticide

31. As the population gets larger, it grows more quickly because the size of each generation of offspring is larger than the generation before, resulting in:

A. multiple growth
B. exponential growth
C. growth density
D. logistic growth

32. The gray-brown haze often found over large cities is:

A. smog
B. particulates
C. greenhouse gases
D. ozone layer

33. Compounds that contribute to the formation of acid rain contain:

A. nitrogen and sulfur
B. ammonia and nitrates
C. carbon dioxide and oxygen
D. calcium and phosphorus

34. The growth phases through which most populations go are represented on:

A. normal curve
B. population curve
C. logistic growth curve
D. exponential growth curve

35. Water lilies do not grow in desert sand because water availability to these plants in a desert is:

A. competition factor
B. logistic growth curve
C. limiting factor
D. carrying capacity

36. The sulfur and nitrogen compounds in smog combine with water to form:

A. acid rain
B. chlorofluorocarbons
C. ozone
D. ammonia

37. Air and water pollution have been reduced by:

A. raising more cattle for food
B. increasing biological magnification
C. using fossil fuels in factories
D. using only unleaded gasoline

38. Which would be least likely to be affected by a density-dependent limiting factor?

A. population with a high immigration rate
B. large, dense population
C. population with a high birth rate
D. small, scattered population

39. Raising cattle and farming rice contribute to air pollution by:

A. releasing the greenhouse gas methane into the atmosphere
B. producing smog which reacts to form dangerous ozone gas
C. producing particulates into the air
D. releasing sulfur compounds that form acid rain

40. For most populations that are growing, as resources become less available, the population:

A. enters a phase of exponential growth
B. reaches carrying capacity
C. increases more rapidly
D. declines rapidly

41. Which term best describes the number of different species in the biosphere or area?

A. species diversity
B. genetic diversity
C. ecosystem diversity
D. biodiversity

42. Which density-dependent factor other than predator/prey relationships affect the populations of moose and wolves on an island?

A. A hurricane for moose and wolves
B. Food availability for the moose and disease for the wolf
C. Extreme temperatures for the moose and flooding for the wolves
D. Parasitic wasps for the wolves and clear-cut forest for the moose

43. How are species diversity and genetic diversity different?

A. Species diversity measures the number of individuals, while genetic diversity is the total variety of species
B. Conservation biology is concerned with species diversity but not with genetic diversity
C. Species diversity is evaluated in ecosystems, while genetic diversity is evaluated in the entire biosphere
D. Species diversity measures the number of species in the biosphere, while genetic diversity measures the variety of genes in the biosphere, including genetic variation within species

44. Which of the following is NOT likely to be a limiting factor on the sea otter population living in the ocean?

A. drought
B. predation
C. disease
D. competition

45. Introduced species can threaten biodiversity because they can:

A. crowd out native species
B. reduce the fertility of native species
C. cause desertification of land
D. cause biological magnification

46. Which of the following is a density-independent limiting factor?

A. parasitism and disease
B. eruption of a volcano
C. predator/prey relationships
D. struggle for food

47. What would reduce competition within a species' population?

A. higher population density
B. fewer resources
C. higher birth rate
D. fewer individuals

48. A major factor that negatively affects biodiversity is:

A. non-renewable resources
B. contour plowing
C. habitat fragmentation
D. biological magnification

49. All the following are threats to biodiversity EXCEPT:

A. habitat fragmentation
B. desertification
C. habitat preservation
D. biological magnification of toxic compounds

50. The graph shows the changes in the mosquito population. What caused the changes seen?

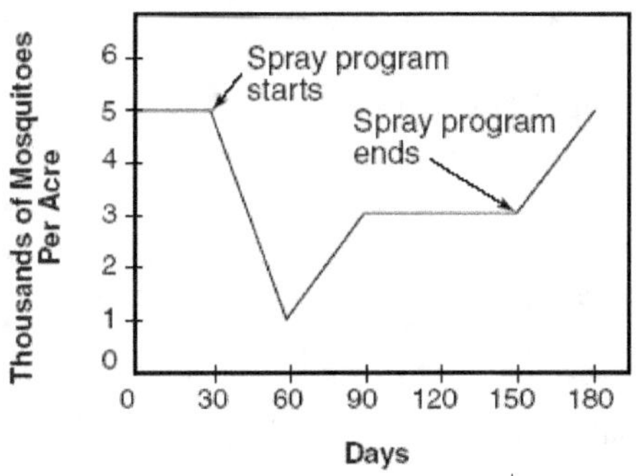

A. density-independent limiting factor
B. density-dependent limiting factor
C. reduction in resources
D. increase in predation

51. Each of the following is a density-dependent limiting factor EXCEPT:

A. crowding
B. disease
C. competition
D. temperature

52. The "hot spot" strategy seeks to protect species in danger of extinction due to:

A. human activity
B. biological magnification
C. captive breeding programs
D. expanding national parks

53. About 500 years ago, the world's population started to:

A. level off
B. grow more rapidly
C. reach carrying capacity
D. decrease

54. Protecting an entire ecosystem ensures that:

A. interactions among many species are preserved
B. governments will set aside land
C. existing parks and reserves will expand
D. captive breeding programs will succeed

55. The goals of biodiversity conservation include all the following EXCEPT:

A. Ensuring that local people benefit from conservation efforts
B. Preserving habitats and ecosystems
C. Introducing exotic species into new environments
D. Protecting individual species

56. Countries in the first stage of demographic transition have:

A. slowly growing population
B. more older people than young people
C. high death rate and a high birth rate
D. high death rate and a low birth rate

57. Demography is the scientific study of:

A. disease
B. human populations
C. parasitism
D. modernized countries

58. The amount of land and water necessary to provide the resources for a person's living and to neutralize that person's waste is that person's:

A. habitat
B. ecological sustainability
C. biodiversity
D. ecological footprint

59. Population density refers to the:

A. number of people in each age group
B. carrying capacity
C. number of individuals per unit area
D. area inhabited by a population

60. The anticipated human population by the year 2050 is about:

A. 780 million
B. 11 trillion
C. 9.8 billion
D. 3.6 billion

61. Demographic transition changes from high birth rates and high death rates to:

A. indefinite growth
B. low birth rates and high death rates
C. low birth rates and low death rates
D. exponential growth

62. Imported plants and animals in Hawaii have:

A. increased crop yields
B. improved soil fertility
C. increased the native bird species
D. caused native species to die out

63. A benefit of monoculture farming practices is:

A. pest resistance of the crops
B. the ability to grow a lot of food
C. the ability to spend less money on fertilizer
D. the use of less water for irrigation

64. An example of a density-independent limiting factor is:

A. disease
B. predation
C. unusual weather
D. competition

65. In countries like India, the human population is growing:

A. logistically
B. linearly
C. exponentially
D. transitionally

66. Most of the worldwide human population is growing exponentially because:

A. human population does not conform to the logistic model
B. the food supply is limitless
C. human populations have not reached their exponential curve
D. most countries have not yet completed the demographic transition

67. Solving an environmental problem is likely when researchers follow basic principles of ecology because:

A. ecology uses scientific research to identify the cause and best practices to solve problems
B. ecologists are good at influencing government officials into changing laws to improve the environment
C. ecological solutions are usually easy to implement and can be done quickly
D. most people worldwide are more interested in saving the environment than in their convenience

68. Which environmental problem can be identified by ecologists through the data in the graph?

A. habitat fragmentation
B. desertification
C. the hole in the ozone layer
D. global warming

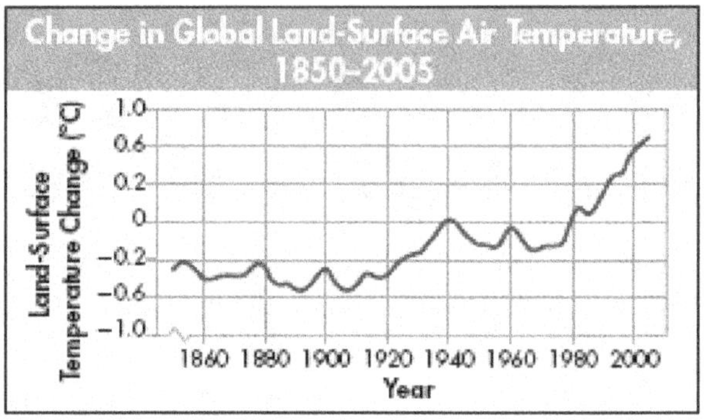

69. The total genetically based variety of organisms in the biosphere is:

A. ecosystem diversity
B. biodiversity
C. genetic diversity
D. species diversity

70. An ecological hot spot is an area where:

A. many habitats and species are at high risk of extinction
B. species diversity is too high
C. habitats show a high amount of biodiversity
D. hunting is encouraged

71. In a climax community, which of these would be a dominant species?

A. shrubs
B. annual grasses
C. mosses
D. deciduous trees

72. Which of the following statements is TRUE for a climax community?

A. It is relatively stable within a given climate
B. It is independent of the environment
C. It consists of only one species of life
D. It is populated mainly by so-called pioneer organisms

Notes for active learning

Notes for active learning

Classification and Diversity

1. The *Canis lupus* is a wolf in the Canidae family. Which statement is accurate about its members?

   A. They may be classified as *lupus* but not *Canis*
   B. They may be classified as *Canis* but not Canidae
   C. More are classified as *lupus* than *Canis*
   D. More are classified as Canidae than *Canis*

2. Two species of the same order must be members of the same:

   A. class
   B. habitat
   C. genus
   D. family

3. Homology serves as the evidence of:

   A. balancing selection
   B. common ancestry
   C. genetic mutation
   D. sexual selection

4. Which statement is correct about orthologous genes?

   A. They are related through speciation
   B. They are house-keeping genes
   C. They are repetitive
   D. They are related via gene duplication within species

5. Which given pair represents homologous structures?

   A. Mouth of a fly and beak of a hummingbird
   B. Wings of a dragonfly and a blue jay
   C. Forelimb of a human and of a dog
   D. Wings of a pigeon and a bat

6. Humans belong to the order:

   A. hominidae
   B. primata
   C. chordata
   D. vertebrata

7. All the following are members of the phylum Chordata EXCEPT:

   A. birds
   B. ants
   C. tunicates
   D. apes

8. All these statements regarding the phylum Echinodermata are true EXCEPT:

   A. echinoderms are heterotrophs
   B. phylum includes crayfish
   C. echinoderms reproduce sexually
   D. phylum includes starfish and sea urchins

9. Two species of the same phylum must be members of the same:

A. order
B. kingdom
C. genus
D. class

10. *Homo sapiens* belong to the phylum:

A. Vertebrata
B. Homo
C. Mammalia
D. Chordata

11. Which of these organisms is a chordate but not a vertebrate?

A. lizard
B. lancelet
C. shark
D. lamprey eel

12. Which represents the levels of complexity at which life is studied, from simplest to most complex?

A. cell, tissue, organ, organism, population, community
B. community, population, organism, organ, tissue, cell
C. cell, organ, tissue, organism, population, community
D. cell, tissue, organ, population, organism, community

13. Which of the following statements is TRUE for the notochord?

A. It is present in chordates during embryological development
B. It is always a vestigial organ in chordates
C. It is present in all adult chordates
D. It is present in all echinoderms

14. All are present in members of the phylum Chordata at some point in their life cycle, EXCEPT:

A. pharyngeal slits
B. tail
C. notochord
D. backbone

15. Which of the following are likely two organisms of the same species?

A. Two South American iguanas which mate in different seasons
B. Two migratory birds nesting on the different Hawaiian Islands
C. Cabbage in South Carolina and Texas mate and produce fertile progeny only in seasons of extreme climate conditions
D. Two fruit flies on the island of Bali with distinct courtship behaviors

16. What is one reason why sponges are classified as animals?

A. Sponges form partnerships with photosynthetic organisms
B. Sponges are autotrophic
C. Their cells have cell walls
D. Sponges are heterotrophic

17. What is a protostome?

A. outermost germ layer
B. animal whose mouth is formed from the blastopore
C. animal whose anus is formed from the blastopore
D. embryo just after fertilization

18. Arthropods have:

A. spiny skin and an internal skeleton
B. flattened body with an external shell
C. segmented body and an exoskeleton
D. soft body and an internal shell

19. A medical student looks at a slice of tissue on an unlabeled microscope slide. The student concludes that the tissue is not from an animal because the cells in the tissue have:

A. cell walls
B. cell membranes
C. nuclei
D. membrane-bound organelles

20. Which of the following is true about annelids?

A. Each body segment contains several pairs of antennae
B. Annelids rely on diffusion to transport oxygen and nutrients to their tissues
C. Annelids have segmented bodies and a true coelom
D. Annelids are acoelomates

21. Ancient chordates are thought to be most closely related to:

A. octopi
B. sea anemones
C. spiders
D. earthworms

22. Which of the following is true about echinoderms?

A. Echinoderms have one pair of antennae and unbranched appendages
B. Adult echinoderms have an exoskeleton
C. Most adult echinoderms exhibit radial symmetry
D. Echinoderm body is divided into a head, a thorax, and an abdomen

23. The simplest animals to have body symmetry are:

A. cnidarians
B. echinoderms
C. sponges
D. algae

24. In chordates with pharyngeal pouches that develop slits leading outside the body for breathing, the adult likely uses:

A. pharynx
B. nose
C. gills
D. lungs

25. In a chordate embryo, nerves branch in intervals from:

A. hollow nerve cord
B. notochord
C. pharyngeal pouches
D. tail

26. Each of the following is an essential function performed by animals, EXCEPT:

A. circulation
B. cephalization
C. excretion
D. respiration

27. Which of the following animals are deuterostomes?

A. mollusks and arthropods
B. echinoderms and chordates
C. annelids and arthropods
D. cnidarians and mollusks

28. In fishes, pharyngeal pouches may develop into:

A. gills
B. tails
C. lungs
D. fins

29. Which of the following is true about corals?

A. The polyp form of corals is restricted to a small larval stage
B. Corals only reproduce by asexual means
C. Corals have only the medusa stage in their life cycles
D. Coral polyps secrete an underlying skeleton of calcium carbonate

30. Amphibians evolved from:

A. jawless fishes
B. lobe-finned fishes
C. ray-finned fishes
D. cartilaginous fishes

31. The notochord is responsible for which function in an embryo?

A. respiration
B. processing nerve signals
C. processing wastes
D. structural support

32. A flexible, supporting structure found only in chordates is the:

A. pharyngeal slits
B. dorsal fin
C. nerve net
D. notochord

33. Which of the following is NOT true about earthworms?

A. Earthworms are hermaphrodites that reproduce sexually
B. Earthworms have a digestive tract that includes a mouth and an anus
C. Earthworms have a pseudocoelom
D. Their tunnels allow for the growth of oxygen-requiring soil bacteria

34. Skeletons of early vertebrates were composed of cartilage instead of bone. Which characteristic does cartilage share with notochords?

A. Oxygen can diffuse through it
B. It contracts
C. It is soft and flexible
D. It is hard and rigid

35. Which of the following do NOT exhibit bilateral symmetry?

A. arthropods
B. cnidarians
C. mollusks
D. annelids

36. Which chordate characteristic is visible on the outside of an adult cat?

A. a tail that extends beyond the anus
B. pharyngeal pouches
C. hollow nerve cord
D. notochord

37. Fewer than 5 percent of animal species have:

A. cell membranes
B. vertebral columns
C. a protostome development pattern
D. eukaryotic cells

38. The eggs of ovoviviparous fish are:

A. released from the female before they are fertilized
B. held inside the female's body as they develop
C. released from the female immediately after being fertilized
D. retained and nourished by the female

39. Which of the following statements about chordates is true?

A. All chordates are vertebrates
B. All chordates have paired appendages
C. All chordates have a notochord
D. All chordates have backbones

40. Which of the following pairs of modern chordate groups are most closely related?

A. birds and crocodilians
B. sharks and the coelacanth
C. hagfishes and lungfishes
D. lampreys and ray-finned fishes

41. The hominoid group of primates includes:

A. anthropoids only
B. apes and hominines
C. New World monkeys and Old-World monkeys
D. lemurs, lorises and anthropoids

42. Fossil evidence indicates how these three vertebrates evolved: bony fish with a jaw, jawless fish, and fish with leglike fins?

A. The jawless fish was the last to evolve
B. The fish with leglike fins evolved before the jawless fish
C. The bony fish evolved before the jawless fish
D. The fish with leglike fins was the last to evolve

43. Jaws and limbs are characteristic of:

A. reptiles only
B. vertebrates only
C. all chordates
D. worms

44. Which structure in the amniotic egg provides nutrients for the embryo as it grows?

A. yolk
B. allantois
C. chorion
D. amnion

45. Which of the following trends in reproduction is evident when following the vertebrate groups?

A. progressively fewer openings in the digestive tract
B. closed circulatory system becomes open
C. external fertilization evolves into internal fertilization
D. endothermy evolves into ectothermy

46. A tail adapted for grasping and holding objects is:

A. prehensile
B. radial
C. bipedal
D. binocular

47. Which of the following is NOT a characteristic of mammals?

A. hair
B. endothermy
C. ability to nourish their young with milk
D. lack of pharyngeal pouches during development

48. The group that includes gibbons and humans but does not include tarsiers is:

A. hominoids
B. hominines
C. primates
D. anthropoids

49. All the following are examples of cartilaginous fishes, EXCEPT:

A. rays
B. sharks
C. hagfishes
D. skates

50. Old World monkeys can be distinguished from New World monkeys by observing:

A. how the monkeys use their tails
B. when the monkeys are most active
C. what the monkeys eat
D. how the monkeys interact with their troop

51. Compared with that of a reptile, a bird's body temperature is:

A. lower and more constant
B. lower and more variable
C. higher and more constant
D. higher and more variable

52. Which of the following internal characteristics would you expect an earthworm to have?

A. prostomium
B. notochord
C. pseudocoelom
D. paired organs

53. Mammals that lay eggs are:

A. reptiles
B. placentals
C. marsupials
D. monotremes

54. The presence of legs or other limbs indicates that the animal is:

A. protostome
B. segmented
C. acoelomate
D. radially symmetrical

55. A primatologist finds a new species of primate in a Madagascar forest. The primate has a long snout and is active at night. The primatologist would likely classify this primate as a:

A. bush baby
B. hominoid
C. anthropoid
D. lemur

56. The vertebrate group with the most complex respiratory system is:

A. birds
B. amphibians
C. mammals
D. reptiles

57. Having a thumb that can move against the other fingers makes it possible for a primate to:

A. judge the locations of tree branches
B. display elaborate social behaviors
C. merge visual images
D. hold objects firmly

58. Which of the following is NOT a characteristic of chordates?

A. pharyngeal pouches
B. vertebrae
C. tail that extends past the anus
D. dorsal, hollow nerve cord

59. An animal that has body parts that extend outward from its center shows:

A. bilateral symmetry
B. radial symmetry
C. segmentation
D. several planes of symmetry

60. Which of the following is a node expected on the cladogram for animals?

A. leg length
B. feather arrangement
C. deuterostome development
D. fur texture

61. Which variation is expected for land vertebrates?

A. Shapes of forelimbs
B. Numbers of germ layers
C. Numbers of limbs
D. Types of symmetry

62. The earliest hominine that belonged to the same genus as modern humans were probably:

A. *Australopithecus afarensis*
B. *Homo ergaster*
C. *Homo habilis*
D. *Homo neanderthalensis*

63. The most important characteristic that separates birds from other chordates is:

A. amniotic egg
B. feathers
C. wings
D. four-chambered heart

64. A double-loop circulatory system is in animals with:

A. two-chambered heart
B. lungs
C. jaws
D. vertebrae

65. Which of the following statements about the cladogram of animals is true?

A. Segmentation evolved more than once in different branches of the cladogram
B. All deuterostomes have cephalization
C. Backbones evolved before segmentation
D. Radial symmetry appears only once in the cladogram

66. All the following are reasons why green algae are classified as plants, EXCEPT:

A. They have genes similar to genes in land plants
B. They have chlorophyll *a*
C. They have cellulose in their cell walls
D. They are single-celled organisms

67. Which of the following statements about green algae is true?

A. Like other plants, they have specialized structures
B. They are multicellular plants
C. Evidence suggests they were the first plants
D. They are found in dry areas on land

Notes for active learning

Evolution and Natural Selection

1. During his trip to the Galapagos Islands in 1835, Charles Darwin observed that:

A. local fossils were not related to contemporary organisms of that region
B. all organisms were related
C. organisms in tropical regions were closely related independent of the region
D. several species of finches varied from island to island

2. In evolutionary terms, organisms belonging to which category would be the most similar?

A. genus
B. family
C. order
D. kingdom

3. What is the source of genetic variation for natural selection?

A. gene duplication
B. mitosis
C. meiosis
D. mutation

4. In modern evolutionary theory, chloroplasts probably descended from:

A. free-living cyanobacteria
B. red algae
C. mitochondria
D. aerobic prokaryote

5. Speciation is the evolution of new, genetically distinct populations from a common ancestral stock by:

I. random mutation II. geographic isolation III. reduction of gene flow

A. I and III only
B. II and III only
C. I and II only
D. I, II, and III

6. Darwin (1809-1882) hypothesized that the mechanism of evolution involves:

A. selective pressure
B. epigenetics
C. natural selection
D. selective breeding

7. The complexity of the organisms that exist on earth today is the result of:

I. multicellularity II. photosynthesis III. eukaryotic cell development

A. I only
B. II only
C. I and II only
D. I, II and III

8. The Archaean Eon contains the oldest known fossil record dating to about:

A. 1.5 billion years
B. 2.0 million years
C. 7,000 years
D. 3.5 billion years

9. The Urey-Miller experiment at the University of Chicago in 1952 demonstrates that:

A. life may have evolved from inorganic precursors
B. humans have evolved from photosynthetic cyanobacteria
C. life existed on prehistoric earth
D. small biological molecules cannot be synthesized from inorganic material

10. To ensure the survival of their species, animals that do not care for their young:

A. have protective coloring
B. produce many offspring
C. lay eggs
D. can live in water and on land

11. Which of the following describes the relationship between nitrogen-fixing bacteria that derive nutrition from plants and plants that benefit from the nitrogen supplied by the bacteria?

A. parasitism
B. commensalism
C. mutualism
D. mimicry

12. Which of the following selection types leads to speciation?

A. sexual selection
B. stabilizing selection
C. directional selection
D. disruptive selection

13. Similarity among the embryos of fish, amphibians, reptiles, and humans is evidence of:

A. genetic drift
B. sexual selection
C. analogous traits
D. common ancestry

14. Selective breeding of soybeans by humans has genetically altered the soybean so that it could not survive in the wild without human intervention. The soybean population is controlled, and most soybean seeds are eaten or spoiled. The relationship between humans and soybeans is best described as:

A. mutualism, because both species benefit
B. parasitism, because humans benefit, and soybeans are harmed
C. commensalism, because there is no benefit to either species
D. commensalism, because humans benefit, and soybeans are neither benefited nor harmed

15. Natural selection leads to:

A. larger population
B. population most adapted to its environment
C. phenotypic diversity
D. broad genetic variation

16. From an evolutionary perspective, which cell property is paramount?

A. passing genetic information to progeny
B. containing a nucleus
C. containing mitochondria
D. interacting with other cells

17. Inherited traits determined by elements of heredity are:

A. crossing over
B. homologous chromosomes
C. locus
D. genes

18. Which statement is correct about the evolutionary process?

A. Organisms develop traits that they need for survival
B. Mutations decrease the rate of evolution
C. Natural selection works on traits that cannot be inherited
D. Natural selection works on existing genetic variation within a population

19. Genetic drift results from:

A. environmental change
B. genetic diversity
C. probability
D. sexual selection

20. The likely result of polygenic inheritance is:

A. human height
B. blood type
C. freckles
D. polydactyly (extra digits)

21. Which of the following is the earliest form of life to evolve on Earth?

A. plants
B. eukaryotes
C. prokaryotes
D. protists

22. If two animals produce viable, fertile offspring under natural conditions, it can be concluded that:

A. for any given allele, they have the same gene
B. their blood types are compatible
C. they have haploid somatic cells
D. they are from the same species

23. The population's total collection of alleles at any one time makes up its:

A. phenotype
B. biodiversity
C. gene pool
D. genotype

24. A pivotal point in Darwin's explanation of evolution is that:

A. biological structures likely inherited are those better suited to the environment through constant use
B. mutations that occur are those that help future generations fit into their environments
C. any trait that confers even a small increase in the probability that its possessor will survive and reproduce will be strongly favored and will spread through the population
D. genes change to help organisms cope with problems encountered within their environments

25. If one allele frequency in a population is 0.7, what is the alternate allele frequency?

A. 0.14
B. 0.21
C. 0.30
D. 0.4

26. Which statement about evolution is CORRECT?

A. Darwin's theory of natural selection relies solely on environmental conditions
B. Natural selection is when random mutations are selected for survival by the environment
C. Darwin's theory of natural selection relies solely on genetic mutation
D. Lamarck's theory of use and disuse adequately describes why giraffes have long necks

27. A significant genetic drift may be prevented by:

A. random mutations
B. large population size
C. small population size
D. genetic variation

28. Which statement applies to a population that survived a bottleneck and recovered to its original size?

A. It is subject to genetic drift
B. It is less likely to get extinct than before the bottleneck
C. It has more genetic variation than before the bottleneck
D. It has less genetic variation than before the bottleneck

29. The difference between the founder effect and a population bottleneck is that the founder effect:

A. only occurs in island populations
B. requires the isolation of a small group from a larger population
C. requires a large gene pool
D. involves sexual selection

30. Which compound was likely unnecessary for the origin of life on Earth?

A. carbon
B. O_2
C. hydrogen
D. H_2O

31. In a community of humans, there is a higher rate of polydactyly (extra fingers or toes) than in the human population. The likely explanation is:

A. bottleneck effect
B. founder effect
C. sexual selection
D. natural selection

32. Gene flow occurs through:

A. random mutations
B. bottleneck effect
C. migration
D. directional selection

33. The first living organisms on Earth derived their energy from:

A. eating dead organisms
B. eating organic molecules
C. the sun
D. eating each other

34. Evolutionary fitness measures:

A. population size
B. reproductive success
C. gene pool size
D. genetic load

35. Which scientist would explain the hawk's lost flying ability with a theory that "Since the hawk stopped using its wings, the wings became smaller, and this acquired trait was passed on to the offspring"?

A. Lamarck
B. de Vries
C. Mendel
D. Darwin

36. Which example is a likely result of directional selection?

A. Increased number of cat breeds
B. Python snakes with different color patterns exhibit different behavior when threatened
C. Female *Drosophila* select usual yellowish-grey colored males over less common yellow-colored males
D. Non-poisonous butterflies evolving color changes to look like poisonous butterflies

37. A genetic variation would be likely decreased by:

A. stabilizing selection
B. balancing selection
C. directional selection
D. sexual selection

38. Asexually reproducing species have a selective disadvantage over sexually reproducing species because sexual reproduction:

A. always decreases an offspring's survival ability
B. decreases the likelihood of mutations
C. creates novel genetic recombination
D. is more energy efficient

39. Which example is a likely result of stabilizing selection?

A. Female *Drosophila* select typical yellowish-grey colored males to mate over less common yellow colored males
B. Increased number of different breeds within one species
C. Non-poisonous butterflies evolving color changes to look like poisonous butterflies
D. Female birds that lay an intermediate number of eggs have the highest reproductive success

40. According to one theory about the origins of life, these molecules were formed by purines, pyrimidines, sugars, and phosphates combining:

A. lipids
B. carbohydrates
C. nucleosides
D. nucleotides

41. Which example is a likely result of sexual selection?

A. Female deer choosing to mate with males with the biggest antlers
B. Python snakes with different color patterns exhibit different behavior when threatened
C. Non-poisonous butterflies evolving color changes that make them look like poisonous butterflies
D. Increased number of different breeds within one species

42. Some fossils from before the Cambrian Explosion are embryos, suggesting that the organisms belong to:

A. reproduce sexually
B. exhibit cephalization
C. reproduce asexually
D. have bilateral symmetry

43. The Cambrian Explosion resulted in the evolution of the first:

A. bacteria
B. land animals
C. dinosaurs and mammals
D. representatives of most animal phyla

44. A primate's ability to hold objects in its hands or feet is an evolutionary development necessary for it to:

A. create elaborate social systems
B. consume food
C. use simple tools
D. walk upright

45. Suppose a paleontologist discovers a fossil skull that he believes might be distantly related to primates. Unlike true primates, the face is not flat, and the eyes do not face entirely forward. The paleontologist would likely conclude that the animal lacked the ability to:

A. manipulate tools
B. judge the location of tree branches
C. form extended family groups
D. grip branches precisely

46. When an animal's environment changes, sexual reproduction improves a species' ability to:

A. increase its numbers rapidly
B. produce genetically identical offspring
C. react to new stimuli
D. adapt to new living conditions

47. Researchers have concluded that Dikika Baby was likely a better climber than modern humans. This evidence suggests that Dikika Baby's relatives may have spent part of their time:

A. hunting in groups
B. waging war on other troops
C. using tools
D. in trees

48. Fossil evidence indicates that *Australopithecus afarensis:*

A. was bipedal
B. appeared later than *Homo ergaster*
C. was primarily a meat-eater
D. had a large brain

49. One way in which reptiles are adapted to fully terrestrial life is:

A. endothermy
B. viviparous development
C. external fertilization
D. amniotic egg

50. Researchers concluded from the leg bones of the fossil Lucy (discovered in 1974) that it was bipedal. Which of the following would indicate that this hominine was bipedal?

A. bowl-shaped pelvis
B. opposable thumbs
C. skull with a flat face
D. broad rib cage

51. Which statement is true of *Homo sapiens?*

A. They replaced *Homo habilis* in the Middle East
B. They became extinct about 1 million years ago
C. They have been Earth's only hominine for the last 24,000 years
D. They evolved after the Cro-Magnons

52. Which of the following was a unique characteristic of Neanderthals?

A. Producing tools from bones and antlers
B. Burying their dead with simple rituals
C. Making sophisticated stone blades
D. Producing cave paintings

53. A small subpopulation of flies with a slightly advantageous modification in the structure was found extinct after a locally-isolated decimating fire. A geneticist would most likely attribute the loss of this advantageous gene to:

A. differential reproduction
B. natural selection
C. Hardy-Weinberg principle
D. genetic drift

Notes for active learning

Notes for active learning

Animal Behavior and Evolution

1. Which of the following behaviors must be learned from another animal?

A. potato washing in macaques
B. suckling of newborn mammals
C. web building in spiders
D. nest building in birds

2. A cat hears a rustling in the leaves that sounds like a mouse digging for food. The cat crouched, becoming still, and focused on the leaves as it waited to pounce. This combination of movements is an example of:

A. behavior
B. stimulus
C. imprinting
D. circadian rhythm

3. For a behavior to evolve under the influence of natural selection, that behavior must be:

A. acquired through learning
B. influenced by genes
C. related to predator avoidance
D. neither adaptive nor harmful

4. If a dog that barks when indoors is always let outside immediately, it learns to bark whenever it wants to go outside. This change in the dog's behavior is an example of:

A. imprinting
B. insight learning
C. operant conditioning
D. classical conditioning

5. A behavior is innate, rather than learned, when:

A. all individuals perform the behavior the same way each time
B. individuals become better at performing the behavior the more they practice it
C. some individuals perform the behavior, and some do not
D. the behavior is different in individuals with different experiences

6. The terms "inborn behavior" and "instinct" have the same meaning as:

A. courtship behavior
B. imprinting
C. learned behavior
D. innate behavior

7. When disturbed, certain moths lift their front wings to expose eyelike markings on their hind wings. This behavior would be most effective against predators that hunt by:

A. touch
B. sight
C. smell
D. sound

8. Which learning occurs when a stimulus produces a particular response because it is associated with a positive or negative experience:

A. trial-and-error learning
B. habituation
C. operant conditioning
D. classical conditioning

9. Certain behaviors are innate in animals because they are essential for:

A. acquiring other behaviors
B. developing a circadian rhythm
C. survival immediately after birth
D. guarding territory

10. Imprinting is a form of behavior that:

A. occurs during a specific time in young animals
B. always involves the sense of sight
C. is restricted to birds
D. is often used in the training of adult animals

11. When people first move into an apartment near railroad tracks, they are awakened when they hear a train. Which of the following responses to the sound of the train would result from habituation?

A. People begin to sleep through the sound of the train over the next few nights
B. People learn that they can cover the sound of the train by sleeping with the radio on
C. People associate the sound of the train with the arrival of the morning newspaper
D. People learn they can sleep between the times that the train travels by their home

12. Aquarium fish often swim to the water's surface when a person approaches. Their behavior has probably formed through:

A. imprinting
B. insight learning
C. instinct
D. classical conditioning

13. What are specific chemical signals that insects use that affect the behavior or development of other individuals of the same species?

A. cilia
B. spiracles
C. pheromones
D. peptidoglycans

14. The ability of salmon to recognize their home stream at spawning time is an example of:

A. communication
B. imprinting
C. competition
D. insight learning

15. Learning occurs whenever:

A. a stimulus does not affect an animal the first time the animal encounters the stimulus
B. an animal leaves a chemical scent on its territory
C. a stimulus causes an animal to change its behavior
D. an animal performs a task perfectly without prior experience

16. The appearance of fireflies at dusk is an example of a circadian rhythm because it:

A. happens daily
B. happens seasonally
C. is related to the phase of the moon
D. is related to the temperature of the air

17. During migration, animals:

A. search for new permanent habitats
B. enter a sleep-like state
C. repeat their daily cycle of behavior
D. conduct seasonal movement

18. The theory stating that helping family members survive increases the probability one's genes will be passed to offspring is:

A. competition
B. classical conditioning
C. kin selection
D. imprinting

19. In winter, bears settle into dens and enter a sleep-like state until spring. This state is:

A. imprinting
B. aggression
C. hibernation
D. migration

20. When a bird learns to press a button to get food, it has learned by:

A. insight learning
B. operant conditioning
C. classical conditioning
D. habituation

21. To survive during winter, when resources are scarce, some animals enter a sleep-like state of:

A. reproductive rhythm
B. territoriality
C. ritual
D. circadian rhythm

22. In some species of balloon flies, males spin balloons of silk and carry them while flying. If a female approaches a male and accepts his balloon, the two will fly off to mate. This behavior is an example of:

A. courtship
B. language
C. aggression
D. territorial defense

23. Members of a society:

A. act independently for each's benefit
B. are usually unrelated to one another
C. belong to at least two species
D. belong to the same species and interact closely with each other

24. Animals that use language are likely those with the greatest capacity for:

A. habituation
B. insight learning
C. behavioral cycles
D. innate behavior

25. An animal can benefit the most by defending a territory if:

A. the animals it defends against do not use the same resources
B. there are more than enough resources in that territory for all competitors
C. that territory has more resources than surrounding areas
D. that territory has many predators

26. An innate behavior:

A. requires reasoning
B. occurs with trial and error
C. requires habituation
D. appears in fully functional form the first time it is performed

27. It is advantageous for grazing mammals to gather in groups because groups:

A. offer greater protection from predation
B. are more difficult for predators to locate than are individuals
C. can migrate more easily than individuals can
D. can make the available food resources last longer

28. Nocturnal animals with a poorly developed sense of smell are likely to communicate by:

A. pheromones
B. chemical signals
C. auditory signals
D. visual displays

29. Many cat species mark their territory by rubbing glands on their faces against surfaces such as tree trunks. This form of communication relies on:

A. chemical messenger
B. defensive display
C. visual signal
D. sound signal

30. When an animal associates a stimulus with a reward or a punishment, it has learned by:

A. imprinting
B. operant conditioning
C. habituation
D. classical conditioning

31. Cephalopods (i.e., squid and octopi) can communicate by changing skin colors and patterns. This communication is an example of:

A. visual signal
B. language
C. sound signal
D. chemical signal

32. Dolphins communicate with one another mainly through:

A. chemical signals
B. pheromones
C. auditory signals
D. visual displays

33. When newly hatched ducklings separated from their mother, follow a person swimming in a pond is:

A. imprinting
B. instrumental conditioning
C. discrimination
D. response to pheromones

Notes for active learning

Cell – the Basic Unit of Life

1. Facilitated transport is differentiated from active transport because:

A. active transport requires a symport
B. facilitated transport displays saturation kinetics
C. active transport displays sigmoidal kinetics
D. active transport requires an energy source

2. The cell is the basic unit of function and reproduction because:

A. subcellular components cannot regenerate whole cells
B. cells can move in space
C. single cells can sometimes produce an entire organism
D. cells can transform energy to do work

3. Which can NOT readily diffuse through the plasma membrane without a transport protein?

I. water
II. small hydrophobic molecules
III. small ions
IV. neutral gas molecules

A. I only
B. I and II only
C. I and III only
D. III and IV only

4. If a membrane-bound vesicle with hydrolytic enzymes is isolated, likely it is a:

A. vacuole
B. lysosome
C. chloroplast
D. phagosome

5. A DNA damage checkpoint arrests cells in:

A. M/G2 transition
B. S/G1 transition
C. G1/S transition
D. anaphase

6. The overall shape of a cell is determined by:

A. cell membrane
B. cytoskeleton
C. nucleus
D. cytosol

7. Which cellular substituent is produced within the nucleus?

A. Golgi apparatus
B. lysosome
C. ribosome
D. rough endoplasmic reticulum

8. Which is the correct sequence during polypeptide synthesis?

A. DNA generates tRNA → tRNA anticodon binds the mRNA codon in the cytoplasm → tRNA is carried by mRNA to the ribosomes, causing amino acids to join in a specific order
B. DNA generates mRNA → mRNA moves to the ribosome → tRNA anticodon binds the mRNA codon, causing amino acids to join in their appropriate order
C. Specific RNA codons cause amino acids to line up in a specific order → tRNA anticodon attaches to mRNA codon → rRNA codon causes a protein to cleave into specific amino acids
D. DNA regenerates mRNA in the nucleus → mRNA moves to the cytoplasm and attaches to the tRNA anticodon → operon regulates the sequence of amino acids in the appropriate order

9. Prokaryotes and eukaryotes contain:

I. a plasma membrane II. ribosomes III. peroxisomes

A. I only
B. II only
C. I and II only
D. II and III only

10. RNA is NOT expected to be in which of the following structures?

A. nucleus
B. mitochondrion
C. vacuole
D. ribosome

11. Mitochondria and chloroplasts are unusual organelles because they:

A. synthesize all their ATP using substrate-level phosphorylation
B. contain cytochrome C oxidase
C. are devoid of heme-containing proteins
D. contain nuclear-encoded and organelle-encoded proteins

12. All the following statements are true about cytoskeleton, EXCEPT that it:

A. is not required for mitosis
B. maintains the cell's shape
C. gives the cell mechanical support
D. is composed of microtubules and microfilaments

13. Inside the cell, several critical events in the cell cycle include:

I. DNA damage repair and replication completion
II. centrosome duplication
III. assembly of the spindle and attachment of the kinetochores to the spindle

A. I and III only
B. I, II and III
C. I and II only
D. II and III only

14. A researcher labeled *Neurospora* mitochondria with a radioactive phosphatidylcholine membrane component and followed cell division by autoradiography in an unlabeled medium, allowing enough time for one cell division. What results did this scientist observe before concluding that pre-existing mitochondria give rise to new mitochondria?

A. Daughter mitochondria are labeled equally
B. Daughter mitochondria are all unlabeled
C. One-fourth of daughter mitochondria are labeled
D. Some of the daughter mitochondria are unlabeled, while some are labeled

15. Which of the following involves the post-translational modification of proteins?

A. peroxisomes
B. vacuoles
C. Golgi complex
D. lysosomes

16. The width of a typical animal cell is closest to:

A. 1 millimeter
B. 20 micrometers
C. 1 micrometer
D. 10 nanometers

17. Which organelle is identified by the sedimentation coefficient – S units (Svedberg units)?

A. peroxisome
B. ribosome
C. mitochondrion
D. nucleolus

18. Which of the following is NOT involved in osmosis?

A. H_2O spontaneously moves from a hypertonic to a hypotonic environment
B. H_2O spontaneously moves from an area of high solvent to low solvent concentration
C. H_2O spontaneously moves from a hypotonic to a hypertonic environment
D. Diffusion of H_2O

19. The smooth ER is involved in:

A. substrate-level phosphorylation
B. exocytosis
C. synthesis of phosphatidylcholine
D. allosteric activation of enzymes

20. Which organelle is in plants but not in animals?

A. ribosomes
B. mitochondria
C. nucleus
D. plastids

21. The rough ER is involved in:

A. oxidative phosphorylation
B. synthesis of plasma membrane proteins
C. endocytosis
D. post-translational modification of enzymes

22. The best definition of active transport is the movement of:

A. solutes across a semipermeable membrane down an electrochemical gradient
B. solutes across a semipermeable membrane up against a concentration gradient
C. substances across a membrane by the Donnan equilibrium
D. solutes via osmosis across a semipermeable membrane from high to low concentration

23. Protein targeting occurs during the synthesis of which proteins?

A. nuclear proteins
B. secreted proteins
C. cytosolic proteins
D. mitochondrial proteins

24. The difference between "free" and "attached" ribosomes is that:

I. Free ribosomes are in the cytoplasm, while attached ribosomes anchor to the endoplasmic reticulum
II. Free ribosomes produce proteins in the cytosol, while attached ribosomes produce proteins inserted into the ER lumen
III. Free ribosomes produce proteins exported from the cell, while attached ribosomes make proteins for mitochondria and chloroplasts

A. I only
B. II only
C. I and II only
D. I, II and III

25. These processes are ATP-dependent, EXCEPT:

A. export of Na^+ from a neuron
B. influx of Ca^{2+} into a muscle cell
C. movement of urea across a cell membrane
D. exocytosis of the neurotransmitter at a nerve terminus

26. Which organelle is most closely associated with exocytosis of newly synthesized secretory protein?

A. peroxisome
B. ribosome
C. lysosome
D. Golgi apparatus

27. Plant membranes are more fluid than animal membranes because plant membranes:

A. contain large amounts of cholesterol
B. have higher amounts of unsaturated fatty acids in the membranes of animals
C. have lower amounts of unsaturated fatty acids in the membranes of animals
D. are only in the inner membrane of the mitochondria

28. Which of the following organelles are enclosed in a double membrane?

I. nucleus II. chloroplast III. mitochondrion

A. I and II only
B. II and III only
C. I and III only
D. I, II and III

29. Proteins are marked and delivered to specific cell locations through:

A. specific protein transport channels
B. regulation signals released by the cell's cytoskeleton
C. post-translational modifications occurring in the Golgi
D. compartmentalization of the rough ER during protein synthesis

30. All the following processes take place in the mitochondrion, EXCEPT:

A. oxidation of pyruvate
B. Krebs cycle
C. electron transport chain
D. glycolysis

31. What is the secretory sequence in the flow of newly-synthesized protein for export from the cell?

A. Golgi → rough ER → smooth ER → plasma membrane
B. Golgi → rough ER → plasma membrane
C. smooth ER → rough ER → Golgi → plasma membrane
D. rough ER → Golgi → plasma membrane

32. Recycling of organelles within the cell is accomplished through autophagy by:

A. mitochondria
B. peroxisomes
C. nucleolus
D. lysosomes

33. Which stage is when human cells with a single unreplicated copy of the genome form?

A. mitosis
B. meiosis II
C. meiosis I
D. interphase

34. Which organelle in the cell is the site of fatty acid, phospholipid, and steroid synthesis?

A. endosome
B. peroxisome
C. rough endoplasmic reticulum
D. smooth endoplasmic reticulum

35. Which eukaryotic organelle is NOT membrane-bound?

A. nucleus
B. plastid
C. centriole
D. chloroplast

36. During which mitotic division do spindle fibers split the centromere and separate the sister chromatids?

A. anaphase
B. telophase
C. prophase
D. metaphase

37. All the following are correct about cyclic AMP (cAMP), EXCEPT:

A. the enzyme that catalyzes the formation of cAMP is in the cytoplasm
B. membrane receptors can activate the enzyme that forms cAMP
C. ATP is the precursor molecule in the formation of cAMP
D. adenylate cyclase is the enzyme that catalyzes the formation of cAMP

38. Which molecule, and its associated protein kinase, ensures a proper progression of cell division?

A. oncogene
B. tumor suppressor
C. cyclin
D. histone

39. Placed in a hypertonic solution, erythrocytes will undergo:

A. crenation
B. expansion
C. plasmolysis and rupture
D. no change

40. Tiny organelles abundant in the liver containing oxidases and detoxify substances (e.g., alcohol and hydrogen peroxide) are:

A. peroxisomes
B. lysosomes
C. endosomes
D. rough endoplasmic reticulum

41. Microtubules can function independently or form protein complexes to produce structures like:

I. flagella
II. actin and myosin filaments in muscle cells
III. mitotic spindle apparatus

A. I, II, and III
B. I and III only
C. II and III only
D. I and II only

42. About a cell that secretes much protein (e.g., a pancreatic exocrine cell), it can be assumed that this cell has:

I. an abundance of rough endoplasmic reticulum
II. a large Golgi apparatus
III. a prominent nucleolus

A. I only
B. I and III only
C. II and III only
D. I, II and III

43. A cell division when each daughter cell receives a chromosome complement identical to that of a parent is:

A. mitosis
B. non-disjunction
C. meiosis
D. replication

44. Which of the following processes is NOT an example of apoptosis?

A. Reabsorption of a tadpole's tail during metamorphosis into a frog
B. Formation of the endometrial lining of the uterus during the menstrual cycle
C. Formation of the synaptic cleft by triggering cell death in brain neuronal cells
D. Formation of fingers in the fetus by the removal of tissue between the digits

45. Cells that utilize large quantities of ATP for active transport (e.g., epithelial cells of the intestine) have:

A. many mitochondria
B. high levels of DNA synthesis
C. high levels of adenylate cyclase
D. polyribosomes

46. Cells respond to peptide hormones through biochemical reactions involving membrane receptors and kinase activation. For cells to respond, the first and second messengers communicate because:

A. peptide hormones pass through the cell membrane and elicit a response
B. the hormone-receptor complex moves into the cytoplasm as a unit
C. hormones alter cellular activities directly through gene expression
D. the G protein acts as a link between the first and second messengers

Notes for active learning

Enzymes and Cellular Metabolism

1. Fermentation yields less energy than aerobic respiration because:

A. it requires a greater expenditure of cellular energy
B. glucose molecules are not completely oxidized
C. it requires more time for ATP production
D. oxaloacetic acid serves as the final H^+ acceptor

2. Metabolism is:

A. consumption of energy
B. release of energy
C. all conversions of matter and energy taking place in an organism
D. production of heat by chemical reactions

3. During alcoholic fermentation, all the following occurs, EXCEPT:

A. release of CO_2
B. oxidation of glyceraldehyde-3-phosphate
C. oxygen is not consumed in the reaction
D. ATP synthesis because of oxidative phosphorylation

4. What is the correct sequence of energy sources used by the body?

A. fats → glucose → other carbohydrates → proteins
B. glucose → other carbohydrates → fats → proteins
C. glucose → other carbohydrates → proteins → fats
D. glucose → fats → proteins → other carbohydrates

5. Enzymes act by:

A. lowering the overall free energy change of the reaction
B. decreasing the distance reactants must diffuse to find each other
C. increasing the activation energy
D. decreasing the activation energy

6. For the following reaction, which statement is TRUE?

ATP + Glucose → Glucose-6-phosphate + ADP

A. reaction results in the formation of a phosphodiester bond
B. reaction is endergonic
C. reaction is part of the Krebs cycle
D. free energy change for the reaction is approx. –4 kcal

7. Which statement is NOT true about the Krebs cycle?

A. Krebs cycle is the single greatest direct source of ATP in the cell
B. Citrate is an intermediate in the Krebs cycle
C. Krebs cycle produces nucleotides such as NADH and $FADH_2$
D. Krebs cycle is linked to glycolysis by pyruvate

8. Coenzymes are:

A. minerals such as Ca^{2+} and Mg^{2+}
B. small inorganic molecules that work with an enzyme to enhance the reaction rate
C. small organic molecules that work with an enzyme to enhance the reaction rate
D. small molecules that do not regulate enzymes

9. The site of the TCA cycle in eukaryotic cells, as opposed to prokaryotes, is:

A. mitochondria
B. endoplasmic reticulum
C. cytosol
D. nucleolus

10. All the following are metabolic waste products, EXCEPT:

A. lactate
B. pyruvate
C. CO_2
D. H_2O

11. A holoenzyme is:

A. inactive enzyme without its cofactor
B. inactive enzyme without its coenzyme
C. active enzyme with its cofactor
D. active enzyme with its coenzyme

12. Which of the following choices represents a correct pairing of aspects for cellular respiration?

A. Krebs cycle – cytoplasm
B. fatty acid degradation – lysosomes
C. electron transport chain – inner mitochondrial membrane
D. glycolysis – inner mitochondrial membrane

13. An apoenzyme is:

A. inactive enzyme without its cofactor
B. inactive enzyme without its inorganic cofactor
C. active enzyme with its cofactor
D. active enzyme with its coenzyme

14. Cofactors are:

I. small inorganic molecules that work with an enzyme to enhance the reaction rate
II. small organic molecules that work with an enzyme to enhance the reaction rate
III. small molecules that regulate enzyme activity

A. I only
B. I and II only
C. II and III only
D. I, II and III

15. All the following statements about glycolysis are true, EXCEPT:

A. end-product can be lactate, ethanol, CO_2, and pyruvate
B. $FADH_2$ is produced during glycolysis
C. one molecule of glucose is converted into two molecules of pyruvate
D. net total of two ATPs are produced

16. Which of the following statements is TRUE for the glycolytic pathway?

A. glucose produces a net of two molecules of ATP and two molecules of NADH
B. glucose produces one molecule of pyruvate
C. O_2 is a reactant for glycolysis
D. glucose is partially reduced

17. Which of these metabolic processes take(s) place in the mitochondria?

I. Krebs cycle   II. glycolysis   III. electron transport chain

A. II only
B. I and III only
C. II and III only
D. I, II and III

18. Which statement best describes the usual relationship of the inhibitor molecule to the allosteric enzyme in feedback inhibition of enzyme activity?

A. The inhibitor is the substrate of the enzyme
B. The inhibitor is the product of the enzyme-catalyzed reaction
C. The inhibitor is the final product of the metabolic pathway
D. The inhibitor is a metabolically unrelated signal molecule

19. All the following statements apply to oxidative phosphorylation, EXCEPT:

A. it can occur under anaerobic conditions
B. it produces two ATPs for each $FADH_2$
C. it involves O_2 as the final electron acceptor
D. it takes place on the inner membrane of the mitochondrion

20. Like other catalysts, enzymes:

I. increase the rate of reactions without affecting ΔG
II. shift the chemical equilibrium from reactants to products
III. do not alter the chemical equilibrium between reactants and products

A. I only
B. I and II only
C. I and III only
D. III only

21. Enzyme activity can be regulated by:

I. zymogen proteolysis
II. changes in substrate concentration
III. post-translational modifications

A. I only
B. II and III only
C. I, II and III
D. I and II only

22. In allosteric regulation, how is enzyme activity affected by binding a small regulatory molecule?

A. It is inhibited
B. It is stimulated
C. Can be stimulated or inhibited
D. Is neither stimulated nor inhibited

23. All biological reactions:

A. are exergonic
B. have an activation energy
C. are endergonic
D. occur without a catalyst

24. In eukaryotes, energy is trapped in a high-energy phosphate group during oxidative phosphorylation in:

A. nucleus
B. mitochondrial matrix
C. inner mitochondrial membrane
D. outer mitochondrial membrane

25. All the following statements about enzymes are true, EXCEPT:

A. They function optimally at a particular temperature
B. They function optimally at a particular pH
C. They may interact with non-protein molecules to achieve biological activity
D. Their activity is not affected by a genetic mutation

26. Allosteric enzymes:

I. are regulated by metabolites that bind at sites other than the active site
II. have a quaternary structure
III. show cooperative binding of substrate

A. I only
B. I and II only
C. II and III only
D. I, II and III

27. The active site of an enzyme is where:

I. prosthetic group is bound
II. proteolysis occurs for zymogens
III. non-competitive inhibitors bind

A. I only
B. II only
C. I and II only
D. II and III only

28. The hydrolysis of ATP → ADP + phosphate allows glucose-6-phosphate to be synthesized from glucose and phosphate because:

A. heat produced from ATP hydrolysis drives the glucose-6-phosphate synthesis
B. enzymatic coupling of these two reactions allows the energy of ATP hydrolysis to drive the synthesis of glucose-6-phosphate
C. energy of glucose phosphorylation drives ATP splitting
D. all the above

Notes for active learning

Microbiology

1. What is the *major* distinction between prokaryotic and eukaryotic cells?

   A. prokaryotic cells do not have DNA, and eukaryotic cells do
   B. prokaryotic cells do not have a nucleolus, but eukaryotic cells do
   C. eukaryotic cells are smaller than prokaryotic cells
   D. prokaryotic cells have not prospered, while eukaryotic cells are evolutionarily advanced

2. Which of these statements describes the actions of penicillin?

   A. it is a reversible competitive inhibitor
   B. it is an irreversible competitive inhibitor
   C. it activates transpeptidase that digests the bacterial cell wall
   D. it is an effective antiviral agent

3. Which of the following statements applies to all viruses?

   A. They have RNA genome
   B. They have DNA genome
   C. They have chromosomes
   D. They cannot replicate outside of the host cell

4. The replica plating technique of Joshua and Esther Lederberg demonstrated that:

   A. mutations are usually beneficial
   B. mutations are usually deleterious
   C. streptomycin caused the formation of streptomycin-resistant bacteria
   D. streptomycin revealed the presence of streptomycin-resistant bacteria

5. Operons:

   A. are a common feature of the eukaryote genome
   B. often coordinate the production of enzymes that function in a single pathway
   C. have multiple translation start and stop sites used by the ribosome
   D. usually undergo alternative splicing

6. Which of the following is TRUE for the life cycle of sexually-reproducing *Neurospora* fungus?

   A. Only mitosis occurs
   B. Fertilization and meiosis are separated
   C. Meiosis quickly follows fertilization
   D. Fertilization immediately follows meiosis

7. Which of the following is in prokaryotic cells?

A. mitochondria
B. chloroplasts
C. nuclei
D. enzymes

8. Which organelle is the site of protein modification and carbohydrate synthesis?

A. Golgi apparatus
B. lysosomes
C. peroxisomes
D. smooth ER

9. Which statement(s) is/are TRUE regarding retrotransposons?

I. They are never between genes in the human genome
II. They comprise close to half of the human genome
III. They can cause mutations by inserting themselves into genes

A. I and II only
B. II and III only
C. I and III only
D. I, II and III

10. Which of the following is TRUE about plasmids?

I. They are small organelles in the bacterial cytoplasm
II. They are transcribed and translated simultaneously
III. They are replicated by bacterial enzymes

A. I only
B. II and III only
C. I and III only
D. I, II, and III

11. The type of bacteria NOT able to grow on minimal media due to mutations affecting metabolism:

A. auxotrophs
B. chemotrophs
C. heterotrophs
D. prototrophs

12. Which of the following statements is TRUE?

A. Endospores are for reproduction
B. Endospores allow a cell to survive environmental changes
C. Endospores are easily stained with Gram stain
D. Cell produces one endospore and keeps growing

13. Most fungi spend the biggest portion of their life cycle as:

A. haploid
B. haploid and diploid
C. diploid
D. polyploid

14. An aerobic bacteria culture that has been exposed to cyanide gas is infected by bacteriophages. However, the replication of viruses does not occur. What is cyanide's mechanism of action?

A. Binding to viral nucleic acid
B. Denaturing bacteriophage enzymes
C. Inhibiting aerobic ATP production
D. Destroying bacteriophage binding sites on the bacterial cell wall

15. Recipient cells acquire genes from free DNA molecules in the surrounding medium by:

A. transformation
B. conjugation
C. transduction
D. recombination

16. Viruses can have a genome that is:

A. single-stranded DNA
B. single-stranded RNA
C. double-stranded RNA
D. all the above

17. Which of the following have a cell wall?

A. protoplasts
B. fungi
C. L forms
D. viruses

18. Which of the following statements about prokaryotic cells is generally FALSE?

A. They have a semirigid cell wall
B. They are motile using flagella
C. They possess 80S ribosomes
D. They reproduce by binary fission

19. All the following events play a role in the life cycle of a typical retrovirus EXCEPT:

A. injection of viral DNA into the host cell
B. integration of viral DNA into the host genome
C. reverse transcriptase gene is transcribed, and mRNA is translated inside the host cell
D. viral DNA incorporated into the host genome may be replicated along with the host DNA

20. All the following are true for viruses EXCEPT:

A. Genetic material may be single-stranded or double-stranded RNA
B. The virus may replicate in a bacterial or eukaryotic host
C. The virus may replicate without a host
D. The protein coat of the virus does not enter a host bacterial cell

21. DNA transfer from a bacterial donor cell to a recipient cell by cell-to-cell contact is:

A. conjugation
B. transformation
C. transduction
D. recombination

22. On the overnight agar plates with *E. coli*, replication of a virus is marked by:

A. no visible change
B. bacterial colonies on the agar surface
C. growth of a smooth layer of bacteria across the plate
D. growth of bacteria across the plate except for small clear patches

23. Which of the following about gram-negative cell walls is FALSE?

A. They protect the cell in a hypotonic environment
B. They have an extra outer layer of lipoproteins, lipopolysaccharides, and phospholipids
C. They are toxic to humans
D. They are sensitive to penicillin

24. All the following may be in a mature virus found outside the host cell EXCEPT:

A. core proteins
B. RNA and DNA
C. protein capsid
D. phospholipid bilayer envelope

25. All the following are correct about *lac* operon EXCEPT:

A. The repressor protein binds to the operator, halting gene expression
B. The promoter is the binding site of RNA polymerase
C. *Lac* operon is in eukaryotes
D. Three structural genes encode functional proteins

26. Phage DNA integrated into the chromosome is:

A. lytic phage
B. specialized transducing phage
C. lysogenic phage
D. prophage

27. Many RNA copies of the retrovirus RNA genome are synthesized by:

A. host cell DNA polymerases
B. reverse transcriptase
C. host cell RNA polymerases
D. host cell ribosomes

28. In the laboratory, *E. coli* are grown at a temperature of 37 °C because:

A. *E. coli* strain is a 37 °C temperature-sensitive mutant
B. *E. coli* reproduce most rapidly at this temperature
C. lower temperatures inhibit conjugation
D. *E. coli* are obligate aerobes

29. Some bacteria can propel themselves through the liquid using:

A. flagellum
B. centriole
C. centrosome
D. peptidoglycan

30. When most viruses infect eukaryotic cells:

A. their capsid does not enter the host cell
B. they replicate independently of the host cell during a lysogenic infection
C. they can enter the cell via endocytosis
D. they do not need to have host-specific proteins to infect the target cell

31. Which organelle(s) is/are NOT in bacteria?

I. peroxisomes
II. nucleolus
III. ribosomes
IV. flagellum

A. I only
B. II only
C. I and II only
D. I, II, and III

32. A bacterial cell carrying a prophage is:

A. lysogen
B. temperate
C. exconjugant
D. transformant

33. All the following are true of prokaryotic translation EXCEPT:

A. mRNA is not spliced before initiation
B. N-terminal amino acid of nascent polypeptides is formylated
C. mRNA chain being translated may not be fully transcribed before translation begins
D. Hydrogen bonds between amino acids and mRNA codons are necessary for translation

34. Prokaryotes are about how many times smaller in diameter than a typical eukaryote?

A. two
B. ten
C. 100
D. 10,000

35. Which of these structures are in prokaryotes?

I. A cell wall containing peptidoglycan
II. A plasma membrane with cholesterol
III. Ribosomes

A. I only
B. II only
C. I and II only
D. I and III only

36. Which of the following is characteristic of viruses?

A. membrane-bound organelles
B. protein coat
C. peptidoglycan cell wall
D. phospholipid bilayer membrane

37. Which of the following statements about a prokaryotic cell is correct?

A. It contains a range of different organelles
B. It uses glycolysis to produce ATP
C. It contains cell walls composed of chitin
D. It has a nuclear membrane that encloses a nucleus

38. The RNA genome of a retrovirus is converted to double-stranded DNA by:

A. host cell DNA polymerases
B. reverse transcriptase
C. host cell RNA polymerases
D. host cell ribosomes

39. Which of these statements applies to a bacteriophage and retrovirus?

A. They are capable of infecting human cells
B. They act as immunosuppressive agents
C. They integrate their genetic material into the genome of the host cell
D. They have genes that encode for reverse transcriptase

40. Which of the following statements about a gram-positive cell wall is FALSE?

A. It maintains the shape of cell
B. It is sensitive to lysozyme
C. It protects the cell in a hypertonic environment
D. It contains teichoic acids

41. Which statement is correct about the lipopolysaccharide layer outside the peptidoglycan cell wall of a Gram-negative bacterium?

A. It allows the bacterium to attach to solid objects
B. It does not contain a phospholipid membrane
C. It protects the bacterium against antibiotics
D. It absorbs and holds Gram stain

42. DNase added to a bacterial cell causes hydrolysis of the cell's DNA, preventing protein synthesis and cell death. Some viruses pretreated with DNase continue to produce new proteins following infection because:

A. viral genome contains multiple copies of their genes
B. viral genome is comprised of RNA
C. viral genome contains multiple reading frames
D. icosahedral protein coat of the virus denatures DNase

43. Integration of the phage DNA into the bacterial chromosome is facilitated by:

A. DNA polymerase encoded in the host genome
B. topoisomerase encoded in the phage
C. site-specific recombinase encoded in the phage
D. site-specific recombinase encoded in the host genome

44. After being digested by restriction enzymes, how is DNA ligated into a plasmid in two different directions?

A. Both ends of a DNA fragment produced by a restriction enzyme are identical if rotated 180°
B. The existing DNA strands serve as primers for DNA polymerase
C. DNA ligase enzymes can link any two pieces of DNA
D. Plasmid DNA is single-stranded, so the ligated strands form double-stranded segments

45. Which statement describes a bacterial cell in a 35% solution of a large polysaccharide (e.g., dextran)?

A. NaCl moves into the cell from a higher to a lower concentration
B. The cell undergoes plasmolysis
C. H_2O moves out of the cell
D. H_2O moves into the cell

46. If the Gram stain method stains a Gram-positive bacterium, it appears?

A. deep purple because of a thicker peptidoglycan cell wall
B. deep purple because of a thinner peptidoglycan cell wall
C. red or pink because of a thicker peptidoglycan cell wall
D. red or pink because of a thinner peptidoglycan cell wall

47. Virulent phage:

I. is capable only of lytic growth
II. can only undergo a process called lysogeny
III. needs several different proteins for incorporation into the bacterial chromosome

A. I and II only
B. III only
C. II only
D. I only

48. Which statement best describes the promoter in an operon?

A. It activates the repressor-inducer complex to permit transcription
B. It is a molecule that inactivates the repressor and turns on the operon
C. It is the binding site for RNA polymerase
D. It is the binding site for the repressor

Notes for active learning

Notes for active learning

Plants: Structure, Function and Reproduction

1. In angiosperms, pollen grains are produced in the:

A. ovule
B. sepal
C. stigma
D. anther

2. In most plants, which organs are adapted to capture sunlight for photosynthesis?

A. leaves
B. flowers
C. roots
D. stems

3. Which is true?

A. Plants take carbon dioxide from the atmosphere and do not use oxygen
B. Most plants get the water they need from the soil
C. Plants have adaptations that minimize light absorption
D. Plants require less water on sunny days

4. In angiosperms, xylem consists of tracheids and:

A. vessel elements
B. parenchyma
C. sieve tube elements
D. albuminous cells

5. In angiosperms, flowers are adaptations for reproduction. Flowers that are pollinated by animals do NOT:

A. produce nectar to attract pollinators
B. pollinate more efficiently than wind-pollinated plants
C. grow larger
D. exist as small structures

6. Without gas exchange, a plant would be unable to:

A. make minerals
B. reproduce
C. make food
D. absorb sunlight

7. Which is NOT part of the female structure of a flower?

A. ovary
B. stigma
C. style
D. filament

8. Living on land requires those plants:

A. have cell walls
B. exchange gases
C. conserve water
D. have photosynthetic pigments

9. If some xylem of a young maple tree were destroyed, it would interfere with the tree's ability to:

A. absorb nutrients from the soil
B. conduct water to the leaves
C. conduct water to the roots
D. conduct sugars to the roots

10. Vascular tissue in plants consists of:

A. epidermis
B. parenchyma and collenchyma
C. meristems
D. xylem and phloem

11. The ancestors of land plants likely evolved from:

A. prokaryotes that carried on photosynthesis
B. a protist that lived on land
C. an organism similar to green algae
D. mosses that lived in water

12. When a plant reproduces vegetatively:

A. only root tissue can be used to produce new offspring
B. offspring are produced by mitosis alone
C. meiosis produces a new gametophyte
D. offspring will differ from the parent

13. Seeds that animals spread usually are contained in:

A. lightweight structures
B. thin coatings that are easily digested
C. unripen ovaries
D. environmentally resistant structures

14. To observe mitosis, which should be examined under a compound microscope?

A. xylem from a tree trunk
B. phloem from the stem of a plant
C. epidermis of a leaf
D. tip of a shoot

15. What would have likely occurred if plants had not begun to live on land?

A. Animals would not have cellular respiration
B. There would be no CO_2 in the atmosphere
C. Animals also would not live on land
D. There would be no green algae in the oceans

16. The vascular cylinder of a root consists of the following:

A. phloem and xylem
B. phloem and ground tissue
C. xylem only
D. phloem only

17. Vegetative reproduction can be a disadvantage when:

A. the parent plant must reproduce quickly
B. the parent plant is not competing with other organisms for resources
C. conditions are favorable for growth
D. conditions in the physical environment suddenly change

18. The layer of cells that encloses the vascular tissue in the central region of a root is:

A. apical meristem
B. epidermis
C. cortex
D. endodermis

19. What is the fundamental difference between a sporophyte and a gametophyte?

A. A sporophyte is in the diploid phase, while a gametophyte is in the haploid phase of the plant life cycle
B. A sporophyte is much smaller than the gametophyte
C. A sporophyte is a reproductive structure, while a gametophyte is not
D. A sporophyte undergoes sexual reproduction, while a gametophyte undergoes asexual reproduction

20. The outer covering of a plant consists of:

A. meristematic tissue
B. dermal tissue
C. vascular tissue
D. ground tissue

21. How have the sporophyte and gametophyte changed as plants have evolved?

A. The gametophyte has become smaller, while the sporophyte has become larger
B. Many more recently evolved plants no longer have a sporophyte
C. The sporophyte has become larger, while the gametophyte has not changed
D. More recently evolved plants are mostly haploid gametophytes

22. A period during which the embryo of a seed is alive but not growing is:

A. germination
B. dormancy
C. dispersal
D. fertilization

23. Which correctly describes seeds and fruits in angiosperms?

A. Fruits provide nutrients for the seed
B. Fruits do not protect the seed, so they are not favored by natural selection
C. Fruits surround no more than one seed apiece
D. Fruits are matured ovaries that contain seeds

24. All are reasons why green algae are classified as plants, EXCEPT:

A. They have genes like genes in land plants
B. They have chlorophyll *a*
C. They have cellulose in their cell walls
D. They are single-celled organisms

25. In bryophytes, which structure is like a root?

A. stalk
B. capsule
C. rhizoid
D. sporophyte

26. A seed that is dispersed to an area far away from the parent plant always faces less:

A. unfavorable conditions for germination
B. favorable conditions for germination
C. chance of self-pollination
D. competition with the parent plants

27. Root pressure:

A. is produced in the vascular cylinder by osmotic pressure
B. is produced within the medulla of the root
C. is produced within the cortex of the stem
D. causes plant roots to increase in size

28. Which statement about green algae is true?

A. Like other plants, they have specialized structures
B. They are multicellular plants
C. Evidence suggests they were the first plants
D. They are found in dry areas on land

29. The plant embryo in a seed begins to grow again during:

A. pollination
B. germination
C. dormancy
D. fertilization

30. A period of dormancy can allow seeds to germinate:

A. in extreme temperatures
B. without water
C. under cold conditions
D. under ideal conditions

31. One of the three main functions of stems is to:

A. store sugars
B. store carbon dioxide
C. facilitate photosynthesis
D. carry nutrients and H_2O between roots and leaves

32. Which hormone stimulates an increase in size, especially in stems and fruits?

A. germination
B. ethylene
C. auxin
D. gibberellin

33. Of the following locations, mosses likely grow in:

A. exposed ground with direct sun
B. shaded ground near a small pond
C. sandy soil near a beach
D. hot and dry soil

34. When environmental conditions are unfavorable, why is it advantageous for green algae to reproduce sexually?

A. Sexual reproduction produces offspring with greater genetic variety and durable zygotes that help green algae survive unfavorable conditions
B. Sexual reproduction produces multicellular algae instead of unicellular algae. Multicellular algae are better able to survive harsh conditions
C. They are multicellular when conditions are unfavorable, which favors sexual reproduction
D. Sexual reproduction produces larger numbers of offspring in less time

35. Many cacti have large stems and no leaves. Which leaves' function is assumed by the stems of these cacti?

A. absorb H_2O from the soil
B. transport materials throughout the plant
C. absorb nutrients from the soil
D. produce food by photosynthesis

36. Which vascular tissue carries solutions of nutrients and carbohydrates produced by photosynthesis?

A. lignin
B. tracheids
C. phloem
D. xylem

37. Unlike roots, stems:

A. may be involved in photosynthesis
B. are protected by epidermal cells
C. have ground tissue
D. transport water

38. During primary growth, a stem:

A. increases in width
B. fills with O_2
C. produces wood
D. increases in length

39. Hormones that stimulate cell elongation and are produced in rapidly growing regions near the tip of plant roots or stems are:

A. cytokinins
B. gibberellins
C. auxins
D. ethylene

40. What might a thin tree ring indicate?

A. decreased production of phloem
B. a year of drought
C. increased production of xylem
D. xylem production in winter

41. What is the phenomenon affecting the growth of the bean shoot in the figure?

A. thigmotropism
B. rapid movement
C. gravitropism
D. phototropism

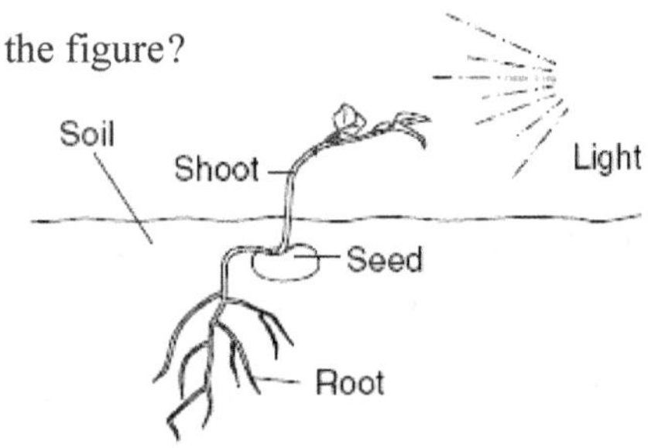

42. Bryophytes must live in moist areas without vascular tissue, and what other reason requires moist areas?

A. Carbon dioxide exchange is more efficient in wet areas
B. Without moisture, rhizoids cannot anchor the plants
C. Bryophytes need extra water for photosynthesis
D. The sperm of bryophytes need water to swim to an egg

43. Which structure includes a plant embryo, a food supply, and a protective covering?

A. Seed
B. Gametophyte
C. Pollen grain
D. Spore

44. When positioned horizontally, a plant's response to gravity is that:

A. all parts of the plant bend upward, away from the force of gravity
B. auxins in the shoot cause the growth of tendrils to pull the shoot upwards
C. auxins in the lower sides of stems cause cell elongation that bends the stem upright
D. the shoot bends to grow downward, toward the force of gravity

45. What is the source of ethylene gas in a plant?

A. stems
B. fruit tissues
C. leaves
D. all the above

46. The stomata of leaves are usually open:

A. at night if a plant has enough water
B. during the day if a plant has enough CO_2
C. during the day if a plant has enough water
D. during the day if a plant has too little water

47. The yellow, orange, and red colors of leaves in the fall are the result of the following:

A. decrease of auxins in the meristems
B. production of anthocyanin and the breakdown of chlorophyll
C. new synthesis of orange and yellow pigments in leaves
D. movement of chlorophyll from the stems to the leaves

48. O_2 and CO_2 move in and out of a leaf through:

A. stomata
B. phloem
C. guard cells
D. palisade mesophyll

49. The growth of plant seedlings is usually:

A. gravitropic and influenced by ethylene
B. phototropic and influenced by ethylene
C. thigmotropic and gravitropic
D. phototropic and gravitropic

50. What is one reason vascular plants are larger than nonvascular plants?

A. Nonvascular plants do not reproduce as quickly as vascular plants do, so they do not grow as large
B. Vascular tissue carries water and nutrients much more efficiently than can be carried by diffusion
C. Vascular plants live in areas where they can get more water than nonvascular plants
D. Nonvascular plants cannot live in as many areas as vascular plants, so they stay small

51. Most of the photosynthesis in plants occurs in:

A. xylem
B. stomata
C. mesophyll tissue
D. guard cells

52. Stomata are only on the lower surface of leaves in many plants. The explanation for this is:

A. water loss would be less on the shaded lower surface than in the direct sun
B. gravity functions in gas exchange
C. stomata are closer to vascular bundles that bring water into the leaf
D. photosynthesis only occurs in the spongy mesophyll near the bottom of the leaf

53. The growth of ivy tendrils that wrap around objects is an example of:

A. photoperiodism
B. chemotropism
C. phototropism
D. thigmotropism

54. Seed-bearing plants differ from all other plants in that:

A. they have true roots, stems, and leaves
B. their gametes do not require water for fertilization
C. they have a gametophyte generation
D. they have only xylem and no phloem

55. Xylem tissue is vital because it:

A. carries out photosynthesis
B. carries carbohydrates to all parts of the plant
C. allows water to diffuse from the roots
D. can conduct water over long distances

56. Corn, broccoli, and cabbage were developed by:

A. auxins
B. selective breeding
C. plant propagation
D. germination

57. In the figure, the water pressure is:

A. high in the guard cells
B. low in the guard cells
C. low in the stoma
D. high in the stoma

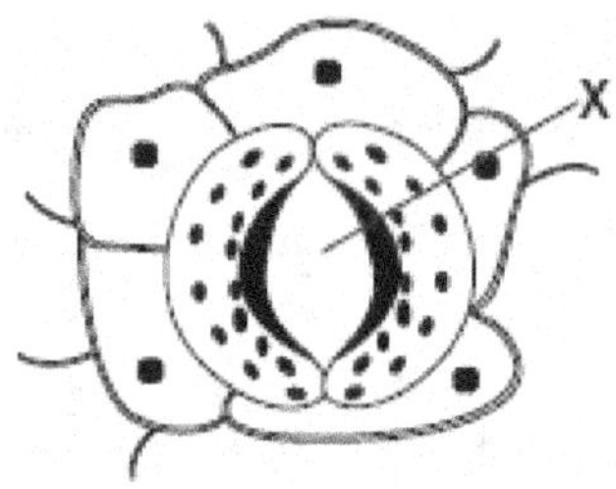

Underside of a leaf

58. Through which plant cell does water move by capillary action?

A. mesophyll cells
B. albuminous cells
C. phloem cells
D. xylem cells

59. Which gymnosperm reproductive structure is haploid?

A. seed
B. cone scale
C. pollen grains
D. seed cone

60. Which BEST describes how people use plants?

A. All medicines come from plants
B. Plants break down decomposing organisms and waste
C. People use only plants for food
D. Plants provide food, raw materials, and several medicines

61. In the figure, which letter indicates the structure where the embryo develops?

A. A
B. B
C. C
D. D

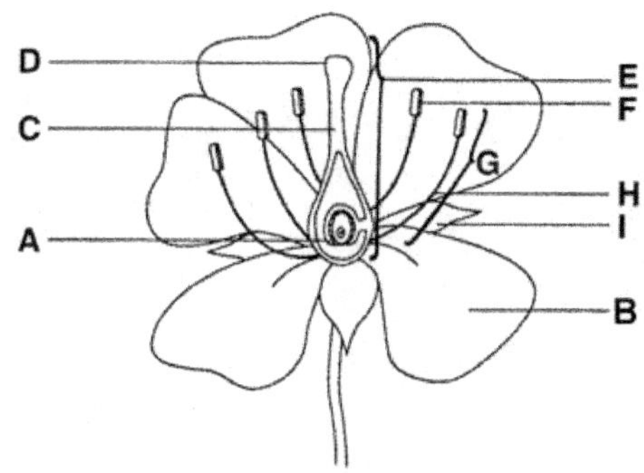

62. The movement of water containing food molecules in the phloem can be explained by

A. root pressure
B. the pressure-flow hypothesis
C. transpiration pull
D. diffusion gradient

63. According to the pressure-flow hypothesis, which statement is NOT true?

A. The movement of H_2O into a nutrient-rich region of the phloem decreases the pressure in that region
B. Phloem can move sugars in either direction to meet the nutritional needs of the plant
C. H_2O is necessary for sugars to move through the phloem
D. H_2O moves from the xylem to the phloem of a plant

64. Angiosperms produce seeds inside protective structures called:

A. petals
B. cones
C. ovaries
D. pollen grains

65. Which correctly relates to an angiosperm structure and what it develops into?

A. embryo → fruit
B. ovule → fruit
C. pollen → grain seed
D. ovary → fruit

66. A pollen grain landing on a stigma of a flower of the same species produces:

A. flower
B. stamen
C. gametophyte
D. pollen tube

67. Flowering plants that grow, flower, produce seeds, and die in one year are:

A. biennials
B. perennials
C. dicots
D. annuals

68. Ground tissue is in a plant's:

A. roots and stems only
B. roots, stems, and leaves
C. stems only
D. stems and leaves only

69. A seed plant is held in the ground by its:

A. phloem
B. epidermis
C. stems
D. roots

Notes for active learning

Notes for active learning

Photosynthesis

1. What mechanism is used by C_4 plants to conserve water?

   A. developing deep roots
   B. performing the Calvin cycle during the day
   C. developing water storages in their leaves and stems
   D. closing the stomata during heat and dryness

2. What mechanism is used by CAM plants to conserve water?

   A. performing the Calvin cycle during the day
   B. developing water storages in their leaves and stems
   C. including carbon dioxide into RuBP
   D. opening their stomata during the night

3. Which of the choices is an autotroph?

   A. beech tree
   B. octopus
   C. zooplankton
   D. coral

4. Which of the following is NOT a true statement about ATP?

   A. ATP provides energy for the mechanical functions of cells
   B. Used ATP is discarded by the cell as waste
   C. ATP consists of ribose, adenine, and three phosphate groups
   D. ADP is produced when ATP releases energy

5. Photosynthesis plays a role in the metabolism of a plant by:

   A. breaking down sugars into H_2O and O_2
   B. converting H_2O into CO_2
   C. consuming CO_2 and synthesizing sugars
   D. converting O_2 into cellulose

6. From the figure, all the following are parts of an ADP molecule EXCEPT:

   A. structure A
   B. structure B
   C. structure C
   D. structure D

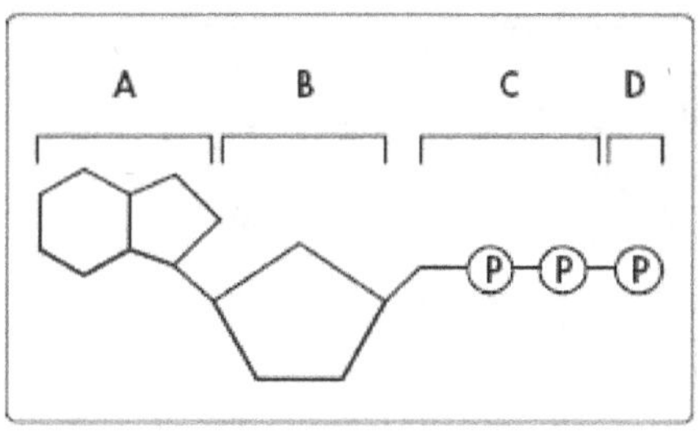

7. Within the inner membrane of a chloroplast, stacks of thylakoids are surrounded by:

A. stroma fluid
B. grana
C. chloroplast
D. thylakoids

8. What happens during photosynthesis?

A. Autotrophs consume carbohydrates
B. Autotrophs produce carbohydrates
C. Heterotrophs consume ATP
D. Heterotrophs produce ATP

9. The role of stomata is to facilitate:

A. gas exchange
B. H_2O release
C. Calvin cycle
D. H_2O uptake

10. Plants gather the sun's energy using:

A. glucose
B. chloroplasts
C. thylakoids
D. pigments

11. Which structure is used by plants to obtain most of their water?

A. Stomata
B. Flowers
C. Roots
D. Chloroplasts

12. Which structure in the image illustrates a single thylakoid?

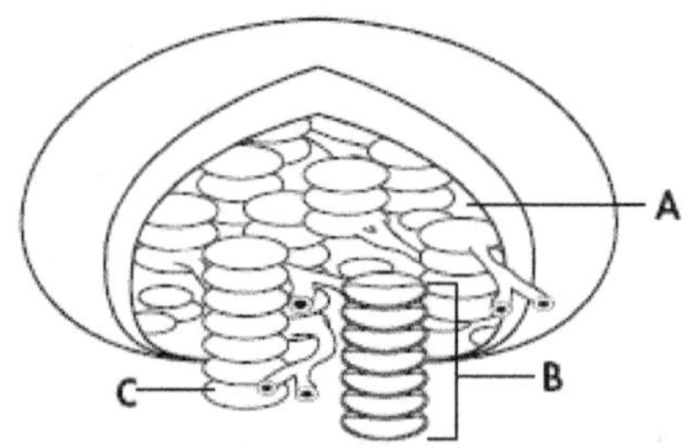

A. Structure A
B. Structure B
C. Structure C
D. Structures A and B

13. Membranous structures in chloroplasts are:

A. organelles
B. thylakoids
C. stomata
D. plasma

14. Chlorophyll is within the chloroplast in:

A. thylakoid space
B. thylakoid membrane
C. stroma
D. ATP

15. Grana is/are:

A. stacks of membranous sacs
B. chloroplast pigments
C. fluid in the chloroplasts
D. the space between the inner and outer membrane of the chloroplasts

16. Which chemical shown in the figure is an electron carrier molecule?

A. $NADP^+$
B. Oxygen
C. H_2O
D. CO_2

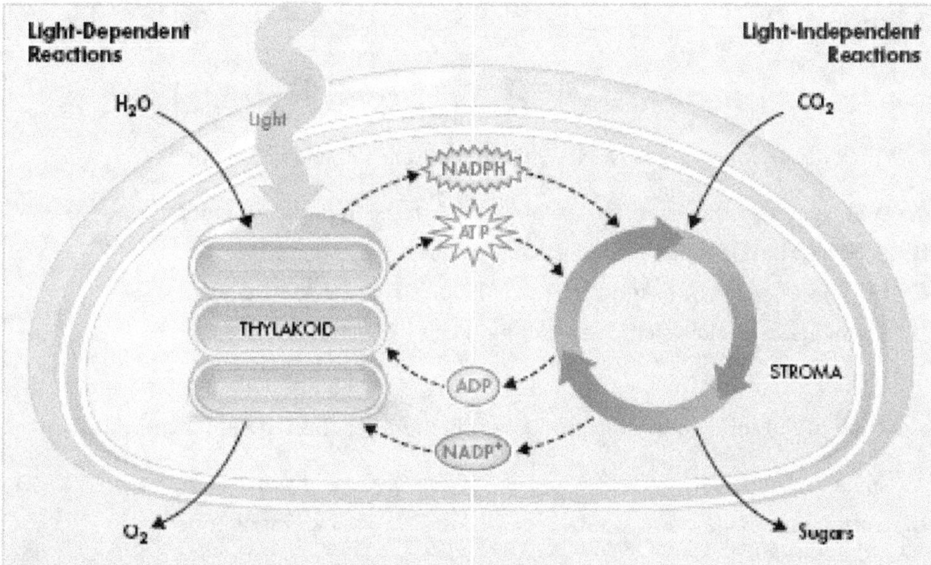

17. Which of the following is likely to result if a shade-tolerant plant would receive a minimal amount of water while receiving its necessary amount of light?

A. Accelerated plant growth
B. Low ATP production
C. Increased output of oxygen
D. Increased consumption of carbon dioxide

18. Why are electron carriers needed for transporting electrons from one part of the chloroplast to another?

A. High-energy electrons get their energy from electron carriers
B. High-energy electrons are not soluble in the cytoplasm
C. High-energy electrons are highly reactive
D. High-energy electrons would be destroyed

19. Which equation describes the reaction of photosynthesis?

A. $6\ O_2 + 6\ CO_2 \rightarrow C_6H_{12}O_6 + 6\ H_2O$
B. $C_6H_{12}O_6 + 6\ H_2O \rightarrow 6\ CO_2 + 6\ O_2$
C. $C_6H_{12}O_6 \rightarrow 6\ CO_2 + 6\ H_2O + 6\ O_2$
D. $6\ H_2O + 6\ CO_2 \rightarrow C_6H_{12}O_6 + 6\ O_2$

20. What is the result of an increase in light intensity that a plant receives?

A. The rate of photosynthesis increases and then levels off
B. The rate of photosynthesis does not change
C. The rate of photosynthesis increases indefinitely with light intensity
D. The rate of photosynthesis decreases indefinitely with light intensity

21. In redox reactions of photosynthesis, electron transfer goes as follows:

A. $H_2O \rightarrow CO_2$
B. $C_6H_{12}O_6 \rightarrow O_2$
C. $O_2 \rightarrow C_6H_{12}O_6$
D. $CO_2 \rightarrow H_2O$

22. Which of the following is NOT a step in the light-dependent reactions?

A. ATP synthase allows H^+ ions to pass through the thylakoid membrane
B. ATP and NADPH are used to produce high-energy sugars
C. High-energy electrons move through the electron transport chain
D. Pigments in photosystem II absorb light

23. Which of the following is the product of the light reaction of photosynthesis?

A. sugar
B. carbon monoxide
C. oxygen
D. carbon dioxide

24. Photosystems I and II are in:

A. cell membrane
B. the Calvin cycle
C. thylakoid membrane
D. stroma

25. Which components from the light reactions are required by the C_3 cycle?

A. NADH and RuBP
B. NADPH and ATP
C. ATP and NADH
D. glucose and $NADP^+$

26. Which pathway is the correct flow of electrons during photosynthesis?

A. Photosystem I → Calvin cycle → $NADP^+$
B. H_2O → $NADP^+$ → Calvin cycle
C. H_2O → Photosystem I → Photosystem II
D. O_2 → ADP → Calvin cycle

27. Light-dependent reactions of photosynthesis produce:

A. H_2O and RuBP
B. ATP and CO_2
C. NADH and ATP
D. NADPH and ATP

28. The Calvin cycle takes place in:

A. chlorophyll molecules
B. thylakoid membranes
C. photosystems
D. stroma

29. The wavelength of which a blue-colored plant reflects light?

A. violet
B. yellow
C. blue
D. lime

30. What are the three parts of an ATP molecule?

A. Adenine, ribose, and three phosphate groups
B. NADH, NADPH, and $FADH_2$
C. Adenine, thylakoid, and a phosphate group
D. Stroma, grana, and chlorophyll

31. When observing a pigmented object, visible color is:

A. wavelength emitted by the object
B. wavelength from the return of excited electrons to their ground state
C. wavelength being reflected by that object
D. wavelength absorbed by the pigment of the object

32. Energy is released from ATP when:

A. a phosphate group is removed
B. ATP is exposed to sunlight
C. adenine binds to ribose
D. a phosphate group is added

33. The wavelength energy least utilized by photosynthesis is:

A. orange
B. blue
C. green
D. cyan

34. Organisms, such as plants that make their food are:

A. symbiotic
B. parasitic
C. autotrophs
D. heterotrophs

35. During the fall foliage, which pigments are responsible for the yellow, red, and orange colors of leaves?

A. carotenoids
B. chlorophyll *b*
C. melanin
D. chlorophyll *a*

36. Which of the following organisms is a heterotroph?

A. sunflower
B. flowering plant
C. alga
D. mushroom

37. In addition to chlorophyll *a*, plants have chlorophyll *b* and carotenoids accessory pigments because:

A. there is not enough chlorophyll *a* produced for the plant's energy needs
B. plants must have leaves of different colors
C. these pigments absorb energy wavelengths that chlorophyll *a* does not
D. these pigments protect plants from UV radiation

38. Plants get the energy they need for photosynthesis by absorbing:

A. chlorophyll *b*
B. energy from the sun
C. high-energy sugars
D. chlorophyll *a*

39. A discrete packet of light is a:

A. photon
B. neutron
C. proton
D. wavelength

40. Most plants appear green because chlorophyll:

A. does not absorb violet light
B. does not absorb green light
C. absorbs violet light
D. absorbs green light

41. In visible light, shorter wavelengths carry:

A. more energy
B. more photons
C. more red color
D. less energy

42. Interconnected sacs of a membrane suspended in a thick fluid are:

A. stroma
B. thylakoids
C. chlorophyll
D. grana

43. When a molecule absorbs a photon, its electrons are raised to a(n):

A. higher state
B. excited state
C. lower state
D. ground state

44. What is the function of $NADP^+$ in photosynthesis?

A. Photosystem
B. Pigment
C. Electron carrier
D. High-energy sugar

45. A molecule releases energy gained from the absorption of a photon through:

I. fluorescence   II. heat   III. light

A. I only
B. II only
C. III only
D. I, II and III

46. Photosynthesis uses sunlight to convert water and carbon dioxide into:

A. oxygen and high-energy sugars
B. ATP and oxygen
C. high-energy sugars and proteins
D. oxygen and carbon

47. Photosystems are in:

A. grana
B. stroma
C. stomata
D. thylakoid membranes

48. The light-dependent reactions take place:

A. within the thylakoid membranes
B. in the outer membrane of the chloroplast
C. in the stroma of the chloroplast
D. within the mitochondria membranes

49. Which compound is in the photosystem's reaction center?

A. chlorophyll *a*
B. rhodopsin
C. ATP
D. ADP

50. What are the products of the light-dependent reactions?

A. CO_2 gas, O_2 gas, and NADPH
B. ATP, CO_2 gas, and NADPH
C. ATP, NADPH, and O_2 gas
D. O_2 gas and glucose

51. Which of the following is the correct matching of molecules and products in the Calvin cycle?

A. ATP + NADPH + carbon dioxide ⇒ sugar
B. light + water + carbon dioxide ⇒ sugar and oxygen
C. ATP + NADPH + carbon dioxide ⇒ sugar and oxygen
D. light + water + carbon dioxide ⇒ sugar

52. Which of the following activities happens within the stroma?

A. Electrons move through the electron transport chain
B. The Calvin cycle produces sugars
C. ATP synthase produces ATP
D. Photosystem I absorbs light

53. Oxygen released by a photosystem comes from:

A. ATP
B. carbon dioxide
C. Chlorophyll *a*
D. water

54. The Calvin cycle is another name for:

A. photosynthesis reaction
B. electron transport chain
C. light-independent reactions
D. light-dependent reactions

55. Electrons for the light reactions originate from:

A. NADPH
B. water
C. light
D. carbon dioxide

56. Which mechanism takes place in both light reactions of photosynthesis and cellular respiration?

A. Electron transport chain
B. Beta oxidation
C. Glycolysis
D. Krebs cycle

57. During photosynthesis, an H^+ ion gradient forms across:

A. mitochondrial inner membrane
B. inner chloroplast membrane
C. thylakoid membrane
D. mitochondrial outer membrane

Notes for active learning

Notes for active learning

Macromolecules in a Biological System

1. The linear sequence of amino acids along a peptide chain determines its:

A. primary structure
B. secondary structure
C. tertiary structure
D. quaternary structure

2. There are [ ] different types of major biomolecules used by humans.

A. a few dozen
B. four
C. several thousand
D. several million

3. Amino acids are linked to one another in a protein by which of the following bonds?

A. amide bonds
B. carboxylate bonds
C. ester bonds
D. amine bonds

4. Insulin is an example of a(n):

A. hormone
B. storage protein
C. structural protein
D. enzyme

5. Which protein structure corresponds to a spiral alpha-helix of amino acids?

A. primary
B. secondary
C. tertiary
D. quaternary

6. Proteins are characterized by:

A. always has a quaternary structure
B. retaining their conformation above 35-40 °C
C. a primary structure formed by covalent linkages
D. composed of a single peptide chain

7. Members of which class of biomolecules is the building block of proteins?

A. fatty acids
B. amino acids
C. glycerols
D. monosaccharides

8. All the bonds stated are apparent in the secondary and tertiary structure of a protein, EXCEPT:

A. electrostatic interactions
B. peptide bonds
C. hydrogen bonding
D. hydrophobic interactions

9. Proteins are polymers. They consist of monomer units which are:

A. keto acids
B. amide
C. amino acids
D. ketones

10. Why might a change in pH cause a protein to denature?

A. The hydrogen bonds between the hydrophobic portions of the protein collapse due to extra protons
B. The disulfide bridges open
C. The functional groups that give the protein its shape becomes protonated or deprotonated
D. The water hardens and causes the protein's shape to change

11. Hydrophobic interactions help stabilize the [ ] structure(s) of a protein.

A. primary
B. secondary
C. secondary and tertiary
D. tertiary and quaternary

12. Collagen is an example of a (an):

A. storage protein
B. transport protein
C. enzyme
D. structural protein

13. The coiling of a chain of amino acids describes protein:

A. primary structure
B. secondary structure
C. tertiary structure
D. quaternary structure

14. The laboratory conditions typically used to hydrolyze a protein are:

A. dilute acid and room temperature
B. dilute base and room temperature
C. concentrated acid and heat
D. concentrated base and heat

15. Which of the following is an essential amino acid?

A. aniline
B. valine
C. glycine
D. serine

16. What is the localized bending and folding of a polypeptide backbone?

A. primary structure
B. secondary structure
C. tertiary structure
D. quaternary structure

17. Which of the following macromolecules are composed of polypeptides?

A. amino acids
B. proteins
C. carbohydrates
D. fats

18. What is the purpose of the plasma membrane?

A. Storing of the genetic material of the cell
B. Retaining water in the cell to prevent it from dehydrating
C. Acting as a cell wall to give the cell structure and support
D. Acting as a boundary, but also letting molecules in and out

19. These fatty acids contain sixteen to eighteen carbons and range from saturated to three double bonds. Which has the lowest melting point?

A. palmitic acid (saturated)
B. oleic acid (one double bond)
C. linoleic acid (two double bonds)
D. linolenic acid (three double bonds)

20. Which statements correctly describe the relationship between fatty acid structure and its melting point?

I. Saturated fatty acids melting points increase gradually with the molecular weights
II. As the number of double bonds in a fatty acid increases, its melting point decreases
III. The presence of a *trans* double bond in the fatty acid has a greater effect on its melting point than does the presence of a *cis* double bond

A. I only
B. II only
C. III only
D. I and II only

21. The hydrocarbon end of a soap molecule is:

A. hydrophilic and attracted to grease
B. hydrophobic and attracted to grease
C. hydrophilic and attracted to water
D. hydrophobic and attracted to water

22. It is important to have cholesterol in one's body because:

A. it breaks down extra fat lipids
B. it serves as the starting material for the biosynthesis of most other steroids
C. it is the starting material for the building of glycogen
D. the brain is made almost entirely of cholesterol

23. Cholesterol belongs to the [ ] group of lipids.

A. prostaglandin
B. triacylglycerol
C. saccharides
D. steroid

24. In chemical terms, soaps are best described as:

A. simple esters of fatty acids
B. mixed esters of fatty acids
C. salts of carboxylic acids
D. long-chain acids

25. The function of cholesterol in a cell membrane is to:

A. act as a precursor to steroid hormones
B. take part in the reactions that produce bile acids
C. maintain structure due to its flat rigid characteristics
D. attract hydrophobic molecules to form solid deposits

26. Which fatty acid is a saturated fatty acid?

A. oleic acid
B. linoleic acid
C. arachidonic acid
D. myristic acid

27. Commercially, liquid vegetable oils are converted to solid fats such as margarine by:

A. oxidation
B. hydration
C. hydrolysis
D. hydrogenation

28. The biochemical roles of lipids are:

A. short-term energy storage, transport of molecules, and structural support
B. storage of excess energy, component of cell membranes, and chemical messengers
C. catalysis, protection against outside invaders, motion
D. component of cell membranes, catalysis, and structural support

29. Which of the following molecules is an omega-3 fatty acid?

A. oleic acid
B. linolenic acid
C. linoleic acid
D. palmitic acid

30. Which of the following is a lipid?

A. lactose
B. aniline
C. nicotine
D. estradiol

31. Which of the following terms best describes the interior of a soap micelle in water?

A. hard
B. saponified
C. hydrophobic
D. hydrophilic

32. Lipids are compounds soluble in:

A. glucose solution
B. organic solvents
C. distilled water
D. normal saline solution

33. Which statement regarding fatty acids is NOT correct? Fatty acids:

A. are always liquids
B. are long-chain carboxylic acids
C. are usually unbranched chains
D. usually have an even number of carbon atoms

34. Unsaturated triacylglycerols are usually [ ] because [ ]?

A. liquids … they have relatively short fatty acid chains
B. liquids … the kinks in their fatty acid chains prevent their fitting closely
C. liquids … they contain impurities from their natural sources
D. solids … they have relatively long fatty acid chains

35. Triacylglycerols are compounds with:

A. cholesterol and other steroids
B. fatty acids and phospholipids
C. fatty acids and glycerol
D. fatty acids and choline

36. Oils are generally [ ] at room temperature and obtained from [ ]:

A. liquids … plants
B. liquids … animals
C. solids … plants
D. solids … animals

37. Lipids are naturally occurring compounds that all:

A. contain fatty acids as structural units
B. are water-insoluble but soluble in nonpolar solvents
C. contain ester groups
D. contain cholesterol

38. Hydrogenation of vegetable oils converts them into which molecule?

A. esters
B. ethers
C. hemiacetals
D. saturated fats

39. Saturated fats are [ ] at room temperature and are obtained from [ ]?

A. liquids; plants
B. liquids; animals
C. solids; plants
D. solids; animals

40. Which of the following is NOT in a lipid wax?

A. saturated fatty acid
B. long-chain alcohol
C. glycerol
D. ester linkage

41. How many fatty acid molecules are needed to produce one fat or oil molecule?

A. 1
B. 1.5
C. 2
D. 3

42. When dietary triglycerides are hydrolyzed, the products are:

A. glycerol and fatty acids
B. carbohydrates
C. amino acids
D. alcohols and lipids

43. How many fatty acids are in a phospholipid molecule?

A. 0
B. 1
C. 2
D. 3

44. Which of the following is NOT a function of lipids?

A. cushioning to prevent injury
B. insulation
C. energy reserve
D. precursor for glucose catabolism

45. Which of the following lipids is an example of a simple lipid?

A. oil
B. wax
C. fat
D. terpene

46. Which of the following statements describes most monosaccharides?

A. They are unsaturated compounds
B. They are rarely monomers in nature
C. They are composed of carbon, hydrogen, and oxygen, with each carbon bound to at least one oxygen
D. They are insoluble

47. The three elements in all carbohydrates are [ ], [ ] and [ ]:

A. nitrogen, oxygen, hydrogen
B. carbon, hydrogen, oxygen
C. carbon, hydrogen, water
D. nitrogen, oxygen, carbon

48. Fructose does not break apart into smaller units because it is a(n):

A. monosaccharide
B. polysaccharide
C. hexose
D. aldose

49. Which of the following molecules is a disaccharide?

A. lactose
B. cellulose
C. amylose
D. glucose

50. What is the primary biological function of the glycogen biomolecule?

A. It is used to synthesize disaccharides
B. It is the building block of proteins
C. It stores glucose in animal cells
D. It is a storage form of sucrose

51. A carbohydrate that gives two molecules when it is completely hydrolyzed is a:

A. polysaccharide
B. starch
C. monosaccharide
D. disaccharide

52. Which group of carbohydrates can NOT be hydrolyzed to give smaller molecules?

A. oligosaccharides
B. trisaccharides
C. disaccharides
D. monosaccharides

53. A carbohydrate can be defined as a molecule with:

A. carbon atoms bonded to water molecules
B. amine and carboxylic acid groups bonded to a carbon skeleton
C. mainly hydrocarbons and soluble in non-polar solvents
D. tan aldehyde or ketone with more than one hydroxyl group

54. Disaccharides are best characterized as:

A. two monosaccharides linked by a nitrogen bond
B. two peptides linked by a hydrogen bond
C. two monosaccharides linked by an oxygen bond
D. two amino acids linked by a peptide bond

55. Which of these molecules has the highest lipid density?

A. very-low-density lipoprotein (VLDL)
B. low-density lipoprotein (LDL)
C. high-density lipoprotein (HDL)
D. chylomicron

56. Which choice is the correct statement about lipids?

A. They are composed of elements C, O, N & H
B. Their secondary structure is composed of α helices and β pleated sheets
C. They are molecules used for long-term energy storage in animals
D. Elements of C:H:O are in the ratio of 1:2:1

57. Which nutrients yield four calories per gram?

A. glucose and proteins
B. fats and glucose
C. proteins and lipids
D. lipids and sugars

58. Which of the following is NOT an end product of digestion?

A. amino acids
B. lactose
C. fructose
D. fatty acids

59. During digestion, fats are catabolized when fatty acids are detached from glycerol, while proteins are digested into amino acids. What do these two processes have in common?

A. Both involve the addition of H_2O to break bonds (hydrolysis)
B. Both occur as intracellular processes in most organisms
C. Both are catalyzed by the same enzyme
D. Both require the presence of hydrochloric acid to lower the pH

60. The two strands of DNA in the double helix are held by:

A. dipole–dipole attractions
B. metallic bonds
C. ionic bonds
D. covalent bonds

61. What happens to DNA when placed into an aqueous solution at physiological pH?

A. Individual DNA molecules repel each other due to the presence of positive charges
B. DNA molecules bind to negatively charged proteins
C. Individual DNA molecules attract each other due to the presence of positive and negative charges
D. Individual DNA molecules repel each other due to the presence of negative charges

62. What is the term for how a DNA molecule synthesizes a complementary single strand of RNA?

A. translation
B. transcription
C. replication
D. duplication

63. Consider the following compounds:

I. amino acid
II. nitrogen-containing base
III. phosphate group
IV. five-carbon sugar

From which of the compounds are the monomers (i.e., nucleotides) of nucleic acids formed?

A. I only
B. I and II only
C. II and IV only
D. II, III and IV only

64. The nucleotide sequence, T–A–G, stands for

A. threonine-alanine-glutamine
B. thymine-adenine-guanine
C. tyrosine-asparagine-glutamic acid
D. thymine-adenine-glutamine

65. The two new DNA molecules formed in replication:

A. contain one parent and one daughter strand
B. contain only the parent DNA strands
C. contain only two new daughter DNA strands
D. are complementary to the original DNA

66. Which of the following is in an RNA nucleotide?

I. phosphoric acid
II. nitrogenous base
III. ribose sugar

A. I only
B. II only
C. I and II only
D. I, II and III

67. If one strand of a DNA double helix has the sequence AGTACTG, what is the sequence of the other strand?

A. GACGTCA
B. AGTACTG
C. GTCATGA
D. TCATGAC

68. The leading role of DNA is to provide instructions on how to build:

I. lipids
II. carbohydrates
III. proteins

A. I only
B. II only
C. III only
D. I and II only

69. Nucleic acids are polymers of [ ] monomers.

A. monosaccharide
B. fatty acid
C. DNA
D. nucleotide

70. During DNA transcription, a guanine base on the template strand codes for which base on the growing RNA strand?

A. guanine
B. thymine
C. adenine
D. cytosine

71. What intermolecular force connects strands of DNA in the double helix?

A. hydrogen bonds
B. ionic bonds
C. amide bonds
D. ester bonds

72. How are codons and anticodons related?

A. Codons are the base pairs on a tRNA that bind to complementary strands of DNA and produce proteins
B. Anticodons are the codons on the mRNA used to bind to DNA
C. Codons start the process of transcription; anticodons end the process
D. Codons and anticodons are complementary base pairs that encode an amino acid

73. Which biological compound is a polymer of sugar, a base, and phosphoric acid?

A. nucleic acid
B. lipid
C. carbohydrate
D. protein

74. What is the term for when a DNA molecule synthesizes an identical molecule of DNA?

A. transcription
B. translation
C. duplication
D. replication

75. The bonds that link the base pairs in the DNA double helix are [ ] bonds?

A. hydrophobic
B. hydrogen
C. peptide
D. ionic

76. Translation is the process whereby:

A. protein is synthesized from DNA
B. protein is synthesized from mRNA
C. DNA is synthesized from DNA
D. DNA is synthesized from mRNA

77. What is the primary structural difference between the sugar of RNA compared to the sugar of DNA?

A. It stabilizes the RNA outside the nucleus
B. It acts as an energy source to produce proteins
C. It allows the RNA to be easily digested by enzymes
D. It keeps the RNA from binding tightly to DNA

78. The double helix of DNA is stabilized mainly by:

A. hydrogen bonds
B. covalent bonds
C. ion–dipole bonds
D. ester bonds

79. Which of the following is the correct listing of DNA's constituents in the order of increasing size?

A. Nucleotide, codon, gene, nucleic acid
B. Nucleic acid, nucleotide, codon, gene
C. Nucleotide, codon, nucleic acid, gene
D. Gene, nucleic acid, nucleotide, codon

80. The number of adenines in a DNA molecule equals the number of thymines because:

A. adenines are paired opposite of guanine in a DNA molecule
B. of the strong attraction between the nucleotides of adenine and thymine
C. the structure of adenine is similar to uracil
D. adenine is paired to cytosine in a DNA molecule

81. What is the sugar component in RNA?

A. fructose
B. galactose
C. glucose
D. ribose

82. What is the process when the DNA double helix unfolds, and each strand serves as a template for synthesizing a new strand?

A. translation
B. replication
C. transcription
D. complementation

83. Which of the following illustrates the direction of flow for protein synthesis?

A. RNA → protein → DNA
B. DNA → protein → RNA
C. RNA → DNA → protein
D. DNA → RNA → protein

84. The three-base sequence in mRNA specifying the amino acid is:

A. rRNA
B. an anticodon
C. a codon
D. tRNA

85. Which of the following is an RNA codon for protein synthesis?

I. GUA   II. CGU   III. ACG

A. I only
B. II only
C. I and II only
D. I, II and III

86. During DNA replication, an adenine base on the template strand codes for which base on the complementary strand?

A. thymine
B. guanine
C. cytosine
D. adenine

87. Which of the following is NOT part of a nucleotide?

A. cyclic nitrogenous base
B. fatty acid
C. phosphate group
D. cyclic sugar

88. Which of the following linkage is in a nucleic acid?

A. phosphate linkage
B. ester linkage
C. glycoside linkage
D. peptide linkage

89. Which of the following codes for an amino acid during protein synthesis?

A. RNA nucleotide
B. RNA trinucleotide
C. DNA nucleotide
D. DNA trinucleotide

90. How does RNA differ from DNA?

A. RNA is double-stranded, while DNA is single-stranded
B. RNA is a polymer of amino acids, while DNA is a polymer of nucleotides
C. RNA contains uracil, while DNA contains thymine
D. In RNA, G pairs with T, while in DNA, G pairs with C

91. The one cyclic amine base that occurs in DNA but not in RNA is:

A. cystine
B. guanine
C. thymine
D. uracil

92. In the synthesis of mRNA, an adenine in the DNA pairs with:

A. guanine
B. thymine
C. uracil
D. adenine

93. Nucleic acids determine the:

A. quantity and type of prions
B. number of mitochondria in a cell
C. sequence of amino acids
D. pH of the cell nucleus

94. DNA is a(n):

A. peptide
B. protein
C. nucleic acid
D. enzyme

95. Which of the following amine bases is NOT in DNA?

A. adenine
B. cytosine
C. guanine
D. uracil

96. In transcription:

A. the mRNA contains the genetic information from DNA
B. uracil pairs with thymine
C. a double helix containing one parent strand and one daughter strand is produced
D. the mRNA produced is identical to the parent DNA

97. The two strands of the double helix of DNA are held by:

A. disulfide bridges
B. ionic bonds
C. hydrogen bonds
D. covalent bonds

98. Which of the following statements is TRUE?

A. Protein function can be altered by modification after synthesis
B. Covalent bonds stabilize the secondary structure of proteins
C. A protein with a single polypeptide subunit has a quaternary structure
D. Integral proteins contain high amounts of acidic amino acids

99. Which of the following statements is TRUE?

A. Disaccharides must contain fructose
B. Most polysaccharides contain ribose
C. Polysaccharides are energy-generating molecules
D. Glucose and fructose have different chemical properties even with the same molecular formula

Notes for active learning

Chromosomes, Genes and DNA

1. DNA and RNA differ because:

A. only DNA contains phosphodiester bonds
B. only RNA contains pyrimidines
C. DNA is in the nucleus, and RNA is in the cytosol
D. RNA is associated with ribosomes, and DNA is associated with histones

2. Tumor-suppressor genes normally control:

A. cell differentiation
B. necrosis
C. cell proliferation or activation of apoptosis
D. sister chromatid separation

3. In the genetic code, deciphered by Nobel laureate Marshall W. Nirenberg, each amino acid is coded for by three nucleotides (codons). How many possible codons exist in nature encoding 20 amino acids in polypeptides?

A. 4
B. 20
C. 27
D. 64

4. Which of the following statements does NOT apply to protein synthesis?

A. The process does not require energy
B. rRNA is required for proper binding of the mRNA message
C. tRNA molecules shuttle amino acids that are assembled into the polypeptide
D. The mRNA is synthesized from 5' → 3'

5. Which stage of cell division is the stage when chromosomes replicate?

A. prophase
B. interphase
C. anaphase
D. metaphase

6. Which of these statements is NOT correct about DNA replication?

A. DNA polymerase synthesizes and proofreads the DNA
B. RNA primers are necessary for the hybridization of the polymerase
C. Ligase relaxes positive supercoils that accumulate as the replication fork opens
D. DNA polymerase adds Okazaki fragments in a 5' → 3' direction

7. If a particular RNA sequence has a cytosine content of 25%, what is its adenine content?

A. 50%
B. cannot be determined
C. 12.5%
D. 25%

8. Which of the following statements is TRUE for the base composition of DNA?

A. In double-stranded DNA, the number of G bases equals the number of T bases
B. In double-stranded DNA, the number of A bases equals the number of T bases
C. In double-stranded DNA, the number of C bases equals the number of T bases
D. In every single strand, the number of A bases equals the number T bases

9. Which of the following is NOT a part of post-translational modification of protein?

A. addition of a 3' poly-A tail
B. phosphorylation
C. methylation
D. glycosylation

10. Which of the following is an exception to the principle of the *central dogma* of molecular biology?

A. yeast
B. retroviruses
C. bread mold
D. skin cells

11. Chromosome regions with few functional genes are:

A. heterochromatin
B. mid-repetitive sequences
C. euchromatin
D. chromatids

12. If a drug inhibits ribosomal RNA synthesis, which of these eukaryotic organelles would be most affected by this drug?

A. Golgi apparatus
B. lysosome
C. mitochondria
D. nucleolus

13. Which of these macromolecules would be repaired rather than degraded?

A. triglyceride
B. polynucleotide
C. polypeptide
D. polysaccharide

14. Which of the following is the correct sequence occurring during polypeptide synthesis?

A. DNA generates tRNA → tRNA anticodon binds to the mRNA codon in the cytoplasm → tRNA is carried by mRNA to the ribosomes, causing amino acids to join in a specific order
B. DNA generates mRNA → mRNA moves to the ribosome → tRNA anticodon binds to the mRNA codon, causing amino acids to join in their appropriate order
C. Specific RNA codons cause amino acids to line up in a specific order → tRNA anticodon attaches to mRNA codon → rRNA codon causes the protein to cleave into specific amino acids
D. DNA regenerates mRNA in the nucleus → mRNA moves to the cytoplasm and attaches to the tRNA anticodon → operon regulates the sequence of amino acids in the appropriate order

15. After the new DNA strands are synthesized, which enzyme is needed to complete the process of DNA replication?

A. ligase
B. primase
C. helicase
D. reverse transcriptase

16. What is the number of double-stranded DNA molecules in a single mouse chromosome immediately after the gametes are formed?

A. 0
B. 1
C. 2
D. 4

17. What distinguishes meiosis from mitosis?

I. Genetic recombination
II. Failure to synthesize DNA between successive cell divisions
III. Separation of homologous chromosomes into distinct cells

A. I only
B. II only
C. I and III only
D. II and III only

18. SRY gene encoding for the testis-determining factor is the master sex-determining gene that resides on:

A. pseudoautosomal region of the Y chromosome
B. short arm of the Y chromosome, but not in the pseudoautosomal region
C. X chromosome
D. pseudoautosomal region of the X chromosome

19. The number of chromosomes contained in a human primary spermatocyte is:

A. 23
B. 46
C. 92
D. 184

20. 47, XXY is a condition known as:

A. Turner syndrome
B. double Y syndrome
C. trisomy X syndrome
D. Klinefelter syndrome

21. Translation, transcription, and replication take place in which phase of the cell cycle:

A. G1
B. G2
C. metaphase
D. S

22. Which type of inheritance has the pattern where an affected male has all affected daughters but no affected sons?

A. X-linked recessive
B. Y-linked
C. X-linked dominant
D. Autosomal recessive

23. Which cell division process results in four genetically different daughter cells that contain one haploid set of chromosomes?

A. interphase
B. meiosis
C. cell division
D. mitosis

24. Which of the following is a type of genetic mutation?

I. insertion
II. frameshift
III. nonsense
IV. missense

A. I and II only
B. I, II and III only
C. II and IV only
D. I, II, III and IV

25. A DNA damage checkpoint arrests cells in:

A. M/G2 transition
B. S/G1 transition
C. G1/S transition
D. anaphase

26. If a segment of a double-stranded DNA has a low ratio of guanine-cytosine (G-C) pairs relative to adenine-thymine (A-T) pairs, it is reasonable to assume that this nucleotide segment:

A. requires less energy to separate the two DNA strands than a comparable segment with a high C-G ratio
B. requires more energy to separate the two DNA strands than a comparable segment with a high C-G ratio
C. contains more cytosine than guanine
D. contains more adenine than thymine

27. A failure in which stage of spermatogenesis produces nondisjunction resulting in a male having an XXY karyotype?

A. prophase I
B. metaphase
C. anaphase I
D. telophase

28. Which of the following stages is when human cells with a single unreplicated copy of the genome are formed?

A. mitosis
B. meiosis II
C. meiosis I
D. interphase

29. In transcription:

A. the mRNA contains the genetic information from DNA
B. uracil pairs with thymine
C. a double helix containing one parent strand and one daughter strand is produced
D. the mRNA produced is identical to the parent DNA

30. The two new DNA molecules formed in replication:

A. contain one parent and one daughter strand
B. both contain only the parent DNA strands
C. both contain only two new daughter DNA strands
D. are complementary to the original DNA

31. The main role of DNA is to provide instructions on how to build:

I. lipids
II. carbohydrates
III. proteins

A. I only
B. II only
C. III only
D. I and II only

32. Nucleic acids are polymers of [ ] monomers.

A. monosaccharide
B. fatty acid
C. DNA
D. nucleotide

33. During DNA transcription, a guanine base on the template strand codes for which base on the growing RNA strand?

A. guanine
B. thymine
C. adenine
D. cytosine

34. What intermolecular force connects strands of DNA in the double helix?

A. hydrogen bonds
B. ionic bonds
C. amide bonds
D. ester bonds

35. What is the term for the process of a DNA molecule synthesizing an identical molecule of DNA?

A. transcription
B. translation
C. duplication
D. replication

36. Translation is the process whereby:

A. protein is synthesized from DNA
B. protein is synthesized from mRNA
C. DNA is synthesized from DNA
D. DNA is synthesized from mRNA

37. Which of the following is the correct listing of DNA's constituents in the order of increasing size?

A. Nucleotide, codon, gene, nucleic acid
B. Nucleic acid, nucleotide, codon, gene
C. Nucleotide, codon, nucleic acid, gene
D. Gene, nucleic acid, nucleotide, codon

38. What is the process in which the DNA double helix unfolds, and each strand serves as a template for the synthesis of a new strand?

A. translation
B. replication
C. transcription
D. complementation

39. Which of the following illustrates the direction of flow for protein synthesis?

A. RNA → protein → DNA
B. DNA → protein → RNA
C. RNA → DNA → protein
D. DNA → RNA → protein

40. The three-base sequence in mRNA specifying the amino acid is called:

A. rRNA
B. an anticodon
C. a codon
D. tRNA

41. DNA is a(n):

A. peptide
B. protein
C. nucleic acid
D. enzyme

42. What is the term for a blotting method where proteins are transferred from a gel to membranes and probed by antibodies to specific proteins?

A. Eastern blotting
B. Western blotting
C. Northern blotting
D. Southern blotting

43. All of the following statements about PCR are correct, EXCEPT:

A. PCR can be used to obtain large quantities of a particular DNA sequence
B. PCR does not require knowledge of the terminal DNA sequences of the region to be amplified
C. PCR uses a DNA polymerase to synthesize DNA
D. PCR uses short synthetic oligonucleotide primers

44. Experiments designed by Avery, McLeod, and McCarty to identify the transforming principle were based on:

I. purifying each of the macromolecule types from a cell-free extract
II. removing each of the macromolecules from a cell, then testing its type
III. selectively destroying the different macromolecules in a cell-free extract

A. I only
B. II only
C. III only
D. I, II and III

45. The Hershey–Chase experiment:

A. proved that DNA replication is semiconservative
B. used ^{32}P to label protein
C. used ^{35}S to label DNA
D. supported the hypothesis that DNA is the transforming molecule

46. The loss of function of p53 protein results in:

A. blockage in the activation of the anaphase-promoting complex
B. increase of contact inhibition
C. elimination of the DNA damage checkpoint
D. activation of apoptosis

Notes for active learning

Heredity

1. Which of the following stages is when human cells with a single unreplicated copy of the genome are formed?

A. mitosis
B. meiosis II
C. meiosis I
D. interphase

2. During which phase of mitotic division do spindle fibers split the centromere and separate the sister chromatids?

A. anaphase
B. telophase
C. prophase
D. metaphase

3. In mice, short hair is dominant over long hair. What can be concluded if a short-haired individual is crossed with a long-haired individual and both long and short-haired offspring?

A. Short-haired individual is homozygous
B. Short-haired individual is heterozygous
C. Long-haired individual is homozygous
D. Long-haired individual is heterozygous

4. Several eye colors are characteristic of *Drosophila melanogaster*. Red eyes are dominant over sepia or white eyes. What percent of offspring of a sepia-eyed fly will have sepia-eyes if that fly mated with a red-eyed fly that was a cross of red-eyed and sepia-eyed parents?

A. 0%
B. 25%
C. 50%
D. 75%

5. All of the following are necessary conditions for the Hardy-Weinberg equilibrium, EXCEPT:

A. forward mutation rate equals backward mutation rate
B. random emigration and immigration
C. large gene pool
D. random mating

6. Mendel's crossing of spherical-seeded pea plants with wrinkled-seeded pea plants resulted in progeny that all had spherical seeds. This indicates that the wrinkled-seed trait is:

A. codominant
B. dominant
C. recessive
D. penetrance

7. The result of mitosis is the production of:

A. two (1N) cells identical to the parent cell
B. two (2N) cells identical to the parent cell
C. four (1N) cells identical to the parent cell
D. four (2N) cells identical to the parent cell

8. The degree of genetic linkage is often measured by the:

A. frequency of nonsense mutations
B. histone distribution
C. frequency of missense mutations
D. probability of crossing over

9. If two species with the AaBbCc genotype reproduce, what is the probability that their progeny have the AABBCC genotype?

A. 1/2
B. 1/4
C. 1/16
D. 1/64

10. What is the pattern of inheritance for a rare recessive allele?

A. Every affected person has an affected parent
B. Unaffected parents can produce children who are affected
C. Unaffected mothers have affected sons and daughters who are carriers
D. Every affected person produces an affected offspring

11. If tall height and brown eye color are dominant, what is the probability for a heterozygous tall, heterozygous brown-eyed mother and a homozygous tall, homozygous blue-eyed father to have a tall child with blue eyes? Note: the genes for eye color and height are unlinked.

A. 3/4
B. 1/8
C. 1/4
D. 1/2

12. Recombination frequencies:

A. arise from completely random genetic exchanges
B. are the same for *cis* and *trans* heterozygotes
C. decrease with distance
D. are the same for all genes

13. How many different gametes can be produced from the genotype *AaBbCc*, assuming independent assortment?

A. 4
B. 6
C. 8
D. 16

14. What is the pattern of inheritance for a rare dominant allele?

A. Every affected person has an affected parent
B. Unaffected parents can produce children who are affected
C. Unaffected mothers have affected sons and daughters who are carriers
D. Every affected person produces an affected offspring

15. People that are homozygous for a recessive autosomal mutation accumulate harmful amounts of lipids. Jane and her parents are not afflicted. However, Jane's sister accumulates lipids. What is the probability that Jane is heterozygous for the mutation?

A. 1/4
B. 1/3
C. 2/3
D. 1/2

16. What is the risk of having a child affected by a disease with autosomal dominant inheritance if the mother and father have one mutant gene for that disease?

A. 0%
B. 25%
C. 50%
D. 75%

17. Which of these statements is/are valid for an autosomal dominant inheritance?

I. A single allele of the mutant gene is needed to exhibit the phenotype
II. transmission to the son by the father is not observed
III. autosomal dominant traits do not skip generations

A. II only
B. I, II and III
C. I only
D. I and III only

18. All of these cell types contain the diploid (2N) number of chromosomes EXCEPT:

A. primary oocyte
B. spermatogonium
C. spermatid
D. zygote

19. The probability that all children in a four-children family will be males is:

A. 1/2
B. 1/4
C. 1/8
D. 1/16

20. Common red-green color blindness is an X-linked trait. When a woman whose father is colorblind has a son with a non-afflicted man, what is the probability that their son will be colorblind?

A. 0
B. 1/4
C. 1/2
D. 3/4

21. Since the gene responsible for color blindness is on the X chromosome, what is the chance that a son of a colorblind man and a woman-carrier will be colorblind?

A. 75%
B. 100%
C. 25%
D. 50%

22. A recessive allele may appear in a phenotype due to:

A. gain-of-function mutation
B. acquired dominance
C. senescence
D. the loss of heterozygosity

23. Mendel concluded that each pea has two units for each characteristic, and each gamete contains one unit. Mendel's "unit" for traits is now referred to as:

A. gene
B. genome
C. codon
D. DNA

24. Which cross must produce all green, smooth peas if green (G) is dominant over yellow (g) and smooth (S) is dominant over wrinkled (s)?

A. GgSs × GGSS
B. GgSS × ggSS
C. Ggss × GGSs
D. GgSs × GgSs

25. What is the probability that a cross between a true-breeding pea plant with a dominant trait and a true-breeding pea plant with a recessive trait will result in all F_1 progeny having the dominant trait?

A. 50%
B. 25%
C. 0%
D. 100%

26. Which of the following methods was NOT used by Mendel to study the genetics of the garden pea?

A. Maintenance of true-breeding lines
B. Cross-pollination
C. Microscopy
D. Production of hybrid plants

27. An unknown inheritance pattern has the following characteristics:

- 25% probability of having a homozygous unaffected child
- 25% probability of having a homozygous affected child
- 50% probability of having a heterozygous child

Which of these Mendel's inheritance patterns best matches the above observations?

A. autosomal recessive
B. autosomal dominant
C. X-linked recessive
D. X-linked dominant

28. Which of the following observations would support the theory of maternal inheritance for the spunky phenotype?

A. Spunky female × wild-type male → progeny all spunky
B. Wild-type female × spunky male → progeny all spunky
C. Wild-type female × spunky male → progeny 1/2 spunky, 1/2 wild-type
D. Spunky female × wild-type male → progeny 1/2 spunky, 1/2 wild-type

29. Hemophilia is a recessive X-linked trait. Knowing that females with Turner's syndrome have a high incidence of hemophilia, it can be concluded that these females have:

A. lost an X and gained a Y
B. lost an X
C. gained an X
D. gained a Y

30. Which of the following is a characteristic that makes an organism unsuitable for genetic studies?

A. Large number of chromosomes
B. Short generation time
C. Ease of cultivation
D. Ability to control crosses

Notes for active learning

Cardiovascular System

1. In a healthy individual, the highest blood pressure would most likely be in the:

A. superior vena cava
B. aorta
C. pulmonary capillaries
D. systemic capillaries

2. Oxygenated blood is pumped by:

A. left ventricle and left atrium
B. left ventricle and right atrium
C. left and right atria
D. left and right ventricles

3. The osmotic pressure at the arterial end of a capillary bed:

I. results in a net outflow of fluid
II. is less than the hydrostatic pressure
III. is greater than the hydrostatic pressure

A. I only
B. II only
C. I and II only
D. I and III only

4. All of these statements about the circulatory system are true, EXCEPT:

A. veins have a strong pulse
B. oxygenated blood is typically transported in arteries
C. mammals have a four-chambered heart
D. the thoracic duct returns lymphatic fluid to the circulatory system

5. Which of the following is the *correct* path of blood flow through the heart?

A. Superior/inferior vena cava → left atrium → left ventricle → pulmonary artery → lungs → → pulmonary vein → right atrium → right ventricle → aorta
B. Superior/inferior vena cava → right atrium → left ventricle → pulmonary artery → lungs → → pulmonary vein → left atrium → right ventricle → aorta
C. Superior/inferior vena cava → right atrium → right ventricle → pulmonary vein → lungs → → pulmonary artery → left atrium → left ventricle → aorta
D. Superior/inferior vena cava → right atrium → right ventricle → pulmonary artery → lungs → → pulmonary vein → left atrium → left ventricle → aorta

6. Which of the following is a protective role of blood?

A. maintenance of pH in body tissue
B. maintenance of body temperature
C. prevention of blood loss
D. maintenance of adequate fluid volume

7. Which blood component is involved in clot formation?

A. erythrocytes
B. macrophages
C. platelets
D. T cells

8. Which of the following is characteristic of a capillary?

I. hydrostatic pressure is higher at the arteriole end than at the venule end
II. osmotic pressure is higher in the blood plasma than in the interstitial fluid
III. hydrostatic pressure results from heart contractions

A. I only
B. I and II only
C. II and III only
D. I, II and III

9. All of the below statements about blood are correct, EXCEPT:

A. mature erythrocytes lack a nucleus
B. blood platelets are involved in the clotting process
C. erythrocytes develop in the adult spleen
D. leukocytes undergo phagocytosis of foreign matter

10. Which of the following is a normal blood flow pathway?

A. inferior vena cava to the left atrium
B. right ventricle to the aorta
C. pulmonary veins to the left ventricle
D. pulmonary veins to the left atrium

11. In an infant born with a congenital heart defect, which of the following most likely results from blood mixing between the right and left ventricles?

A. recurrent fever
B. hemoglobin deficiency
C. poor oxygenation of the tissues
D. low blood pressure

12. Which statement is accurate regarding blood plasma?

A. It is over 90% water
B. It contains about 20 dissolved components
C. It is the same as the serum but without the clotting proteins
D. The main protein component is hemoglobin

13. Which of these vessels has the highest partial pressure of oxygen in a healthy person?

A. aorta
B. coronary veins
C. superior vena cava
D. pulmonary arteries

14. Which of these structures would be reached LAST by a tracer substance injected into the superior vena cava?

A. tricuspid valve
B. pulmonary veins
C. left ventricle
D. right ventricle

15. In the mammalian heart, semi-lunar valves are found:

A. where the blood goes from the atria to ventricles
B. on the right side of the heart, only
C. where the pulmonary veins attach to the heart
D. where blood leaves via the aorta and pulmonary arteries

16. Which statement is correct about the atrioventricular (AV) node?

A. It regulates the contraction rhythm of cardiac cells
B. It delays the contraction of the heart ventricles
C. It is the parasympathetic ganglion located in the left atrium of the heart
D. It conducts action potentials from the vagus nerve to the heart

17. Stenosis is a condition whereby the valve's opening is narrowed and results in decreased blood flow through the valve. If diagnosed with stenosis of the mitral valve, where would a patient experience the highest blood pressure increase?

A. right atrium
B. aorta
C. left ventricle
D. left atrium

18. Which of the following would result from a significant loss of blood due to a hemorrhage?

A. No change in blood pressure but a slower heart rate
B. No change in blood pressure but a change in respiration
C. Lowering of blood pressure due to change in cardiac output
D. Rise in blood pressure due to change in cardiac output

19. What is the correct order for the cardiac conduction pathway?

A. Purkinje fibers → SA node→ AV node → bundle of His
B. SA node → bundle of His → Purkinje fibers → AV node
C. SA node → AV node → bundle of His → Purkinje fibers
D. AV node → SA node → bundle of His → Purkinje fibers

20. Which of these heart structures has blood with the highest O_2 content?

A. right atrium
B. left ventricle
C. pulmonary artery
D. thoracic duct

21. Which of these heart structures has blood with the lowest O_2 content?

A. thoracic duct
B. left ventricle
C. pulmonary artery
D. inferior vena cava

22. In which of the following vessels has the highest protein concentration?

A. proximal tubule
B. renal artery
C. afferent arteriole
D. vasa recta

23. Which blood cell is responsible for coagulation?

A. leukocyte
B. erythrocyte
C. lymphocyte
D. platelet

24. Which of the following is NOT correct about the hepatic portal system?

A. It branches from the inferior vena cava
B. It consists of a vein connecting two capillary beds
C. Its major vessels are the superior mesenteric, inferior mesenteric, and splenic veins
D. It carries nutrients to the liver for processing

25. Which of the following is valid according to Starling's capillary physiology hypothesis?

A. Filtration occurs where the hydrostatic pressure is less than the osmotic pressure
B. Ultrafiltrate is returned to the bloodstream by the lymphatic system
C. Hydrostatic pressure is lower at the arteriolar end
D. Osmotic pressure drives fluid out of the capillary

26. Which of the following is correct about the hemoglobin molecule of the red blood cell?

A. It does not bind CO_2
B. It has a higher O_2 affinity in adults compared with fetal hemoglobin
C. It does not bind CO
D. It exhibits positive cooperative binding for O_2

27. All of the following are functions of blood, EXCEPT:

A. transport of salts to maintain blood volume
B. transport of hormones to their target organs
C. transport of metabolic wastes from cells
D. delivery of O_2 to body cells

28. All of the following statements are true regarding blood, EXCEPT:

A. It carries body cells to injured areas for repair
B. Its pH usually is between 7.34 and 7.45
C. It contains albumins to regulate osmolarity
D. It is denser and more viscous than water

29. When inhaled, carbon monoxide (CO) can be lethal because it:

A. irritates the pleura
B. blocks electron transport within the cytochrome system
C. binds with a strong affinity to hemoglobin
D. is insoluble in the bloodstream

30. The site of O_2 absorption in the capillaries of the lungs is:

A. alveoli
B. pleura
C. bronchi
D. bronchioles

31. Which of the following statements is valid for cardiac muscle?

A. It is multi-nucleated
B. It is not striated
C. It is under involuntary control
D. It is innervated by the somatic motor nervous system

Notes for active learning

Immune System

1. A person lacking gamma globulins would have:

A. diabetes
B. hemophilia
C. severe allergies
D. low resistance to infection

2. Humoral immunity depends on the function of:

A. erythrocytes
B. cytotoxic T cells
C. immunoglobulins
D. albumin

3. Which cell type produces antibodies?

A. neurons
B. T cells
C. B cells
D. macrophages

4. Inflammatory responses may include which of the following?

A. clotting proteins migrating away from the site of infection
B. increased activity of phagocytes in an inflamed area
C. reduced permeability of blood vessels to conserve plasma
D. release of substances that decrease the blood supply to an inflamed area

5. Which innate immune response is exhibited by a patient initially exposed to a pathogen?

A. T killer cells
B. neutrophils
C. T helper cells
D. B lymphocytes

6. Which of the following would be impaired due to HIV infecting and killing helper $CD4^+$ T lymphocytes?

I. humoral immunity
II. cell-mediated immunity
III. non-specific immunity

A. I only
B. I and II only
C. I and III only
D. II and III only

7. What are antigens?

A. proteins in the blood that cause foreign blood cells to clump
B. proteins embedded in B cell membranes
C. proteins that consist of two light and two heavy polypeptide chains
D. foreign molecules that trigger the production of antibodies

8. Which cells are responsible for secreting virus-neutralizing antibodies?

A. T cells
B. plasma cells
C. macrophages
D. thymus cells

9. All of the following is a T cell type, EXCEPT:

A. memory
B. regulatory
C. cytotoxic
D. antigenic

10. B lymphocytes develop immunocompetence in:

A. lymph nodes
B. bone marrow
C. the spleen
D. the thymus

11. The inflammatory response is involved in all of the following processes, EXCEPT:

A. disposing of cellular debris and pathogens
B. setting the stage for repair processes
C. preventing the spread of the injurious agent to nearby tissue
D. replacing injured tissues with connective tissue

12. B cells are dormant in lymph nodes and other lymphoid tissues until activated by specific lymphocyte antigens. When activated, B cells produce:

A. cytotoxic T cells
B. lymphokines
C. antibodies
D. macrophages

13. All of the following are examples of innate immunity, EXCEPT:

A. phagocytotic cells
B. digestive enzymes and stomach acid
C. memory B cells
D. skin as a physical barrier to antigens

14. Which statement is true regarding interferons?

A. Interferon produced against one virus would not protect cells against another virus
B. They act by increasing the rate of cell division
C. They interfere with viral replication within cells
D. They are used in nasal sprays for the common cold

15. White blood cells at the scene of a wound:

A. produce the blood clot
B. secrete scar tissue fibers
C. stimulate the epidermal cells to divide
D. engulf microbes and cell debris

16. A patient with AB blood type is the *universal recipient* because this patient's blood:

A. contains neither antibody
B. contains both antibodies
C. is the most common blood type
D. is the least common blood type

17. A patient with O type blood is the *universal donor* because this patient's blood:

A. contains neither antigen
B. contains both antigens
C. is the least common
D. is the most common

18. Antibodies that function against a particular foreign substance are released by:

A. lymph nodes
B. natural killer cells
C. T lymphocytes
D. plasma cells

19. Which statement about lymphocytes is correct?

A. T cells are the precursors of B cells
B. T cells are the only form of lymphocytes in lymphoid tissue
C. The two main types are T cells and macrophages
D. B cells differentiate into plasma cells that secrete antibodies into the blood

20. What is the function of lymph nodes?

A. breaking down hemoglobin
B. filtering of lymph
C. facilitating the absorption of amino acids
D. increasing glucose concentrations in blood

21. All of the following are functions of the lymphatic system, EXCEPT:

A. regulation of erythropoiesis
B. destruction and removal of foreign particles
C. removal of proteins from interstitial spaces
D. absorption of lipids from the small intestine

22. Antibodies act by:

A. aiding in phagocytosis of antigens
B. binding to antigens via the variable portion
C. inhibiting stem cell production in the bone marrow
D. binding to plasma cells and marking them for destruction via phagocytosis

23. The correct sequence of events in phagocytosis is:

A. ingestion → adherence → chemotaxis → digestion → killing
B. chemotaxis → adherence → ingestion → digestion → killing
C. chemotaxis → ingestion → digestion → adherence → killing
D. adherence → digestion → killing → ingestion → chemotaxis

24. The reason for *swollen glands* in the neck of a patient with a cold is that:

A. blood pools in the neck to keep it warm

B. lymph nodes swell during infection as white blood cells proliferate within them

C. inflammation is initiated by the infection that causes fluid to be drained from the area

D. fever initiates a general expansion of the tissues of the head and neck

Notes for active learning

Notes for active learning

Central Nervous System

1. The central nervous system determines the strength of a stimulus by the:

A. amplitude of action potentials
B. frequency of action potentials
C. type of stimulus receptor
D. origin of the stimulus

2. The blood-brain barrier is effective against:

A. metabolic waste such as urea
B. nutrients such as glucose
C. alcohol
D. anesthetics

3. Which structure of the brain controls the breathing rate?

A. cerebrum
B. cerebellum
C. medulla oblongata
D. hypothalamus

4. The partial decussation (i.e., crossing) of axons in the optic chiasm results in:

A. both sides of the visual field being represented together in each hemisphere of the primary visual cortex
B. representation of the left side of the visual field remaining ipsilateral
C. the left side of the visual field is represented in the right hemisphere of the primary visual cortex
D. the left side of the visual field is represented in the left hemisphere of the primary visual cortex

5. Which is characteristic of an accident victim who sustained isolated damage to the cerebellum?

A. loss of muscular coordination
B. loss of speech
C. loss of voluntary muscle contraction
D. loss of sensation in the extremities

6. All these matches of brain structure to function are correct, EXCEPT:

A. reticular activating system – sensory processing
B. hypothalamus – appetite
C. medulla oblongata – basic emotional drives
D. cerebral cortex – higher intellectual function

7. Which is NOT a location where the white matter would be found?

A. pyramidal tracts
B. outer portion of the spinal cord
C. corpus callosum
D. cerebral cortex

8. Which function would be significantly impaired by destroying the cerebellum?

A. thermoregulation
B. coordinated movement
C. sense of smell
D. urine formation

9. Detection of sound involves air pressure waves that are converted into neural signals in the:

A. semicircular canals
B. cochlea
C. tympanic membrane
D. retina

10. The brain stem consists of:

A. midbrain
B. pons, medulla, midbrain, and cerebellum
C. pons, medulla, and midbrain
D. cerebrum, pons, and medulla

11. Which portion of the brain does nerve cells that control thermoregulation concentrate in?

A. medulla
B. hypothalamus
C. cerebellum
D. cerebrum

12. All the following are in normal cerebrospinal fluid, EXCEPT:

A. potassium
B. protein
C. glucose
D. red blood cells

13. Which area of the brain is where the primary auditory cortex is located?

A. temporal lobe
B. parietal lobe
C. prefrontal lobe
D. frontal lobe

14. Cell bodies of the sensory neurons of the spinal nerves are in the:

A. sympathetic ganglia
B. thalamus
C. ventral root ganglia of the spinal cord
D. dorsal root ganglia of the spinal cord

15. The white matter of the spinal cord contains:

A. myelinated nerve fibers
B. neither myelinated nor unmyelinated nerve fibers
C. myelinated and unmyelinated nerve fibers
D. unmyelinated nerve fibers

16. Which statement is true regarding Broca's area?

A. It is considered a motor speech area
B. It serves for the recognition of complex objects
C. It is in the right hemisphere
D. It serves for auditory acuity

17. Which structures contain hair cells that detect motion?

I. the skin II. the organ of Corti III. the semicircular canals

A. I only
B. III only
C. II and III only
D. I, II and III

18. The part of the cerebral cortex involved in cognition, personality, intellect, and recall is:

A. the limbic association area
B. combined primary somatosensory cortex and somatosensory association
C. the prefrontal cortex
D. the cortex posterior association area

19. Nerves that carry impulses towards the central nervous system are:

A. motor nerves
B. mixed nerves
C. afferent nerves
D. efferent nerves

20. Which is true regarding regeneration within the central nervous system?

A. It is prevented due to growth-inhibiting proteins of oligodendrocytes
B. It is promoted by growth inhibitors and glial scars
C. It is more successful than with the PNS
D. It is promoted by glial cells

Notes for active learning

Reproductive System

1. Which event is essential for the menstrual cycle?

I. adrenal medulla releases norepinephrine
II. FSH stimulates ovarian follicle development
III. progesterone stimulates the formation of the endometrial lining

A. II only
B. III only
C. I and II only
D. II and III only

2. What number of double-stranded DNA molecules are in a single mouse chromosome immediately after gametes form?

A. 0
B. 1
C. 2
D. 4

3. All the following about human gamete production is true, EXCEPT:

A. meiosis in females produces four egg cells
B. sperm develop in the seminiferous tubules within the testes
C. eggs develop in the ovarian follicles within the ovaries
D. FSH stimulates gamete production in males and females

4. Common red-green color blindness is an X-linked trait. When a woman whose father is colorblind has a son with a non-afflicted man, what is the probability that their son is colorblind?

A. 0
B. 1/4
C. 1/2
D. 3/4

5. At birth, a woman possesses a finite number of ova. In oogenesis, the meiotic division is arrested at which stage until she reaches menarche?

A. ovum
B. oogonium
C. primary oocytes
D. secondary oocytes

6. The likely gamete cell to be produced from meiosis in the seminiferous tubules is:

A. diploid 2° spermatocytes
B. haploid 1° spermatocytes
C. haploid spermatids
D. diploid spermatids

7. A human cell after the first meiotic division is:

A. 2N and 2 chromatids
B. 2N and 4 chromatids
C. 1N and 2 chromatids
D. 1N and 1 chromatid

8. All these statements regarding the menstrual cycle are true, EXCEPT:

A. Graafian follicle, under the influence of LH, undergoes ovulation
B. follicle secretes estrogen as it develops
C. corpus luteum develops from the remains of the post-ovulatory Graafian follicle
D. posterior pituitary secretes FSH and LH

9. Which is the initial site of spermatogenesis?

A. seminiferous tubules
B. seminal vesicles
C. vas deferens
D. epididymis

10. What distinguishes meiosis from mitosis?

I. Genetic recombination
II. Failure to synthesize DNA between successive cell divisions
III. Separation of homologous chromosomes into distinct cells

A. I only
B. II only
C. I and III only
D. II and III only

11. Oocytes within the primordial follicles of the ovary are arrested in:

A. interphase
B. prophase II of meiosis
C. prophase I of meiosis
D. prophase of mitosis

12. Which event occurs first?

A. formation of corpus luteum
B. rupture of the Graafian follicle
C. secretion of estrogen
D. release of progesterone

13. SRY gene encoding for the testis-determining factor is the master sex-determining gene that resides on:

A. pseudoautosomal region of the Y chromosome
B. short arm of the Y chromosome, but not in the pseudoautosomal region
C. X chromosome
D. pseudoautosomal region of the X chromosome

14. The number of chromosomes contained in the primary human spermatocyte is:

A. 23
B. 46
C. 92
D. 184

15. Which process does NOT contribute to genetic variation?

A. Random segregation of homologous chromosomes during meiosis
B. Random segregation of chromatids during mitosis
C. Recombination
D. Mutation

16. Which statement correctly describes the role of LH in the menstrual cycle?

A. stimulates the ovary to increase LH secretions
B. inhibits secretions of GnRH
C. stimulates the development of the endometrium for implantation of the zygote
D. induces the corpus luteum to secrete estrogen and progesterone

17. What likely causes the testes not fully to descend into the scrotum due to abnormal testicular development?

A. cortisol deficiency
B. testosterone deficiency
C. excess LH
D. excess estrogen

18. Testosterone is synthesized primarily by:

A. sperm cells
B. hypothalamus
C. Leydig cells
D. anterior pituitary gland

19. In human females, secondary oocytes do not complete meiosis II until:

A. menarche
B. menstruation
C. fertilization
D. menopause

20. 47, XXY is a condition known as:

A. Turner syndrome
B. double Y syndrome
C. trisomy X syndrome
D. Klinefelter syndrome

21. Which condition is the LEAST likely cause of male infertility?

A. acrosomal enzymes denaturation
B. immotility of cilia
C. abnormal mitochondria
D. testosterone deficiency

22. The surgical removal of the seminiferous tubules would likely cause:

A. sterility, because sperm would not be produced
B. sterility, because sperm would not be able to exit the body
C. reduced the volume of semen
D. enhanced fertilization potency of sperm

23. All these cell types contain the diploid (2N) number of chromosomes EXCEPT:

A. primary oocyte
B. spermatogonium
C. spermatid
D. zygote

24. The probability that all children in a four-children family will be males is:

A. 1/2
B. 1/4
C. 1/8
D. 1/16

25. Which results from scars in women's reproductive system with a high risk of infections (e.g., chlamydia)?

A. elevated levels of estrogen
B. decreased ovulation
C. infertility
D. reduced gamete production

26. The surgical removal of the seminal vesicles would likely cause:

A. sterility, because sperm would not be produced
B. sterility, because sperm would not be able to exit the body
C. reduced volume of semen
D. enhanced fertilization potency of sperm

27. Unequal cytoplasm division is characteristic of:

A. binary fission of bacteria
B. mitosis of a kidney cell
C. production of a sperm
D. production of an ovum

28. All the following are clinical manifestations of Kartagener's syndrome, resulting from defective dynein that causes paralysis of cilia and flagella, EXCEPT:

A. chronic respiratory disorders
B. cessation of ovulation
C. male infertility
D. ectopic pregnancy

29. The primary difference between estrous and menstrual cycles is that:

A. in the estrous cycle, the endometrium is shed and reabsorbed by the uterus, while in the menstrual cycle the shed endometrium is excreted
B. behavioral changes during estrous cycles are much less apparent than those during menstrual cycles
C. season and climate have less pronounced effects on the estrous cycle than they do on menstrual cycles
D. copulation commonly occurs across the estrous cycle, whereas in menstrual cycles, copulation only occurs during the period surrounding ovulation

30. Translation, transcription, and replication occur in which cell cycle phase?

A. G1
B. G2
C. metaphase
D. S

31. Progesterone is primarily secreted by the:

A. primary oocyte
B. hypothalamus
C. corpus luteum
D. anterior pituitary gland

32. All statements are true for a typical human gamete, EXCEPT it:

A. originates via meiosis from a somatic cell
B. contains genetic material that has undergone recombination
C. contains a haploid number of genes
D. always contains an X or Y chromosome

33. All statements are correct regarding cleavage in human embryos, EXCEPT:

A. blastomeres are genetically identical to the zygote
B. holoblastic cleavage occurs in one portion of the egg
C. morula is a solid mass of cells produced via cleavage of the zygote
D. size of the embryo remains constant throughout cleavage of the zygote

34. One function of estrogen is to:

A. induce the ruptured follicle to develop into the corpus luteum
B. stimulate testosterone synthesis in males
C. maintain female secondary sex characteristics
D. promote the development and release of the follicle

35. Which inheritance has a pattern where an affected male has all affected daughters but no affected sons?

A. X-linked recessive
B. Y-linked
C. X-linked dominant
D. Autosomal recessive

36. Increasing estrogen levels during the female menstrual cycle triggers which feedback mechanism?

A. positive, which stimulates LH secretion by the anterior pituitary
B. positive, which stimulates FSH secretion by the anterior pituitary
C. negative, which stimulates the uterine lining to be shed
D. negative, which inhibits progesterone secretion by the anterior pituitary

37. Which cell division results in four genetically different haploid daughters?

A. interphase
B. meiosis
C. cell division
D. mitosis

38. Crossing over occurs during:

A. anaphase I
B. telophase I
C. prophase I
D. metaphase I

39. The epididymis functions to:

A. synthesize and release testosterone
B. store sperm until they are released during ejaculation
C. initiate the menstrual cycle by secreting FSH and LH
D. provide a conduit for the ovum as it moves from the ovary into the uterus

40. A genetically important event of crossing over occurs during:

A. metaphase II
B. telophase I
C. anaphase I
D. prophase I

41. Which does NOT describe events occurring in prophase I of meiosis?

A. chromosomal migration
B. genetic recombination
C. formation of a chiasma
D. spindle apparatus formation

42. The difference between spermatogenesis and oogenesis is:

A. spermatogenesis produces 4 1N sperms, while oogenesis produces 1 egg cell and polar bodies
B. spermatogenesis produces gametes, while oogenesis does not produce gametes
C. oogenesis is a mitotic process, while spermatogenesis is a meiotic process
D. spermatogenesis is a mitotic process, while oogenesis is a meiotic process

43. What is the purpose of the cilia covering the inner linings of the Fallopian tubes?

A. Preventing polyspermy by immobilizing additional incoming sperm after fusion
B. Facilitating movement of the ovum towards the uterus
C. Removing particulate matter that becomes trapped in the mucus layer
D. Protecting the ovum from pH fluctuations

44. How many Barr bodies are in the white blood cells of a 48, XXYY individual?

A. 0
B. 1
C. 2
D. 3

45. Polar bodies are the products of:

A. meiosis in females
B. meiosis in males
C. mitosis in females
D. mitosis in males

46. Which endocrine structure initiates the production of testosterone?

A. adrenal medulla
B. hypothalamus
C. pancreas
D. anterior pituitary

47. Turner syndrome results from:

A. monosomy of X chromosome
B. absence of Y chromosome
C. presence of an extra Y chromosome
D. trisomy of X chromosome

48. Decreasing progesterone during the luteal phase of the menstrual cycle results from:

A. increased secretion of estrogen in the follicle followed by the menstruation phase
B. degeneration of the corpus luteum in the ovary
C. increased secretion of LH, which produces the luteal surge and onset of ovulation
D. thickening of the endometrial lining in preparation for implantation of the zygote

49. Vasectomy prevents the:

A. movement of sperm along the vas deferens
B. transmission of sexually transmitted disease
C. production of sperm via spermatogenesis
D. production of semen in the seminal vesicles

Fetal Development

1. Which of the following statements is true regarding respiratory exchange during fetal life?

A. Respiratory exchanges occur through the placenta
B. Since lungs develop later in gestation, a fetus does not need a mechanism for respiratory exchange
C. The respiratory exchange is made through the ductus arteriosus
D. The respiratory exchange is not necessary

2. When each cell in an early stage of embryonic development still has the ability to develop into a complete organism, it is known as:

A. gastrulation
B. blastulation
C. determinate cleavage
D. indeterminate cleavage

3. Mutations in *Drosophila*, transforming one body segment into another, effect:

A. maternal-effect genes
B. homeotic genes
C. execution genes
D. segmentation genes

4. Which structure originates from embryonic ectoderm?

A. connective tissue
B. hair
C. rib cartilage
D. epithelium of the digestive system

5. The placenta is a vital metabolic organ made from a contribution by the mother and fetus. The portion of the placenta contributed by the fetus is:

A. amnion
B. yolk sac
C. umbilicus
D. chorion

6. What occurs when an embryo lacks synthesis of human chorionic gonadotropin (hCG)?

A. The embryo does not support the maintenance of the corpus luteum
B. The embryo increases the production of progesterone
C. The embryo develops immunotolerance
D. The placenta forms prematurely

7. Which germ layer gives rise to smooth muscle?

A. ectoderm
B. epidermis
C. mesoderm
D. endoderm

8. In early embryogenesis, the most critical morphological change in the development of cellular layers is:

A. epigenesis
B. gastrulation
C. conversion of morula to blastula
D. acrosomal reaction

9. During the first eight weeks of development, all the following events occur, EXCEPT:

A. myelination of the spinal cord
B. presence of all body systems
C. formation of a functional cardiovascular system
D. beginning of ossification

10. Which structures derive from the ectoderm germ layers?

A. epidermis and neurons
B. heart, kidneys, and blood vessels
C. nails, epidermis, and blood vessels
D. epidermis and adrenal cortex

11. A diagram of the blastoderm that identifies regions from which specific adult structures are derived is a:

A. phylogenetic tree
B. pedigree diagram
C. fate map
D. linkage map

12. Which tissue is the precursor of long bones in the embryo?

A. hyaline cartilage
B. fibrocartilage
C. dense fibrous connective tissue
D. elastic cartilage

13. What changes are observed in cells as development proceeds?

A. Cytoskeletal elements involved in forming the mitotic spindle
B. Energy requirements of each cell
C. Genetic information that was duplicated with each round of cell division
D. The composition of polypeptides within the cytoplasm

14. During fertilization, what is the earliest event in the process?

A. sperm contacts the cortical granules around the egg
B. sperm nucleic acid enters the egg's cytoplasm to form a pronucleus
C. acrosome releases hydrolytic enzymes
D. sperm contacts the vitelline membrane around the egg

15. The embryonic ectoderm layer gives rise to the following, EXCEPT:

A. eyes
B. fingernails
C. integument
D. blood vessels

16. All the following are correct matches of a fetal structure with what it becomes at birth, EXCEPT:

A. ductus venosus—ligamentum venosum
B. umbilical arteries—medial umbilical ligament
C. foramen ovale—fossa ovalis
D. ductus arteriosus—ligamentum teres

17. All the following statements regarding gastrulation are true, EXCEPT:

A. the primitive gut that results from gastrulation is the archenteron
B. for amphibians, gastrulation is initiated at the gray crescent
C. after gastrulation, the embryo consists of two germ layers of endoderm and ectoderm
D. amphibian blastula cells migrate during gastrulation through an invagination known as the blastopore

18. A homeobox is a:

A. protein involved in the control of meiosis
B. transcriptional activator of other genes
C. sequence that encodes for a DNA binding motif
D. DNA binding site

19. After the gastrulation's completion, the embryo undergoes:

A. cleavage
B. blastulation
C. blastocoel formation
D. neurulation

20. Shortly after implantation:

A. trophoblast forms two distinct layers
B. embryo undergoes gastrulation within 3 days
C. maternal blood sinuses bathe the inner cell mass
D. myometrial cells cover and seal off the blastocyst

21. What is the anatomical connection between the placenta and embryo?

A. chorion
B. umbilical cord
C. endometrium
D. corpus luteum

22. Which statement correctly illustrates the principle of induction during vertebrate development?

A. Neural tube develops into the brain, the spinal cord, and the nervous system
B. Secretion of TSH stimulates the release of thyroxine hormone
C. Presence of a notochord beneath the ectoderm results in the formation of a neural tube
D. Neurons synapse with other neurons via neurotransmitters

23. The acrosomal reaction by the sperm is:

A. a transient voltage change across the vitelline envelope
B. the jelly coat blocking penetration by multiple sperms
C. the consumption of yolk protein
D. hydrolytic enzymes degrading the plasma membrane

24. Which primary germ layer gives rise to the cardiovascular system, bones, and skeletal muscles?

A. endoderm
B. blastula
C. mesoderm
D. ectoderm

25. The dorsal surface cells of the inner cell mass form:

A. notochord
B. primitive streak
C. one of the fetal membranes
D. embryonic disc

26. All the following occurs in a newborn immediately after birth, EXCEPT:

A. the infant completely stops producing fetal hemoglobin
B. resistance in the pulmonary arteries decreases
C. pressure in the left atrium increases
D. pressure in the inferior vena cava and the right atrium increase

27. Homeobox sequences are in:

A. introns
B. exons
C. 3'-untranslated regions
D. 5'-untranslated regions

28. After 30 hours of incubation, the ectoderm tissue within the gastrula differentiates into different specific tissues, which supports the conclusion that:

I. cells become endoderm or mesoderm
II. cells contain a different genome from their parental cells
III. gene expression is altered

A. III only
B. I and III only
C. I only
D. I and II only

29. What is the function of the yolk sac in humans?

A. forms into the placenta
B. secretes progesterone in the fetus
C. gives rise to blood cells and gamete-forming cells
D. stores embryonic wastes

30. The trophoblast is primarily responsible for forming:

A. placental tissue
B. lining of the endometrium
C. allantois
D. archenteron

31. Certain lesions to the mesodermal embryonic primary germ layer may stimulate the development of *spina bifida*, a congenital fissure in the lower vertebrae. Besides the spinal column, what structures would likely be affected by such lesions?

I. intestinal epithelium
II. skin and hair
III. blood vessels
IV. muscles

A. I and II only
B. II and III only
C. III and IV only
D. I, III and IV only

32. The homeotic genes encode for:

A. repressor proteins
B. transcriptional activator proteins
C. helicase proteins
D. single-strand binding proteins

33. Which of the following cell groups gives rise to muscles in a frog embryo?

A. neural tube
B. ectoderm
C. endoderm
D. mesoderm

34. Which of the following is NOT involved in the implantation of the blastocyst?

A. Settling of the blastocyst onto the prepared uterine lining
B. Adherence of the trophoblast cells to the uterine lining
C. Inner cell mass giving rise to the primitive streak
D. Endometrium proteolytic enzymes produced by the trophoblast cells

35. During labor, which hormone stimulates contractions of uterine smooth muscle?

A. oxytocin
B. prolactin
C. luteinizing hormone
D. hCG

36. In a chick embryo, specific ectoderm cells give rise to wing feathers, while others develop into thigh feathers or feet claws. The ectodermal cells that develop into wing feathers were transplanted to an area that develops into thigh feathers or feet claws. It was observed that the transplanted cells developed into claws. Which of the following best explains the experimental results?

A. ectoderm cells possess positional information
B. destiny of the cells was already determined
C. ectoderm cells can develop into any tissue
D. the underlying mesoderm induced cells

37. The *slow block* to polyspermy is due to:

A. transient voltage change across the membrane
B. jelly coat blocking sperm penetration
C. consumption of yolk protein
D. formation of the fertilization envelope

38. Polyspermy in humans results in:

A. mitotic insufficiency
B. interruption of meiosis
C. nonviable zygote
D. multiple births

39. Mesoderm gives rise to:

A. intestinal mucosa
B. nerves
C. heart
D. lung epithelium

40. Genomic imprinting is:

A. inactivation of a gene by interruption of its coding sequence
B. organization of molecules in the cytoplasm to provide positional information
C. DNA modification in gametogenesis that affects gene expression in the zygote
D. suppression of a mutant phenotype because of a mutation in a different gene

41. Comparing a developing frog embryo and an adult, cells of which organism have a greater translation rate?

A. Adult, because ribosomal production is more efficient in a mature organism
B. Adult, because a mature organism has more complex metabolic requirements
C. Embryo, because a developing organism requires more protein production than an adult
D. Embryo, because ribosomal production is not yet under regulatory control by DNA

42. Which structure is the first to form during fertilization in humans?

A. blastula
B. morula
C. zygote
D. ectoderm

43. It is impossible for sperm to be functional (i.e., able to fertilize the egg) until after:

I. they undergo capacitation
II. the tail disappears
III. they have been in the uterus for several days
IV. they become spermatids

A. I only
B. II and IV only
C. I and IV only
D. I and III only

44. Where is the likely deformity if a teratogen affects endoderm development shortly after gastrulation?

A. nervous system
B. lens of the eye
C. liver
D. skeleton

45. Mutations that cause cells to undergo developmental fates of other cell types are:

A. heterochronic mutations
B. loss-of-function mutants
C. homeotic mutations
D. execution mutations

46. Each of the following is true for the cells of an early gastrula's eye field, EXCEPT that they are:

A. terminally differentiated
B. competent
C. derived from the ectoderm layer
D. capable of becoming other ectoderm structures

47. What is the role of proteases and acrosin enzymes in reproduction?

A. They degrade the nucleus of the egg and allow the sperm to enter
B. They degrade the protective barriers around the egg and allow the sperm to penetrate
C. They direct the sperm to the egg through chemotaxis messengers
D. They neutralize the mucous secretions of the uterine mucosa

48. When does the human blastocyst implant in the uterine wall?

A. about a week past fertilization
B. at blastulation
C. a few hours past fertilization
D. at primary germ formation

49. Which primary layer develops into the retina of the eye?

I. endoderm II. ectoderm III. mesoderm

A. I only
B. II only
C. III only
D. II and III

50. What changes must occur in a newborn's cardiovascular system after the infant takes its first breath?

A. Ductus arteriosus constricts and is converted to the ligamentum arteriosum
B. The urinary system is activated at birth
C. Ductus venosus is severed from the umbilical cord, and visceral blood enters the vena cava
D. Foramen ovale between the atria of the fetal heart closes at the moment of birth

51. Early activation of the mammalian zygote nucleus may be necessary because:

A. most developmental decisions occur under the influence of the paternal genome
B. mammalian oocytes are too small to store molecules needed to support the cleavage divisions
C. in gametogenesis, specific genes undergo imprinting
D. mammals do not have maternal-effect genes

52. In reptiles, the aquatic environment necessary for the embryonic development of amphibians is replaced by:

A. amniotic fluid
B. humid atmospheric conditions
C. shells which prevent the escape of gas
D. intrauterine development

53. What would be expected to form from a portion of cells destined to become the heart if it was excised from an early gastrula and placed in a culture medium?

A. undifferentiated mesoderm
B. undifferentiated ectoderm
C. differentiated endoderm
D. undifferentiated endoderm

54. When can an ultrasound determine the sex of the fetus during gestation?

A. at the midpoint of the first trimester
B. about 18 weeks after fertilization
C. at the end of the first trimester
D. at the midpoint of the second trimester

55. How long is the egg viable and able to be fertilized after ovulation?

A. 36-72 hours
B. a full week
C. 12-24 hours
D. 24-36 hours

56. Which stage of embryonic development has a hollow ball of cells surrounding a fluid-filled center?

A. 3-layered gastrula
B. blastula
C. morula
D. zygote

57. Which statement describes the acrosome of a sperm cell?

A. It contains the nucleic acid
B. It contains hydrolytic enzymes which are released when the sperm encounters the jelly coat of the egg
C. It fuses with the egg's cortical granules
D. It functions to prevent polyspermy

58. Which statement is correct about fertilization?

A. Most sperm cells are protected and remain viable once inside the uterus
B. If estrogen is present, the pathway through the cervical opening is blocked from sperm entry
C. The vagina's acidic environment destroys millions of sperm cells
D. Spermatozoa remain viable for about 72 hours in the female reproductive tract

59. All the following statements are correct regarding cleavage in human embryos, EXCEPT:

A. blastomeres are genetically identical to the zygote
B. holoblastic cleavage occurs in one portion of the egg
C. morula is a solid mass of cells produced via cleavage of the zygote
D. size of the embryo remains constant throughout the initial cleavage of the zygote

Notes for active learning

Florida Biology EOC

Answer Keys
and
Detailed Explanations

Answer Keys

Chapter 1. Ecosystems, Biomes, Energy Flow and Nutrient Cycles

1: A	11: D	21: C	31: B	41: B	51: C
2: B	12: D	22: D	32: C	42: C	52: D
3: A	13: A	23: C	33: B	43: D	53: D
4: C	14: B	24: C	34: A	44: A	54: A
5: A	15: A	25: B	35: D	45: C	55: B
6: D	16: A	26: B	36: C	46: B	56: B
7: D	17: A	27: B	37: C	47: C	57: D
8: C	18: B	28: A	38: B	48: A	58: D
9: C	19: C	29: D	39: A	49: D	59: A
10: B	20: D	30: A	40: B	50: A	60: D

Chapter 2. Populations, Communities and Conservation Biology

1: D	11: A	21: D	31: B	41: A	51: D	61: C
2: A	12: B	22: A	32: A	42: B	52: A	62: D
3: B	13: C	23: D	33: A	43: D	53: B	63: B
4: D	14: D	24: A	34: C	44: A	54: A	64: C
5: C	15: A	25: B	35: C	45: A	55: C	65: C
6: C	16: C	26: C	36: A	46: B	56: C	66: D
7: A	17: A	27: D	37: D	47: D	57: B	67: A
8: B	18: C	28: C	38: D	48: C	58: D	68: D
9: C	19: B	29: D	39: A	49: C	59: C	69: B
10: A	20: D	30: B	40: B	50: A	60: C	70: A
						71: D
						72: A

Chapter 3. Classification and Diversity

1: D	11: B	21: B	31: D	41: B	51: C	61: A
2: A	12: A	22: C	32: D	42: D	52: D	62: C
3: B	13: A	23: A	33: C	43: B	53: D	63: B
4: A	14: D	24: C	34: C	44: A	54: B	64: B
5: C	15: B	25: A	35: B	45: C	55: D	65: A
6: B	16: D	26: B	36: A	46: A	56: A	66: D
7: B	17: B	27: B	37: B	47: D	57: D	67: C
8: B	18: C	28: A	38: B	48: A	58: B	
9: B	19: A	29: D	39: C	49: C	59: B	
10: D	20: C	30: B	40: A	50: A	60: C	

Chapter 4. Evolution and Natural Selection

1: D	11: C	21: C	31: B	41: A	51: C
2: A	12: D	22: D	32: C	42: A	52: B
3: D	13: D	23: C	33: C	43: D	53: D
4: A	14: A	24: C	34: B	44: C	
5: D	15: B	25: C	35: A	45: B	
6: C	16: A	26: B	36: D	46: D	
7: D	17: D	27: B	37: A	47: D	
8: D	18: D	28: D	38: C	48: A	
9: A	19: C	29: B	39: D	49: D	
10: B	20: A	30: B	40: D	50: A	

Chapter 5. Animal Behavior and Evolution

1: A	11: A	21: A	31: A
2: A	12: D	22: A	32: C
3: B	13: C	23: D	33: A
4: C	14: B	24: B	
5: A	15: C	25: C	
6: D	16: A	26: D	
7: B	17: D	27: A	
8: D	18: C	28: C	
9: C	19: C	29: A	
10: A	20: B	30: B	

Chapter 6. Cell – the Basic Unit of Life

1: D	11: D	21: B	31: D	41: B
2: A	12: A	22: B	32: D	42: D
3: C	13: B	23: B	33: B	43: A
4: B	14: A	24: C	34: D	44: B
5: C	15: C	25: C	35: C	45: A
6: B	16: B	26: D	36: A	46: D
7: C	17: B	27: B	37: A	
8: B	18: A	28: D	38: C	
9: C	19: C	29: C	39: A	
10: C	20: D	30: D	40: A	

Chapter 7. Enzymes and Cellular Metabolism

1: B	11: C	21: C
2: C	12: C	22: C
3: D	13: A	23: B
4: B	14: D	24: C
5: D	15: B	25: D
6: A	16: A	26: D
7: A	17: B	27: C
8: C	18: C	28: B
9: A	19: A	
10: B	20: C	

Chapter 8. Microbiology

1: B	11: A	21: A	31: C	41: C
2: B	12: B	22: D	32: A	42: B
3: D	13: A	23: D	33: D	43: C
4: D	14: C	24: B	34: B	44: A
5: B	15: A	25: C	35: D	45: C
6: C	16: D	26: D	36: B	46: A
7: D	17: B	27: C	37: B	47: D
8: A	18: C	28: B	38: B	48: C
9: B	19: A	29: A	39: C	
10: B	20: C	30: C	40: C	

Chapter 9. Plants: Structure, Function and Reproduction

1: D	11: C	21: A	31: D	41: D	51: C	61: A
2: A	12: B	22: B	32: D	42: D	52: A	62: B
3: B	13: D	23: D	33: B	43: A	53: D	63: A
4: A	14: D	24: D	34: A	44: C	54: B	64: C
5: D	15: C	25: C	35: D	45: D	55: D	65: D
6: C	16: A	26: D	36: C	46: C	56: B	66: D
7: D	17: D	27: A	37: A	47: B	57: A	67: D
8: C	18: D	28: C	38: D	48: A	58: D	68: B
9: B	19: A	29: B	39: C	49: D	59: C	69: D
10: D	20: B	30: D	40: B	50: B	60: D	

Chapter 10. Photosynthesis

1: D	11: C	21: A	31: C	41: A	51: A
2: D	12: C	22: B	32: A	42: B	52: B
3: A	13: B	23: C	33: C	43: B	53: D
4: B	14: B	24: C	34: C	44: C	54: C
5: C	15: A	25: B	35: A	45: D	55: B
6: D	16: A	26: B	36: D	46: A	56: A
7: A	17: B	27: D	37: C	47: D	57: C
8: B	18: C	28: D	38: B	48: A	
9: A	19: D	29: C	39: A	49: A	
10: D	20: A	30: A	40: B	50: C	

Chapter 11. Macromolecules in a Biological System

1: A	21: B	41: D	61: D	81: D
2: B	22: B	42: A	62: B	82: B
3: A	23: D	43: C	63: D	83: D
4: A	24: C	44: D	64: B	84: C
5: B	25: C	45: D	65: A	85: D
6: C	26: D	46: C	66: D	86: A
7: B	27: D	47: B	67: D	87: B
8: B	28: B	48: A	68: C	88: A
9: C	29: B	49: A	69: D	89: B
10: C	30: D	50: C	70: D	90: C
11: D	31: C	51: D	71: A	91: C
12: D	32: B	52: D	72: D	92: C
13: B	33: A	53: D	73: A	93: C
14: C	34: B	54: C	74: D	94: C
15: B	35: C	55: D	75: B	95: D
16: B	36: A	56: C	76: B	96: A
17: B	37: B	57: A	77: C	97: C
18: D	38: D	58: B	78: A	98: A
19: D	39: D	59: A	79: A	99: D
20: D	40: C	60: A	80: B	

Chapter 12. Chromosomes, Genes and DNA

1: D	11: A	21: D	31: C	41: C
2: C	12: D	22: C	32: D	42: B
3: D	13: B	23: B	33: D	43: B
4: A	14: B	24: D	34: A	44: C
5: B	15: B	25: C	35: D	45: D
6: C	16: B	26: A	36: B	46: C
7: B	17: D	27: C	37: A	
8: B	18: B	28: B	38: B	
9: A	19: B	29: A	39: D	
10: B	20: D	30: A	40: C	

Chapter 13. Heredity

1: B	11: D	21: D
2: A	12: A	22: D
3: B	13: C	23: A
4: C	14: A	24: A
5: B	15: C	25: D
6: C	16: D	26: C
7: B	17: D	27: A
8: D	18: C	28: A
9: D	19: D	29: B
10: B	20: C	30: A

Chapter 14. Cardiovascular System

1: B	11: C	21: C	31: C
2: A	12: A	22: C	
3: C	13: A	23: D	
4: A	14: C	24: A	
5: D	15: D	25: B	
6: C	16: B	26: D	
7: C	17: D	27: A	
8: D	18: C	28: A	
9: C	19: C	29: C	
10: D	20: B	30: A	

Chapter 15. Immune System

1: D	11: D	21: A
2: C	12: C	22: B
3: C	13: C	23: B
4: B	14: C	24: B
5: B	15: D	
6: B	16: A	
7: D	17: A	
8: B	18: D	
9: D	19: D	
10: B	20: B	

Chapter 16. Central Nervous System

1: B	11: B
2: A	12: D
3: C	13: A
4: C	14: D
5: A	15: A
6: C	16: A
7: D	17: C
8: B	18: C
9: B	19: C
10: C	20: A

Chapter 17. Reproductive System

1: D	11: C	21: B	31: C	41: A
2: B	12: C	22: A	32: A	42: A
3: A	13: B	23: C	33: B	43: B
4: C	14: B	24: D	34: C	44: B
5: C	15: B	25: C	35: C	45: A
6: C	16: D	26: C	36: A	46: B
7: C	17: B	27: D	37: B	47: A
8: D	18: C	28: B	38: C	48: B
9: A	19: C	29: A	39: B	49: A
10: D	20: D	30: D	40: D	

Chapter 18. Fetal Development

1: A	11: C	21: B	31: C	41: C	51: C
2: D	12: A	22: C	32: B	42: C	52: A
3: B	13: D	23: D	33: D	43: A	53: A
4: B	14: C	24: C	34: C	44: C	54: C
5: D	15: D	25: B	35: A	45: C	55: C
6: A	16: D	26: D	36: D	46: A	56: B
7: C	17: C	27: B	37: D	47: B	57: B
8: B	18: C	28: A	38: C	48: A	58: C
9: A	19: D	29: C	39: C	49: B	59: B
10: A	20: A	30: A	40: C	50: A	

Ecosystems, Biomes, Energy Flow and Nutrient Cycles – Detailed Explanations

1. A is correct.

Ecosystem is a community of organisms (e.g., plants, animals, microbes) with nonliving (i.e., *abiotic*) components of their environment (e.g., air, water, minerals). It is the network of interactions among organisms and between organisms and their environment.

Nutrient cycles and *energy flows* link *biotic* (i.e., living) and *abiotic* (i.e., nonliving) components.

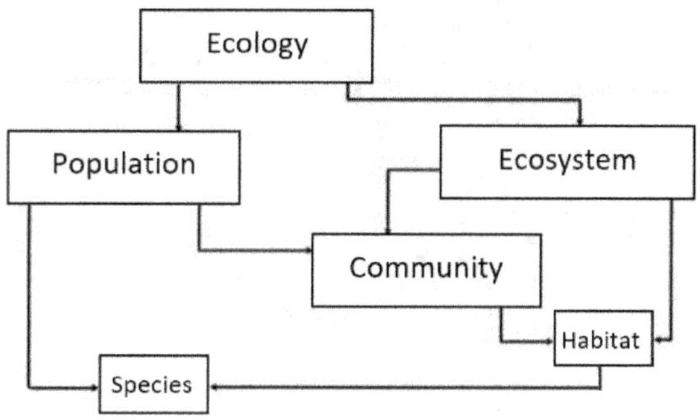

Relationship between organisms and their environment. Population, ecosystem, community, habitat, and species relationship are highlighted.

Ecology is the study of relationships between living organisms and between their environment.

Population is a group of organisms of the same species living in the same area.

Community is a group of populations living and interacting in an area.

Habitat is the environment in which a species lives.

Species are a group of organisms that can breed and produce fertile offspring.

2. B is correct.

Weather is the immediate conditions (e.g., hourly, daily, weekly), while *climate* is the average conditions.

3. A is correct.

Symbiosis is the close and often long-term interaction between different species.

Historically, some biologists proposed symbiosis as *persistent mutualisms* (i.e., all organisms benefit).

Biologists and ecologists now propose symbiosis as *persistent biological interactions*; mutualistic (+/+), commensalism (+/0), or parasitic (+/–).

4. C is correct.

Climate zones are distinct in an east-west direction around Earth and are classified by climate parameters.

Climate zones are primarily determined by *variations in temperature* related to *latitude* and *altitude.*

Angle of the sun's rays contributes to differences in climate zones.

5. A is correct.

Greenhouse effect is a natural phenomenon that maintains Earth's temperature range.

Greenhouse gases on Earth are water vapor (36–70%), CO_2 (9–26%), methane (4–9%), and ozone (3–7%).

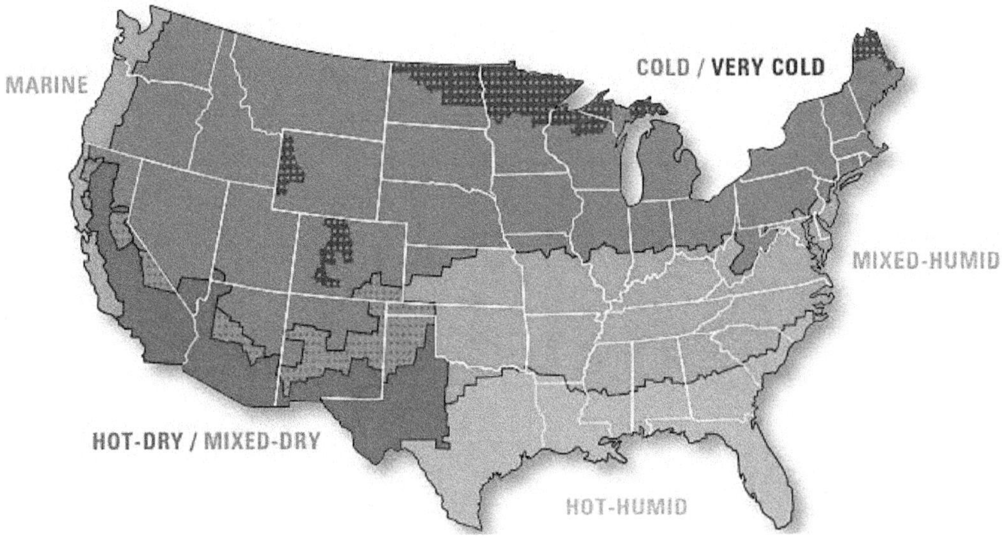

U.S. climate zones

6. D is correct.

Convection is movement caused by hotter (i.e., less dense) air rising and colder (i.e., denser) air sinking under the influence of gravity, consequently resulting in heat transfer.

Global wind patterns in the atmosphere are created by *convection* (i.e., air movement based on density).

When *warm air rises* and cools in a low-pressure zone, it may not hold the water it contains as vapor. Some water vapor condenses to form clouds or precipitation.

When *cool air descends*, it warms.

7. D is correct.

Wetlands ecosystems have saturated ground surfaces either permanently or temporarily.

8. C is correct.

Niche describes when each species has separate, unique physical and environmental conditions.

Biological aspects of an organism's niche are *biotic factors* (i.e., living) required for survival.

Abiotic factors (i.e., nonliving) include soil, sunlight, water, and minerals.

Ecological niche is how organisms (or populations) respond to the distribution of resources and competitors (e.g., growing with abundant resources while predators are scarce) and how it alters those factors (e.g., limiting access by other organisms, being a food source for predators and a consumer of prey).

9. C is correct.

Niche is when each species has separate, unique physical and environmental conditions.

Niche is the range of *physical and biological conditions* in which an organism lives and how it obtains resources for survival and reproduction.

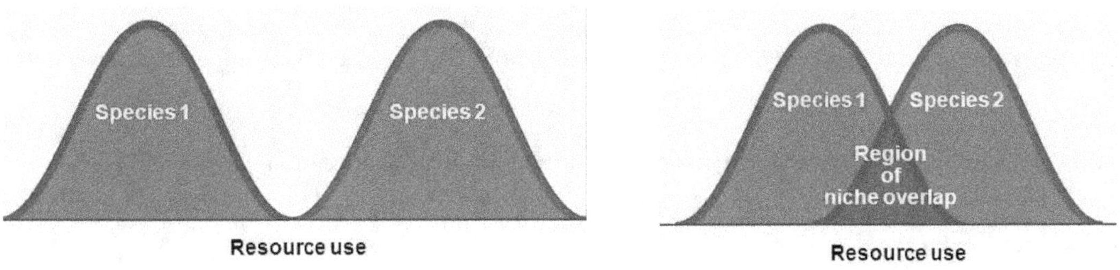

Resource partitioning results in competition for scarce resources when niche regions overlap

10. B is correct.

Competitive exclusion principle (*Gause's law of competitive exclusion* or *Gause's law*) states that two species competing for the same resources cannot coexist when ecological factors are constant.

Domination occurs when one species has even a slight advantage. One competitor overcomes another, leading to extinction or behavioral shift toward a different ecological niche.

11. D is correct.

Primary producers (i.e., *autotrophs*) produce energy from inorganic compounds in an ecosystem. Primary producers often use *photosynthesis* (e.g., plants, cyanobacteria).

Autotrophs (e.g., plants) support the ecosystem and feed heterotrophs.

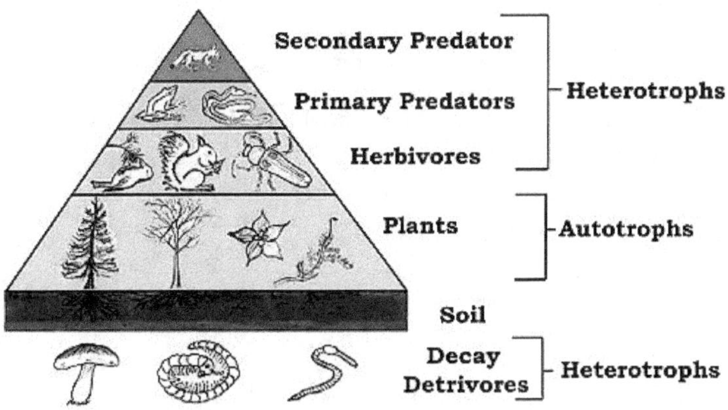

Trophic levels as positions on a food web. Autotrophs support the ecosystem and feed heterotrophs.

12. D is correct.

Primary producers (i.e., *autotrophs*) produce energy from inorganic compounds in an ecosystem. Primary producers often use *photosynthesis* (e.g., plants, cyanobacteria). Plants are primary producers.

Autotrophs support the ecosystem and feed heterotrophs. Autotrophs create organic compounds using energy from the sun or inorganic compounds. They are divided into photoautotrophs and chemoautotrophs.

Photoautotrophs (e.g., algae, plants, cyanobacteria) use photosynthesis, converting solar energy into organic compounds.

Chemoautotrophs are bacteria that oxidize inorganic compounds such as ammonia, nitrite, and sulfide to generate organic compounds. They are rare and are typically found in caves, hydrothermal ocean vents, and other environments lacking light.

Archaea (single-cell prokaryotes) may produce biomass from oxidizing inorganic compounds (i.e., chemoautotrophs). For example, processes in *deep ocean hydrothermal vents*.

Decomposers are fungi and other organisms that produce biomass from oxidizing organic materials. They absorb and metabolize nutrients (i.e., saprotrophic nutrition from dead or decaying material).

13. A is correct.

Open ocean waters are separated from cold, nutrient-rich interior water by *density differences* restricting water mixing, reducing nutrient supply, which becomes the limiting factor for productivity.

Open oceans have negligible nutrients from land and little upwelling to supply nutrients from deep oceans.

14. B is correct.

Habitat is an ecological area inhabited by a species. It is the natural environment in which an organism lives or the physical environment encompassing a population.

Niche describes when each species has separate, unique physical and environmental conditions.

Niche is the range of *physical and biological conditions* in which an organism lives and how it obtains resources for survival and reproduction.

Niche breadth is the range of habitat by an organism.

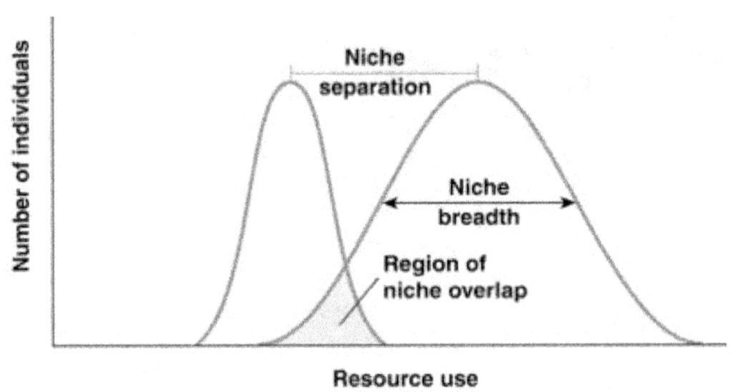

Niches may overlap and introduce competition for scarce resources

15. A is correct.

Food web is a graphical model depicting linked *food chains by* feeding relationships in an ecosystem.

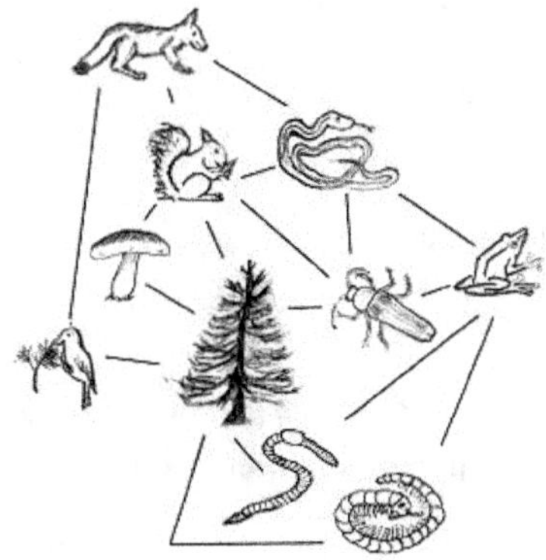

Food web and energy flow relationship examples

Food chain is a *linear succession* whereby another eats each species.

16. A is correct.

Mutualism is when two organisms of different species exist in a relationship, each benefit.

B: *parasitism* is a non-mutual symbiotic relationship between species, where one species (i.e., the parasite) benefits at the expense of the other (i.e., the host).

C: *commensalism is* when two organisms of different species exist in a relationship, each benefit.

D: *omnivorism* is a human diet consuming meat, eggs, dairy, and produce traced to an organic farm.

17. A is correct.

Predation is a biological interaction when a predator (i.e., a hunting organism) feeds on prey.

A predator may (or may not kill) its prey before feeding, but the act of predation most often results in the prey's death and the eventual absorption of the prey's tissue through consumption.

Given the alternative, *predators* (e.g., lions, tigers, wolves) may engage in *scavenging* feeding.

Mutualism (+/+) is a symbiosis of *persistent biological interactions.*

18. B is correct.

Biomass is the mass (i.e., a body of matter without defined shape) of living organisms in an area (or ecosystem). Biomass can include microorganisms, plants, or animals.

The mass can be expressed as the average mass per unit area or as the total mass in the community. How biomass is measured depends on why it is being measured.

Species biomass is the mass of one or more species.

Community biomass is the mass of the species in the community.

Trophic levels show a succession of the flow of *food energy* and *feeding relationships*. It is a food chain or ecological pyramid occupied by groups with similar feeding modes.

Trophic level 1	Producers
Trophic level 2	Primary consumers
Trophic level 3	Secondary consumers
Trophic level 4	Tertiary consumers

Food chain is the hierarchy in which organisms in an ecosystem are grouped into trophic (nutritional) levels.

Ecological pyramid represents the biomass or the energy flow in an ecosystem.

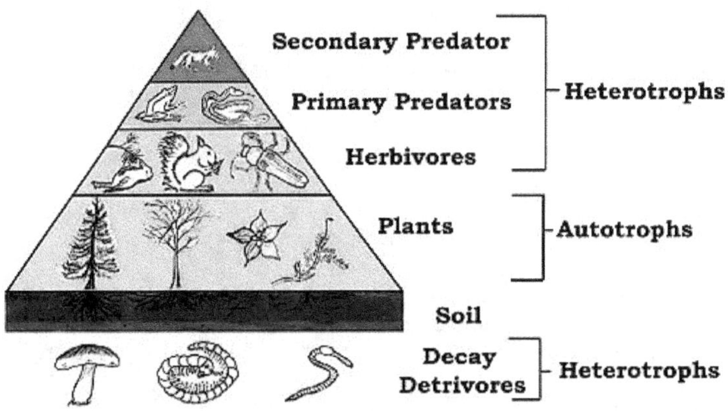

Trophic levels as positions on a food web. Autotrophs support the ecosystem and feed heterotrophs.

19. C is correct.

Herbivory is an animal anatomically and physiologically adapted to eating plants. As a result of their plant diet, herbivorous animals typically have mouthparts adapted to rasping or grinding.

For example, foliage or marine algae is the main component of herbivores.

20. D is correct.

Parasitism (+/–) is a non-mutual symbiotic relationship between species, where one species (i.e., the parasite) benefits at the expense of the other (i.e., the host).

A: *synnecrosis* is a rare symbiosis when the interaction between species is detrimental to both. It is short-lived because the interaction eventually causes death. Evolution selects against synnecrosis and is uncommon.

B: *predation* is a biological interaction when a predator (i.e., hunting organism) feeds on prey.

Predators may not kill their prey before feeding, but predation often results in the prey's death and eventual absorption of their tissue through consumption. The remains of prey may become a food source for scavengers.

Scavengers feed on dead organisms that they did not kill.

Organisms (e.g., lions, tigers, wolves) considered predators may engage in scavenging feeding behavior when given the alternative.

C: *mutualism* (+/+) is when two organisms of different species exist in a relationship, each benefit.

21. C is correct.

Omnivores consume meat, eggs, dairy, and produce.

Autotrophs support the ecosystem and feed heterotrophs. Autotrophs create organic compounds using energy from the sun or inorganic compounds. They divide into photoautotrophs and chemoautotrophs.

Photoautotrophs (e.g., algae, plants, cyanobacteria) use photosynthesis to convert solar energy into organic compounds.

Chemotrophs are organisms that obtain energy by the oxidation of electron donors in their environments. These molecules can be organic or inorganic.

Chemotrophs can be either *autotrophic* or *heterotrophic.*

Herbivores are animals anatomically and physiologically adapted to eating plants. As a result of their plant diet, herbivorous animals typically have mouthparts adapted to rasping or grinding.

22. D is correct.

Ecosystems include food production, solar energy (photosynthesis), and O_2 production (e.g., photosynthesis). They are the network of interactions among organisms and between organisms and their environment.

23. C is correct.

Autotrophs support the ecosystem and feed heterotrophs. Autotrophs create organic compounds using energy from the sun or inorganic compounds. They divide into *photoautotrophs* and *chemoautotrophs*.

Photoautotrophs (e.g., algae, plants, cyanobacteria) use photosynthesis, converting solar energy into organic compounds.

Heterotrophs obtain nutrients and energy by consuming organic substances.

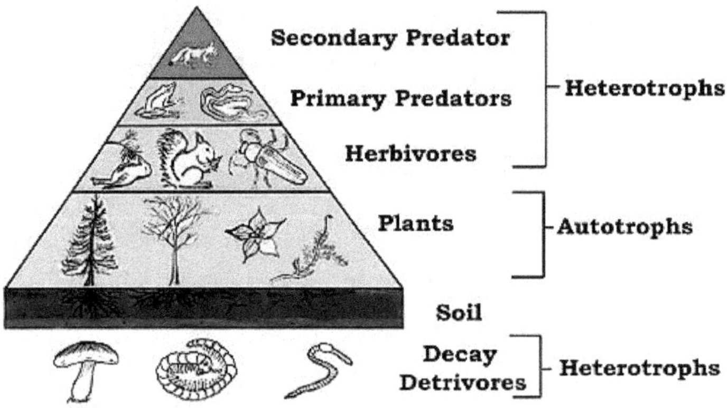

Trophic levels as positions on a food web. Autotrophs support the ecosystem and feed heterotrophs.

24. C is correct.

Primary succession is one of two biological and ecological succession of plant life, occurring in an environment devoid of vegetation and usually lacking soil (e.g., lava flow). It is the gradual growth of an ecosystem over a longer period.

Secondary succession occurs on a substrate that supports vegetation before an ecological disturbance from less cataclysmic events (e.g., floods, hurricanes, tornadoes), which destroy plant life.

25. B is correct.

Heterotrophs obtain nutrients and energy by consuming organic substances.

Autotrophs are *producers*.

Heterotrophs are *consumers* and rely on producers.

26. B is correct.

Primary succession is one of two biological and ecological succession of plant life, occurring in an environment devoid of vegetation and usually lacking soil (e.g., lava flow).

It is the gradual growth of an ecosystem over a longer period.

Secondary succession occurs on a substrate that supports vegetation before an ecological disturbance from less cataclysmic events (e.g., floods, hurricanes, tornadoes), which destroy plant life.

27. B is correct.

The snail is a typical detritivore.

Detritivores are heterotrophs that obtain nutrition from decomposing (waste or detritus) plants, dead organisms, and feces and contribute to decomposition and the nutrient cycles. Detritivores are not to be confused with other decomposers such as bacteria, fungi, and protists, which cannot ingest minute lumps of matter.

Decomposers absorb and metabolize nutrients (i.e., saprotrophic nutrition from dead or decaying material).

28. A is correct.

Matter recycles through the biosphere because biological systems do not deplete it but transform it.

29. D is correct.

Clearing and farming are physical disturbances to the ecosystem.

Climax community is a biological community of plants, animals, and fungi that has reached a *steady-state* through ecological succession (i.e., vegetation in an area over time).

Equilibrium occurs because the climax community has species best adapted to average conditions in that area.

30. A is correct.

Ecosystems include food production, solar energy (photosynthesis), and O_2 production (e.g., photosynthesis). They are the network of interactions among organisms and between organisms and their environment.

Biomes are geographically and climatically defined as contiguous with similar conditions, such as communities of plants, animals, and soil organisms.

31. B is correct.

Tundra is a biome where low temperatures and short growing seasons hinder tree growth. The vegetation includes dwarf shrubs, grasses, mosses, and lichens.

Desert is a barren area of land where little precipitation occurs, and consequently, living conditions are hostile to plant and animal life.

32. C is correct.

The bird is the carnivore, while the insect is the consumer (heterotroph).

33. B is correct.

Tundra is a biome where low temperatures and short growing seasons hinder tree growth. The vegetation includes dwarf shrubs, grasses, mosses, and lichens.

Desert is a barren area of land where little precipitation occurs, and consequently, living conditions are hostile to plant and animal life.

Boreal forest (or *taiga*) is a biome characterized by coniferous (i.e., softwood) forests mainly consisting of pines, spruces, and larches and is the world's largest terrestrial biome.

34. A is correct.

Decomposers are fungi and other organisms that produce biomass from oxidizing organic materials. They absorb and metabolize nutrients (i.e., saprotrophic nutrition from dead or decaying material).

Scavengers feed on dead organisms that they did *not* kill. Often, the remains of the prey become a food source for scavengers.

35. D is correct.

Climate is the average conditions, *while the weather* is the immediate conditions (e.g., hourly, daily, weekly).

36. C is correct.

Nitrogen fixation is when atmospheric nitrogen (N_2) is converted into ammonium (NH_4). It is necessary for life because nitrogen is required for nucleotides of DNA and RNA and amino acids of proteins.

Diazotrophs are prokaryotic (i.e., bacteria and archaea) microorganisms that can fix nitrogen.

Some higher plants and animals (e.g., termites) have symbiotic relationships with diazotrophs.

37. C is correct.

Primary productivity is the rate at which producers (autotrophs) produce organic matter.

38. B is correct.

Biomes are climatically and geographically defined as *contiguous areas* with *similar climatic conditions* such as communities of animals, plants, and soil organisms and are often called ecosystems.

Factors such as plant structures define biomes (e.g., shrubs, trees, and grasses), leaf types (e.g., broadleaf and needleleaf), plant spacing (e.g., forest, woodland, savanna), and climate.

Biomes, unlike ecozones, are not defined by genetic, taxonomic, or historical similarities.

Biomes are often identified by patterns of ecological succession and climax vegetation.

39. A is correct.

North Pole and the South Pole are not classified into major biomes.

40. B is correct.

Predation is a biological interaction when a predator (i.e., a hunting organism) feeds on prey. A predator may (or may not kill) its prey before feeding on them, but the act of predation most often results in the prey's death and the eventual absorption of the prey's tissue through consumption.

Scavengers feed on dead organisms that they did not kill. Often, the remains of the prey become a food source for scavengers.

Organisms (e.g., lions, tigers, wolves) considered predators may engage in scavenging feeding behavior when given the alternative.

41. B is correct.

Aphotic zone is the area of a lake or ocean with little or no sunlight, the depths beyond which less than 1 percent of sunlight penetrates. Below the photic zone is the dark aphotic zone, where photosynthesis cannot occur.

Benthic zone is the lowest ecological zone in water bodies and usually involves sediments on the seafloor. These sediments play an essential role in providing nutrients for the organisms that live in the benthic zone.

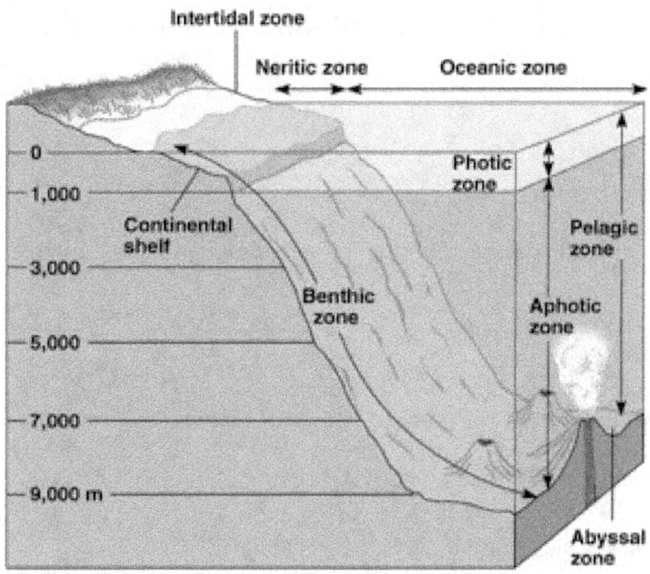

Deep ocean zones include photic, pelagic, aphotic, and abyssal zones

Phytoplankton (microalgae), like terrestrial plants, contain chlorophyll and require sunlight to live and grow.

Most phytoplankton is buoyant and floats in the ocean's upper layers, where sunlight penetrates the water. Phytoplankton is the base of many aquatic food webs whereby shrimp, jellyfish, and whales feed on them.

Zooplankton is heterotrophic (sometimes detritivores) plankton – organisms drifting in oceans, seas, and freshwater bodies. Individual zooplankton is usually microscopic, but some (e.g., jellyfish) are larger and visible to the unaided eye.

42. C is correct.

Transpiration is when water moves in a plant and evaporates from aerial parts, such as leaves, stems, and flowers.

43. D is correct.

In aquatic ecosystems, *nitrogen* and *phosphorus* are the most important, often in short supply relative to the needs of plants, algae, and microbes.

Elements like iron, manganese, and copper are needed in small amounts.

44. A is correct.

Phytoplankton (microalgae), like terrestrial plants, contain chlorophyll and require sunlight to live and grow.

Most phytoplankton is buoyant and floats in the ocean's upper layers, where sunlight penetrates the water. Phytoplankton is the base of many aquatic food webs whereby shrimp, jellyfish, and whales feed on them.

Zooplankton is heterotrophic (sometimes detritivores) plankton – organisms drifting in oceans, seas, and freshwater bodies. Individual zooplankton is usually microscopic, but some (e.g., jellyfish) are larger and visible to the unaided eye.

45. C is correct.

Energy flows in one direction while nutrients recycle.

Nutrient cycles and *energy flows* link *biotic* (i.e., living) and *abiotic* (i.e., nonliving) components.

46. B is correct.

Boreal forest (or *taiga*) is a biome characterized by coniferous (i.e., softwood) forests mainly consisting of pines, spruces, and larches and is the world's largest terrestrial biome.

47. C is correct.

Freshwater ecosystems naturally share resources between habitats.

At lower elevations, rivers and streams bring salts and nutrients from the mountains to lakes, ponds, and wetlands. Eventually, they bring those nutrients to the ocean.

48. A is correct.

Wetlands are areas filled with water most of the year. It might not always be wet.

Ponds and *lakes* are usually kept filled with water from many sources. They receive more water than they give off through evaporation.

Lake and ponds are distinguished by *depth* and *surface area.*

Lakes usually are much deeper than ponds and have a larger surface area.

Ponds have water in the photic zone, meaning ponds are shallow enough to allow sunlight to reach the bottom.

49. D is correct.

Estuaries are partly enclosed coastal bodies of brackish water (i.e., saline levels of 0.05 to 3% between fresh and saltwater).

Estuaries form when rivers or streams flow in and have a free connection to the open sea. They form a transition zone between the river and maritime environments and have marine influences (e.g., tides, waves, and saline water) and river influences (e.g., freshwater and sediment).

The inflows of seawater and freshwater provide nutrients in the water column and sediment, making *estuaries* among the most *productive natural habitats.*

50. A is correct.

Aphotic zone is the area of a lake or ocean with little or no sunlight, the depths beyond which less than 1 percent of sunlight penetrates. Below the photic zone is the dark aphotic zone, where photosynthesis cannot occur.

51. C is correct.

Abiotic factors (i.e., nonliving) include soil, sunlight, water, and minerals.

Biotic factors are living components affecting other organisms. Biotic factors need food and metabolic energy for proper growth.

52. D is correct.

Estuaries are partly enclosed coastal bodies of brackish water (i.e., saline levels of 0.05 to 3% between fresh and saltwater).

Estuaries form when rivers or streams flow in and have a free connection to the open sea. They form a *transition zone* between rivers and maritime environments and have marine influences (e.g., tides, waves, and saline water) and river influences (e.g., freshwater and sediment).

The inflows of seawater and freshwater provide nutrients in the water column and sediment, making estuaries among the most *productive natural habitats*.

53. D is correct.

Scavengers feed on dead organisms that they did not kill. Often, the remains of the prey become a food source for scavengers.

Detritivores are heterotrophs that obtain nutrition from decomposing (waste or detritus) plants, dead organisms, and feces and contribute to decomposition and the nutrient cycles.

Detritivores are not to be confused with other decomposers such as bacteria, fungi, and protists, which cannot ingest minute lumps of matter.

Decomposers absorb and metabolize nutrients (i.e., saprotrophic nutrition from dead or decaying material).

54. A is correct.

Open ocean (or *pelagic zone*) is the area of the ocean outside of coastal areas.

The open ocean lies beyond the continental shelf: extending from the Arctic to the Antarctic. It is from the surface to the deepest parts of the ocean and encompasses the entire water column. It accounts for 64% of the ocean and 45% of Earth.

55. B is correct.

A: *paleontology* studies prehistoric life and includes the study of fossils to determine organisms' evolution.

C: *microbiology* is the scientific study of microscopic organisms, unicellular (i.e., single-cell), multicellular (i.e., cell colony), or acellular (i.e., lacking cells).

Microbiology encompasses sub-disciplines of *mycology* (i.e., fungi), *virology*, *parasitology*, and *bacteriology*.

D: *entomology* is the study of insects (a branch of arthropodology), a branch of zoology.

56. B is correct.

Species make up *populations* that comprise *communities*.

57. D is correct.

Photic zone (sunlight zone) is the depth of the water in a lake or ocean exposed to sufficient sunlight for photosynthesis. It extends from the surface to a depth where light intensity falls to one percent.

The thickness of the photic zone depends on the extent of intensity loss in the water column. It can vary from a few inches in highly turbid lakes (i.e., seasonal effects) to around 600 feet in the ocean.

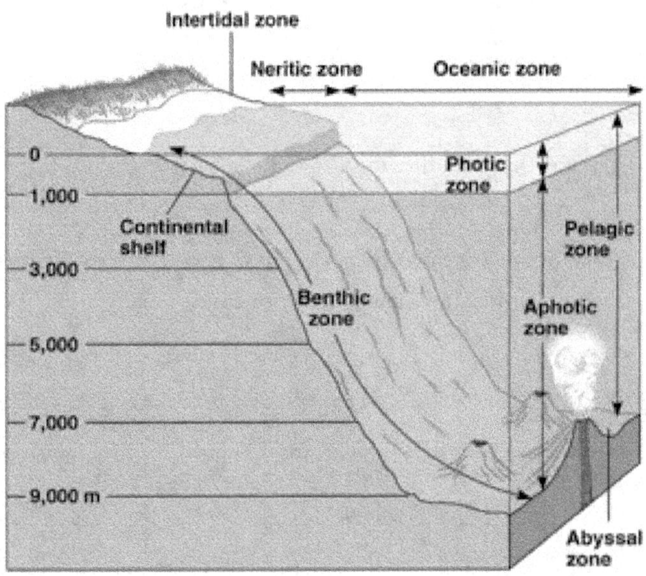

Deep ocean zones include photic, pelagic, aphotic, and abyssal zones

58. D is correct.

Mutualism is when organisms of different species exist in a relationship where each benefit (+/+ relationship).

Symbiosis includes *mutualistic*, *parasitic*, or *commensal* relationships. Mutualism is one *type* of symbiosis.

The body of most lichens is different from those of either the fungus or alga growing separately.

The fungus surrounds the algal cells, often enclosing them within complex fungal tissues.

continued...

Algal cells are photosynthetic and reduce atmospheric CO_2 into organic carbon sugars to feed both symbionts.

Both partners gain water and mineral nutrients mainly from the atmosphere through rain and dust.

The fungal partner protects the alga by retaining water, serving as a larger capture area for mineral nutrients, and in some cases, providing minerals obtained from the substrate.

If a cyanobacterium is another symbiont in addition to green alga (tripartite lichens), they can fix atmospheric nitrogen, complementing the activities of the green alga.

A: *nematodes* (i.e., roundworms) are unlike cnidarians or flatworms because nematodes have tubular digestive systems with openings at each end.

Nematodes are obligate aquatic organisms consisting of cylindrical unsegmented worms. They are free-living saprophytes or parasites.

B: *bread mold* is a *saprophyte* that feeds on dead and decaying material.

C: t*apeworms* are *parasites* extracting nutrition (they benefit) from the host while the host is depleted (host harmed) of nutrients.

This is a +/– relationship.

59. A is correct.

Heterotrophic organisms obtain nutrients and energy by consuming organic substances.

Autotrophic organisms synthesize food from inorganic environmental substances (e.g., light, chemical energy).

Autotrophs are the basis of the ecosystem and feed heterotrophs. Autotrophs create organic compounds using energy from the sun or inorganic compounds. They can be divided into photoautotrophs and chemoautotrophs.

Photoautotrophs (e.g., algae, plants, cyanobacteria) use photosynthesis to convert solar energy into organic compounds.

Chemoautotrophs are bacteria that oxidize inorganic compounds such as ammonia, nitrite, and sulfide to generate organic compounds. They are rare and are typically found in caves, hydrothermal ocean vents, and other environments lacking light.

60. D is correct.

Extracellular digestion allows for the digestion of food outside of the cell, permitting the ingestion of large pieces of food. Intracellular digestion is limited to particles small enough to be taken into the individual cells.

Complex animals typically have digestive tract enzymes that act on the food material.

Notes for active learning

Populations, Communities and Conservation Biology – Detailed Explanations

1. D is correct.

Sustainable development minimizes environmental impact. Sustainable buildings preserve precious natural resources and improve the quality of life. Buildings should consider design, construction, or operation, reducing or eliminating negative impacts and positively impacting climate and natural environment.

Several features can make a development *green*, including:

Efficient use of energy, water, and other resources; renewable energy, such as solar energy

Pollution and waste reduction measures and enabling re-use and recycling

Good indoor environmental air quality

Use of materials that are non-toxic, ethical, and sustainable

Consideration of the environment in design, construction, and operation

Design enables adaptation to a changing environment

2. A is correct.

Community is a population of species occupying the same geographical area. In ecology, a community is an association of populations of two or more species occupying the same geographical area simultaneously.

Ecosystem is a community of organisms (e.g., plants, animals, microbes) with nonliving (i.e., *abiotic*) components of their environment (e.g., air, water, minerals). Ecosystems are the network of interactions among and between organisms and their environment.

Habitat is an *ecological area* inhabited by a species. It is the natural environment in which an organism lives or the physical environment encompassing a population.

3. B is correct.

Population density measures population per unit area.

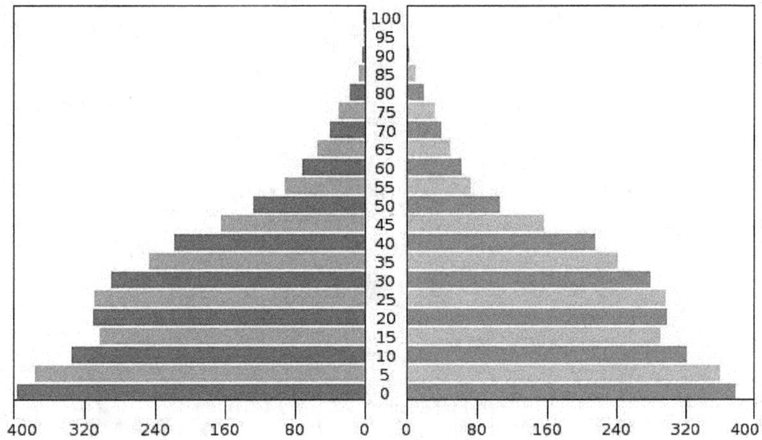

Population density plots age vs. population (in thousands) with male (left) and female comparison

4. D is correct.

Sustainable development meets the needs of people without compromising the ability of future generations to meet their needs for natural resources and environmental quality. Preventing long-term harm to the environment ensures that people can benefit from its use in the future.

5. C is correct.

Range (or *distribution*) of a population is the geographical area.

Dispersion is the variation in local *density* within a range.

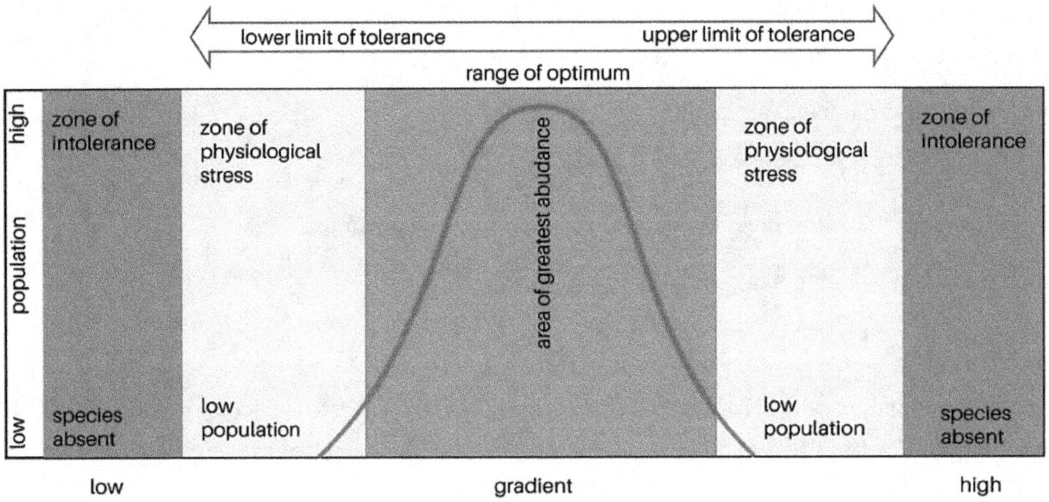

Range of tolerance plots population vs. bifurcated range of optimum stress

6. C is correct.

Observation involves noting characteristics *without* engagement or intervention by the researcher.

7. A is correct.

Demography uses statistics such as births, deaths, income, the incidence of disease, and education, illustrating the changing structure of human populations.

Population growth rate is how the number of people changes in a given period; the flow of people in and out, births, and deaths affect the number of individuals within a population.

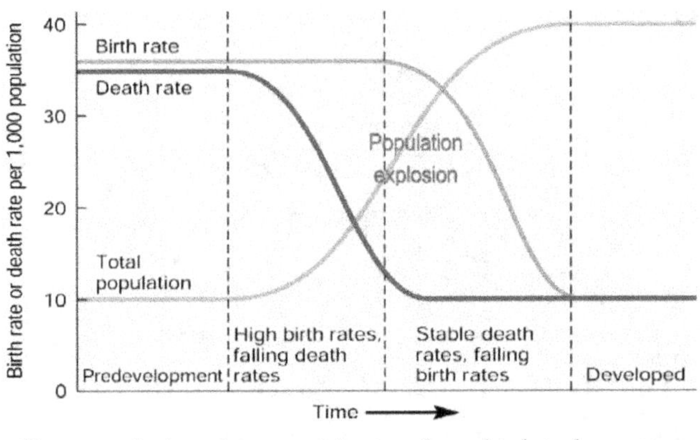

Birth rate and death rate affect population size transitioning from high to low rates with increasing population

8. B is correct.

Population density measures population per unit area (e.g., people per square mile).

9. C is correct.

Biotic factors are living components affecting other organisms.

For example, animals consume other organisms and organic food. Biotic factors need food and metabolic energy for proper growth.

Biotic factors include human influence.

10. A is correct.

Coal, natural gas, oil, and energy are derived from fossil fuels, including minerals (e.g., copper, magnesium). Non-renewable resources are not replenished at a sufficient rate for sustainable economic extraction.

Non-renewable resources require tens of thousands to millions of years and specific conditions to form.

Renewable resources replenish over time through biological reproduction or other naturally recurring processes. They are integral to Earth's natural environment and the largest components of its ecosphere.

Renewable resources may be the source of power for renewable energy. However, renewal and sustainability will not be sustainable when the resource consumption rate exceeds the renewal rate.

11. A is correct.

Ecological models trace the interaction between and interdependence of factors within and across all population levels. It highlights people's interactions with physical and environmental factors.

12. B is correct.

The Dust Bowl was a period of severe dust storms that significantly damaged the ecology and agriculture of the American and Canadian prairies during the 1930s; severe drought and a failure to apply dryland farming methods to prevent the aeolian processes (wind erosion) caused the phenomenon.

Crops began to fail with the onset of drought in 1931, exposing the bare, over-plowed farmland. It began to blow away without deep-rooted prairie grasses to hold the soil in place. The eroding soil led to massive dust storms and economic devastation—especially in the Southern Plains.

13. C is correct.

Immigration is movement *into* an area.

Carrying capacity is the maximum population size that the environment can sustain based on resources. If a population grows larger than the carrying capacity, species may die due to a lack of environmental resources.

Emigration is movement *out* of an area.

14. D is correct.

Animal training is not a method used by ecologists to study the environment.

15. A is correct.

Slower growth rate means fewer new individuals in a population. Increased birth rate and decreased death rates lead to a rise in growth rate, whereas decreased birth rate slows down the population growth rate.

	High stationary	**Early expanding**	**Late expanding**	**Low stationary**	**Declining**
	Stage 1	Stage 2	Stage 3	Stage 4	Stage 5
Birth rate	High	High	Falling	Low	Rising again
Death rate	High	Falls rapidly	Falls slowly	Low	Low
Natural output	Stable or slow increase	Very rapid increase	Increase slows down	Falling and then stable	Stable or slow increase

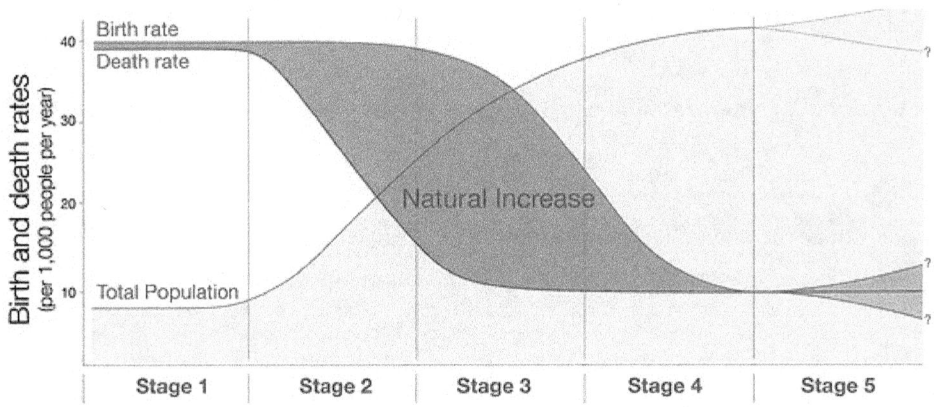

Five stages of demographic transition transitioning from high to low rates with increasing population

16. C is correct.

Desertification is land degradation when a relatively dry land region becomes increasingly arid, typically losing its bodies of water and vegetation, and wildlife. It is caused by factors such as climate change and human activities. The immediate cause is the removal of most vegetation.

Desertification is driven by drought, climatic shifts, tillage for agriculture, overgrazing, deforestation for fuel, and harvesting of construction materials. Desertification is a global ecological and environmental problem.

B: *monoculture* is the agricultural practice of growing a single crop for many consecutive years. It is widely used in industrial agriculture, and its implementation has allowed for large harvests from minimal labor.

However, monocultures quickly spread pests and diseases whereby a uniform crop is susceptible to a pathogen.

continued...

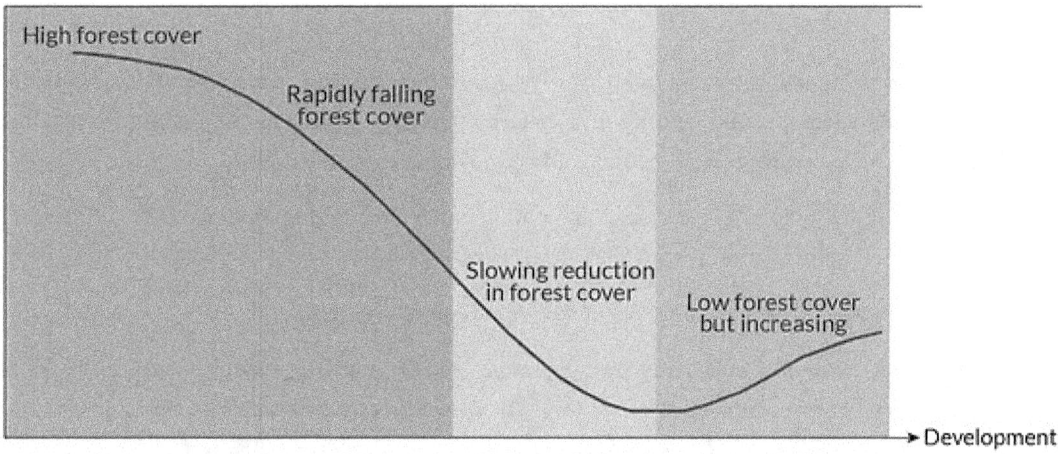

Four stages of forest transition model plotting forest cover vs. development

Stage 1 Pre-transition: high forest cover. No or slow loss of forest. (left panel on the graph)

Stage 2 Early transition: deforestation rates increasing.

Stage 3 Late transition: low forest cover but a slowing rate of deforestation

Stage 4 Post-transition: low forest cover but increasing through reforestation. (right panel on the graph)

17. A is correct.

Exponential growth is a process that increases quantity over time. The graph exhibits a steep curve up.

Animals require resources (e.g., food, water, biotic factors) for survival. The number of microorganisms in culture increases exponentially until an essential nutrient is exhausted, which impedes the organisms' growth.

Typically the first organism divides into two, who split to form four, who split to form eight, etc.

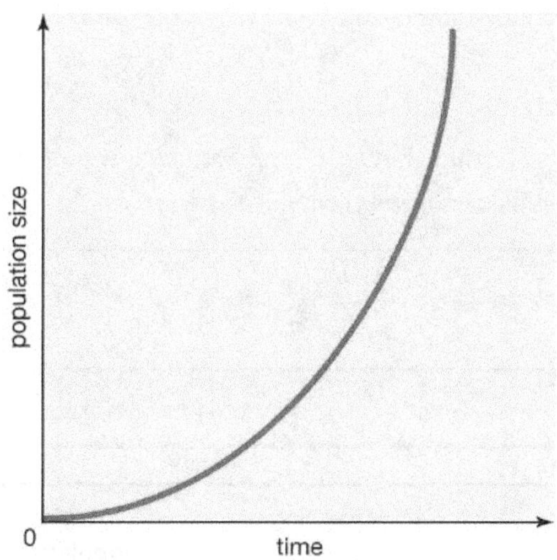

Exponential growth plotting population size vs. time with a steep positive slope indicating increasing numbers

18. C is correct.

Desertification is land degradation when a relatively dry land region becomes increasingly arid, typically losing its bodies of water and vegetation, and wildlife. It is caused by factors such as climate change and human activities. The immediate cause of desertification is the removal of most vegetation.

Desertification is driven by factors (alone or in combination) such as drought, climatic shifts, tillage for agriculture, overgrazing, deforestation for fuel, and harvesting of construction materials.

Desertification is a significant global ecological and environmental problem.

B: *monoculture* is the agricultural practice of growing a *single crop* for many consecutive years. It is widely used in industrial agriculture, and its implementation has allowed for large harvests from minimal labor.

However, monocultures quickly *spread pests and diseases* whereby a uniform crop is susceptible to a pathogen.

Stage 1 Pre-transition: high forest cover. No or slow loss of forest.

Stage 2 Early transition: deforestation rates increasing.

Stage 3 Late transition: low forest cover but a slowing rate of deforestation

Stage 4 Post-transition: low forest cover but increasing through reforestation.

19. B is correct.

Decreased birth rate and *emigration* (i.e., outflow) cause population size to *decrease.*

20. D is correct.

Soil erosion is the gradual wearing away of a field's top layer of soil by natural elements such as water and wind and farming activities.

Irrigation controls the amount of water released by supplying it at regular intervals for farming, which is an effective way of maintaining the quality of the soil.

21. D is correct.

Demographic transition is the shift from high birth and death rates to low birth and death rates as a country develops from a pre-industrial to an industrialized economic system.

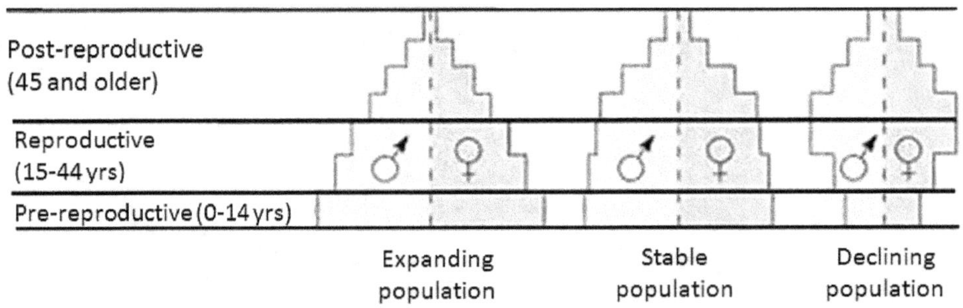

Age structure diagrams: pyramid (left) for expanding, bell-shaped (center) for stable, urn-shaped (right) for declining population profiles

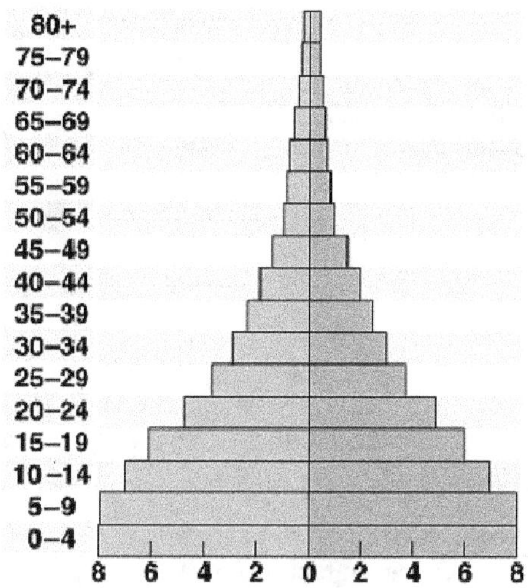

An expansive population pyramid plots age vs. percent. Males on the left. The graph is skewed towards younger members when the demographic transition is incomplete

22. A is correct.

Births and *immigration* (inflows) *increase* population size.

23. D is correct.

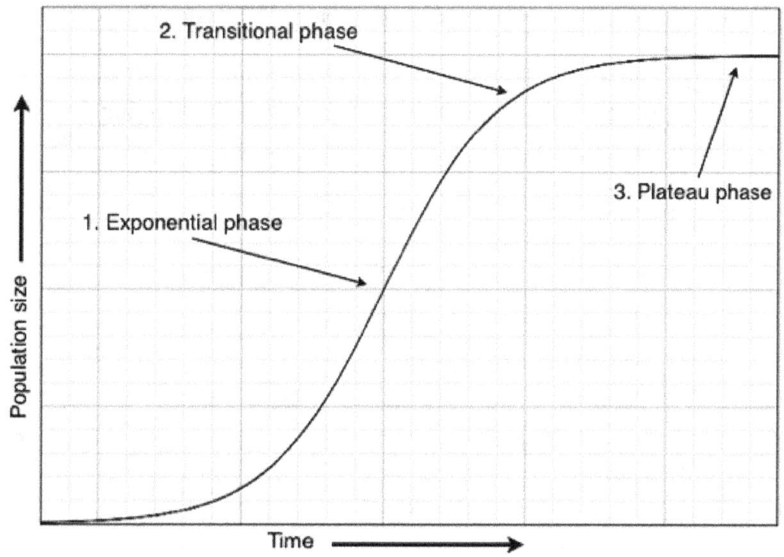

Sigmoidal (S)-shaped logistic growth curve of population growth with environmental resistance. The exponential phase has an accelerating growth rate. The exponential phase is marked by a steep positive slope, transitional phase with a slightly positive slope, and plateau phase with flat curve profiles

Exponential growth is a process that increases quantity over time. The graph exhibits a steep curve up.

Logistic curve (or *function*) is a common *sigmoid function* (S-shape) concerning population growth.

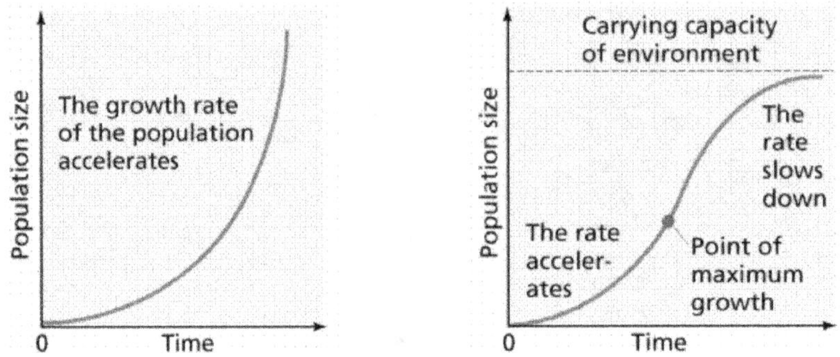

Exponential (unrestricted) growth (left) has accelerating growth patterns and logistic (restricted) growth (right) where the rate decreases as environmental carrying capacity is achieved

Animals require resources (e.g., food, water, biotic factors) for survival.

The number of microorganisms in culture increases exponentially until an essential nutrient is exhausted, which impedes the organisms' growth.

24. A is correct.

Using predators and parasites for pest management is a sustainable resource in ecology.

25. B is correct.

Exponential growth is a process that increases quantity over time. The graph exhibits a steep curve up.

Animals require resources (e.g., food, water, biotic factors) for survival. The number of microorganisms in culture increases exponentially until an essential nutrient is exhausted, which impedes the organisms' growth. Typically the first organism divides into two, who split to form four, who split to form eight, etc.

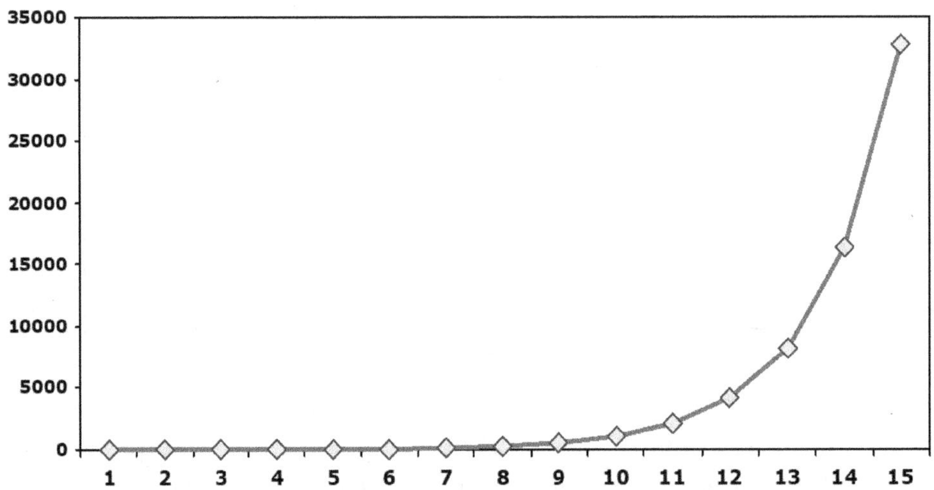

Exponential growth with population vs. generation plotted shows accelerated growth rates

26. C is correct.

DDT (or d*ichlorodiphenyltrichloroethane*) was a modern synthetic pesticide used in the United States (beginning in the 1940s) to control insects on food crops and in buildings for pest control.

DDT was discontinued in 1972 because of the harmful effects on humans and wildlife.

27. D is correct.

Renewable resources can be regenerated or replenished. A renewable resource is a natural resource that can replenish over time through biological reproduction or other naturally recurring processes.

Renewable resources are a part of Earth's natural environment and the largest components of its ecosphere.

Renewable resources may be the source of power for renewable energy.

Renewal and sustainability are not ensured if the renewable resource consumption exceeds its renewal rate.

28. C is correct.

Density-dependent limiting factors affect the size or growth due to variation in population density. Dense populations are more strongly affected than less crowded ones.

Density-dependent limiting factors include food availability, disease, living space, predation, and migration.

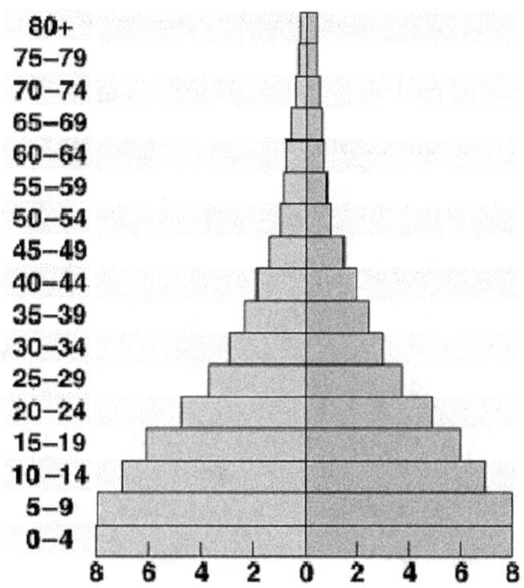

An expansive population pyramid plots age vs. percent. Males on the left. The graph is skewed towards younger members when the demographic transition is incomplete

29. D is correct.

Carrying capacity is the maximum population size that the environment can sustain indefinitely based on the available resources. If a population grows larger than the carrying capacity, species may die due to a lack of environmental resources.

30. B is correct.

Biological magnification is any concentration of a toxin, such as pesticides, in the tissues of tolerant organisms at successively higher levels in a food chain.

Biomagnification is when the chemical concentration in an organism exceeds the concentration of its food when the primary exposure route occurs from the diet. Higher-level predators (e.g., fish, birds, marine mammals) build up greater dangerous amounts of toxic materials than animals lower on the food chain.

DDT was banned in 1972 as a suspected carcinogen (i.e., cancer causing agent).

31. B is correct.

Exponential growth is a process that increases quantity over time. The graph exhibits a steep curve up.

Animals require resources (e.g., food, water, biotic factors) for survival. The number of microorganisms in culture increases exponentially until an essential nutrient is exhausted, which impedes the organisms' growth.

Typically the first organism divides into two, who split to form four, who split to form eight, etc.

32. A is correct.

Photochemical smog is air pollution resulting from the interaction of sunlight with certain chemicals in the atmosphere. One of the primary components of photochemical smog is *ozone.*

Smog is air pollution that reduces visibility and first labeled in the early 1900s to describe a mix of smoke and fog. Smog was common in industrial areas and remains a familiar sight in cities today. The smoke usually came from burning coal.

The atmospheric pollutants or gases that form smog are released into the air when fuels are burnt. When sunlight and its heat react with these gases and fine particles in the atmosphere, smog is formed.

Particulate matter is a mixture of tiny particles and liquid droplets. Particle pollution comprises several components, including acids (nitrates and sulfates), organic chemicals, metals, and soil or dust particles.

Greenhouse gas absorbs infrared radiation (net heat energy) emitted from Earth's surface and reradiates it back to Earth's surface, thus contributing to the greenhouse effect. Carbon dioxide, methane, and water vapor are the most important greenhouse gases.

Ozone layer is a region in the Earth's stratosphere that contains high concentrations of ozone (O_3) and protects the Earth from the sun's harmful ultraviolet radiations.

33. A is correct.

Acid rain is any form of precipitation when acidic components (e.g., sulfur dioxide, nitrogen oxides) fall to the ground from the atmosphere. These compounds rise into the atmosphere, reacting with water and oxygen and falling to the ground as precipitation.

34. C is correct.

Logistic curve (or *function*) is a common *sigmoid function* (S-shape) concerning population growth. The initial growth stage is approximately exponential. As saturation begins, growth slows, and at maturity, growth stops.

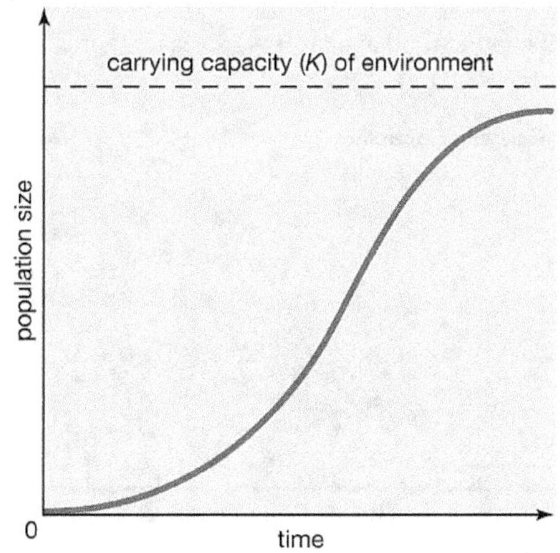

Logistic growth plotting population size vs. time where the rate decreases at carrying capacity

Exponential growth is a process that increases quantity over time. The graph exhibits a steep curve up.

35. C is correct.

Limiting factors in an ecosystem are typically food, water, habitat, and mate. The availability of these factors will affect the carrying capacity of an environment. As the population increases, food demand increases as well. Since food is a limited resource, organisms will begin competing for it.

36. A is correct.

Acid rain is a chemical reaction with sulfur dioxide, and nitrogen oxides are released into the air. These compounds rise into the atmosphere, reacting with water and oxygen and falling to the ground as precipitation.

37. D is correct.

The vapors given off when gasoline evaporates and the substances produced when gasoline is burned (carbon monoxide, nitrogen oxides, particulates, and unburned hydrocarbons) contribute to air pollution.

Burning gasoline produces carbon dioxide (CO_2), a greenhouse gas.

Lead gasoline (prohibited in 1996) releases suspended particles into the air. Motor-vehicle emissions have been reduced by banning lead gasoline for motor vehicles.

However, lead is used in general-aviation gasoline for piston-engine aircraft. Lead poisoning causes immense societal harm: brain damage, chronic illness, lowered IQ, and elevated mortality.

38. D is correct.

Density-dependent limiting factors affect the size or growth due to variation in population density.

Dense populations are more strongly affected than less crowded ones.

Density-dependent limiting factors include food availability, disease, living space, predation, and migration.

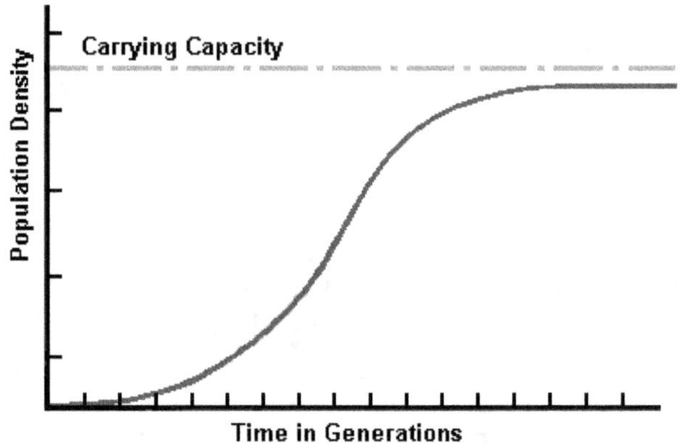

Logistic growth plotting population size vs. time where the rate decreases at carrying capacity

39. A is correct.

Particulate matter is a mixture of tiny particles and liquid droplets. Particle pollution comprises several components, including acids (nitrates and sulfates), organic chemicals, metals, and soil or dust particles.

40. B is correct.

Carrying capacity is the maximum population size that the environment can sustain indefinitely based on the available resources. If a population grows larger than the carrying capacity, species may die due to a lack of environmental resources.

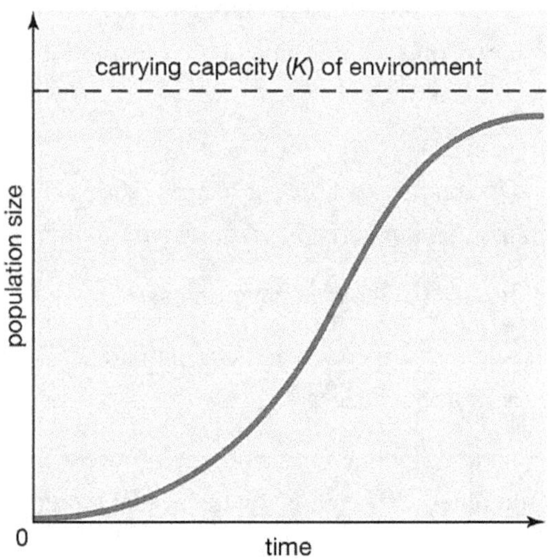

Logistic growth plotting population size vs. time where the rate decreases at carrying capacity

41. A is correct.

Species diversity is the adequate number of different species represented in a collection of individuals.

B: *genetic diversity* refers to the total number of genetic characteristics in the genetic makeup of a species.

It is distinguished from genetic variability, which describes the tendency of genetic characteristics to vary.

Genetic diversity serves as a way for populations to adapt to changing environments.

C: *ecosystem diversity* refers to the diversity of a place at the level of ecosystems (i.e., biotic and abiotic factors).

D: *biodiversity* refers to variation in species rather than ecosystems.

42. B is correct.

Density-dependent limiting factors affect the size or growth due to variation in population density. Dense populations are more strongly affected than less crowded ones.

Density-dependent limiting factors include food availability, disease, living space, predation, and migration.

43. D is correct.

Species diversity is the adequate number of different species represented in a collection of individuals.

B: *genetic diversity* refers to the total number of genetic characteristics in the genetic makeup of a species.

Genetic diversity is distinguished from *genetic variability*, which describes the tendency of genetic characteristics to vary. It serves as a way for populations to adapt to changing environments.

44. A is correct.

Drought is a prolonged period of abnormally low rainfall, leading to a water shortage. This limiting factor does not apply to marine animals.

45. A is correct.

Introduced species (or exotic species) is an organism that is not native to the place or area where it is considered introduced. Instead, it has been accidentally or deliberately transported to the new location by human activity.

Introduced species may be predators or ecological consumers and crowd out native species.

46. B is correct.

Density-independent limiting factors are *not* influenced by population size.

Factors include weather, climate, and natural disasters. All species populations in the ecosystem will be similarly affected, regardless of population size.

Density-dependent limiting factors affect the size or growth due to variation in population density. Dense populations are more strongly affected than less crowded ones.

Density-dependent limiting factors include food availability, disease, living space, predation, and migration.

47. D is correct.

Higher birth rates and lower death rates increase the number of individuals, thus increasing competition.

Fewer resources and higher population density lead to increased competition.

A decrease in population size can reduce competition.

48. C is correct.

Habitat is the natural environment in which an organism lives or the physical environment encompassing a population. It is an ecological area inhabited by a species.

Habitat fragmentation divides large habitats into smaller, isolated patches due to *habitat loss* by human activity and natural causes. This negatively affects biodiversity and reduces the amount of suitable habitat available for certain species.

49. C is correct.

Habitat is an ecological area inhabited by a species. It is the natural environment in which an organism lives or the physical environment encompassing a population.

50. A is correct.

Density-independent limiting factors are *not* influenced by population size. Factors include weather, climate, and natural disasters. Species populations in the ecosystem are similarly affected, regardless of population size.

The spray program affected the mosquito population as a density-independent limiting factor.

51. D is correct.

Density-dependent limiting factors affect the size or growth due to variation in population density. Dense populations are more strongly affected than less crowded ones. Density-dependent limiting factors include food availability, disease, living space, predation, and migration.

Density-independent limiting factors are *not* influenced by population size. Factors include weather, climate, and natural disasters. All species populations in the ecosystem will be similarly affected, regardless of population size.

52. A is correct.

Hot spots are regions of species found in a single geographic region (i.e., endemism). Ecological hot spots tend to occur in tropical environments where species richness and biodiversity are higher than in ecosystems closer to the poles.

For example, biodiversity hotspots are forest habitats as they constantly face destruction and degradation due to illegal logging, pollution, and deforestation.

53. B is correct.

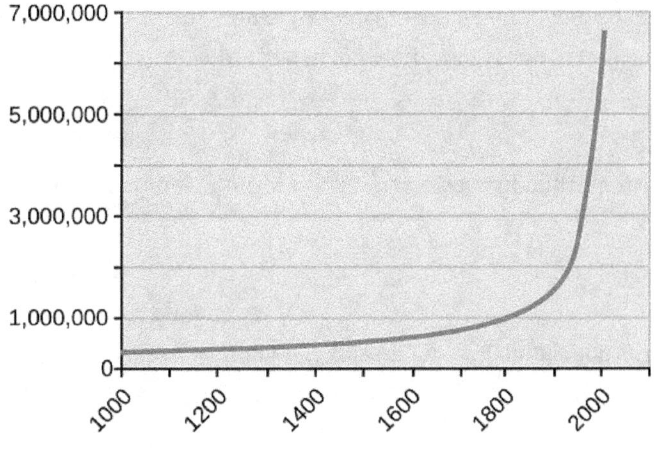

World population from 1000 AD. Graph plots population (in thousands) vs. year

54. A is correct.

Ecosystems include interactions between species. Protecting the entire ecosystem preserves this equilibrium.

55. C is correct.

Biodiversity conservation protects and preserves the wealth and variety of species, habitats, ecosystems, and genetic diversity.

Introduced species (or exotic species) is an organism not native to the place or area where it is considered introduced. Instead, it has been accidentally or deliberately transported to the new location by human activity. Introduced species may be predators or ecological consumers and crowd out native species.

56. C is correct.

Demographic transition is the shift from high birth and high death rates to low birth and low death rates as a country develops from a pre-industrial to an industrialized economic system.

	High stationary	**Early expanding**	**Late expanding**	**Low stationary**	**Declining**
	Stage 1	Stage 2	Stage 3	Stage 4	Stage 5
Birth rate	High	High	Falling	Low	Rising again
Death rate	High	Falls rapidly	Falls slowly	Low	Low
Natural output	Stable or slow increase	Very rapid increase	Increase slows down	Falling and then stable	Stable or slow increase

Five stages of demographic transition transitioning from high to low rates with increasing population

57. B is correct.

Demography is the statistical study of human populations. Demographers use census data, surveys, and statistical models to analyze populations' size, movement, and structure.

58. D is correct.

Ecological footprint is environmental impact expressed as the amount of land required to sustain the use of natural resources.

59. C is correct.

Population density measures population per unit area (e.g., people per square mile).

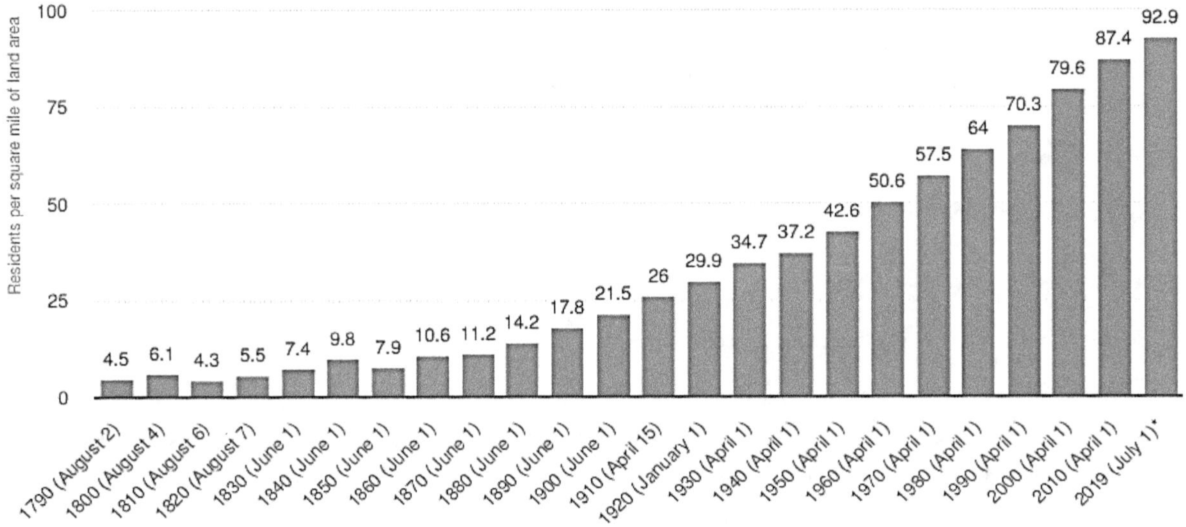

Population density of the United States from 1790 to 2019 in residents per square mile of land area

60. C is correct.

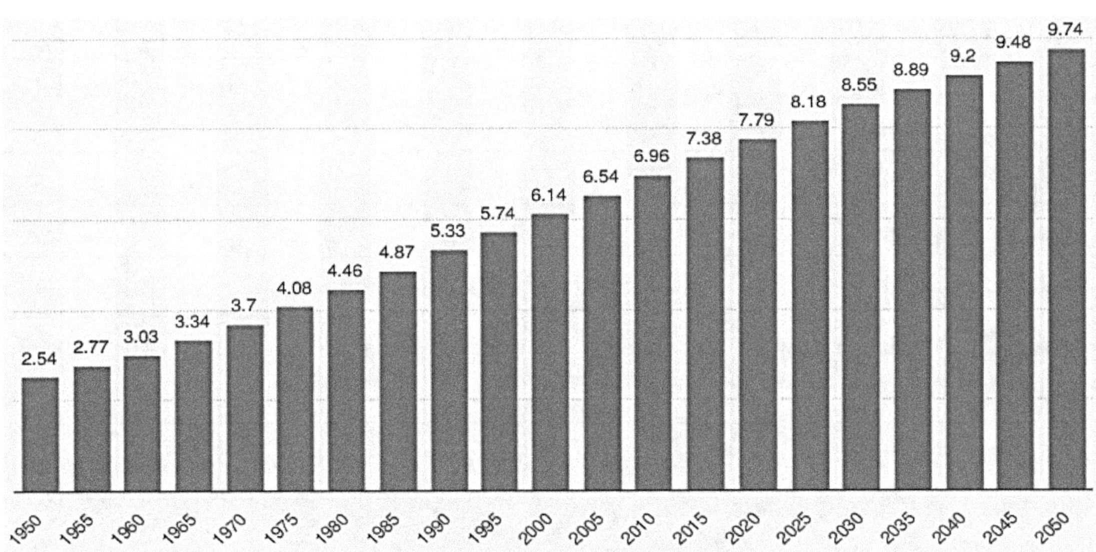

World population from 1950 to 2020 in billion

61. C is correct.

Demographic transition is the shift from high birth and high death rates to low birth and low death rates as a country develops from a pre-industrial to an industrialized economic system.

62. D is correct.

Introduced species (or exotic species) is an organism that is not native to the place or area where it is considered introduced. Instead, it has been accidentally or deliberately transported to the new location by human activity. Introduced species may be predators or ecological consumers and crowd out native species.

63. B is correct.

Monoculture farming grows the same crop or plant in a field year after year. This requires standardized planting, maintenance, and harvesting, resulting in greater yields and lower costs but can lead to quicker buildup and *spread of pests and diseases*.

64. C is correct.

Density-independent limiting factors are *not* influenced by population size. Factors include weather, climate, and natural disasters. All species populations in the ecosystem will be similarly affected, regardless of population size.

Predation is a biological interaction when a predator (i.e., a hunting organism) feeds on prey.

Exponential growth is a process that increases quantity over time. The graph exhibits a steep curve up.

65. C is correct.

Exponential growth is a process that increases quantity over time. The graph exhibits a steep curve up.

66. D is correct.

Demographic transition is the shift from high birth and high death rates to low birth and low death rates as a country develops from a pre-industrial to an industrialized economic system.

	High stationary	**Early expanding**	**Late expanding**	**Low stationary**	**Declining**
	Stage 1	Stage 2	Stage 3	Stage 4	Stage 5
Birth rate	High	High	Falling	Low	Rising again
Death rate	High	Falls rapidly	Falls slowly	Low	Low
Natural output	Stable or slow increase	Very rapid increase	Increase slows down	Falling and then stable	Stable or slow increase

Five stages of demographic transition transitioning from high to low rates with increasing population

continued...

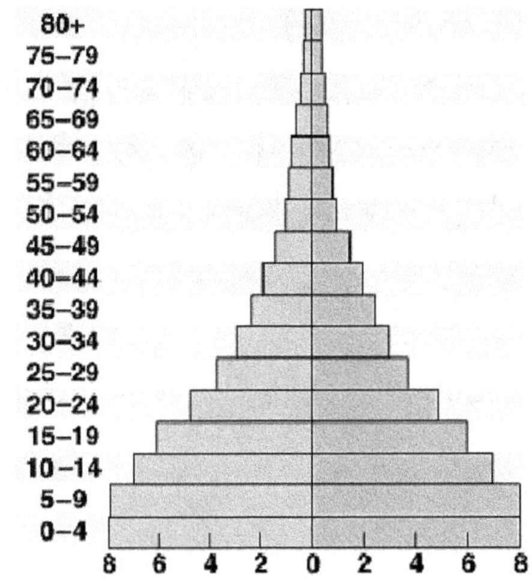

An expansive population pyramid plots age vs. percent. Males on the left. The graph is skewed towards younger members when the demographic transition is incomplete

67. A is correct.

Ecology uses scientific research to identify the cause and best practices to solve problems.

68. D is correct.

Global warming refers to the rise in the average temperature of the Earth's climate system.

The graph in the question shows that the land-surface air temperature has increased over the last century.

69. B is correct.

Biodiversity is the biological variety and variability of life on Earth. Biodiversity measures variation at the genetic, species, and ecosystem levels. Terrestrial biodiversity is usually greater near the equator, resulting from the warm climate and high primary productivity.

Genetic diversity refers to the total number of genetic characteristics in the genetic makeup of a species.

Genetic variability is the tendency of genetic characteristics to vary. It serves as a way for populations to adapt to changing environments.

70. A is correct.

Hot spots are regions of species found in a single geographic region (i.e., endemism). These regions are sensitive to human intervention, and many habitats and species are at a high risk of extinction.

Ecological hot spots tend to occur in tropical environments where species richness and biodiversity are higher than in ecosystems closer to the poles.

For example, biodiversity hotspots are forest habitats as they constantly face destruction and degradation due to illegal logging, pollution, and deforestation.

71. D is correct.

Ecological succession in a rocky, barren area to a final climax community is:

> lichen → mosses → annual grasses → perennial grasses → shrubs → deciduous trees (e.g., thick shade trees such as oak and hemlock).

72. A is correct.

Climax community is when a biological community of plants, animals, and fungi reaches a steady-state through *ecological succession* (i.e., vegetation development in an area over time).

Equilibrium is when the climax community is composed of species best adapted to average conditions.

For example, the climax community is grasslands in the Midwest, while deciduous forests are in the northeast.

B: *climax community* is mainly dependent on *environmental factors*.

Factors such as temperature, soil, and amount of rainfall determine which organisms survive and thrive.

C: *biomes* are geographically and climatically defined as contiguous with similar conditions, such as communities of plants, animals, and soil organisms.

Many species live in a biome.

D: *pioneer species* are hardy species and the first to colonize previously disrupted ecosystems,

Pioneer species are the earliest species in a biome and initially colonize it (e.g., lichen on rocks). They begin a chain of ecological succession leading to a biodiverse steady-state ecosystem.

Notes for active learning

Classification and Diversity – Detailed Explanations

1. D is correct.

Mnemonic: **D**arn **K**ing **P**hillip **C**ame **O**ver **F**or **G**ood **S**oup:

Domain → Kingdom → Phylum → Class → Order → Family → Genus → Species

Canis lupus indicates that the wolf's genus is *Canis,* and the species is *lupus*.

Family is more inclusive than genus or species, so any member of the species *lupus* is in the genus *Canis* of the Canidae family.

2. A is correct.

A single class encompasses several orders.

Mnemonic: **D**arn **K**ing **P**hillip **C**ame **O**ver **F**or **G**ood **S**oup:

Domain → kingdom → phylum → class → **order** → family → genus → species

3. B is correct.

Homology serves as evidence of common ancestry. Homology is a relationship between structures or DNA derived from a common ancestor.

Homologous traits are traced by descent from a common ancestor. Analogous organs have similar functions in two taxa that were not in the last common ancestor but rather evolved separately.

Homologous sequences are *orthologous* if they descend from the same ancestral sequence separated by a speciation event. When a species diverges into two separate species, the copies of a single gene in the two resulting species are orthologous.

Orthology is strictly defined by ancestry.

*Orthologs (*i.e., orthologous genes) are genes in different species that originated by vertical descent from a single gene of the last common ancestor.

Due to gene duplication and genome rearrangement, the ancestry of genes is difficult to ascertain. The strongest evidence that two similar genes are orthologous is usually a phylogenetic analysis of the gene lineage. Orthologous genes often, but not always, have the same function.

4. A is correct.

Homology is a relationship between structures or DNA derived from a common ancestor.

Homologous traits descend from a common ancestor (e.g., the forelimb of a human and dog).

Analogous organs, the opposite of homologous organs, have similar functions in two taxa, not in the last common ancestor but instead evolved separately.

continued...

Homologous sequences are *orthologous* if they descend from the same ancestral sequence separated by a speciation event. When a species diverges into two separate species, the copies of a single gene in the two resulting species are orthologous.

Orthologues (i.e., orthologous genes) are genes in different species that originated by *vertical descent* from a single gene of the last common ancestor.

Orthology is strictly defined by ancestry. Due to gene duplication and genome rearrangement, the ancestry of genes is difficult to ascertain. The strongest evidence that two similar genes are orthologous is through phylogenetic analysis of gene lineage.

Orthologous genes often, but not always, have the same function.

	Analogous leg	Analogous flipper
Homologous mammals	 Cat leg	 Whale flipper
Homologous insects	 Praying mantis leg	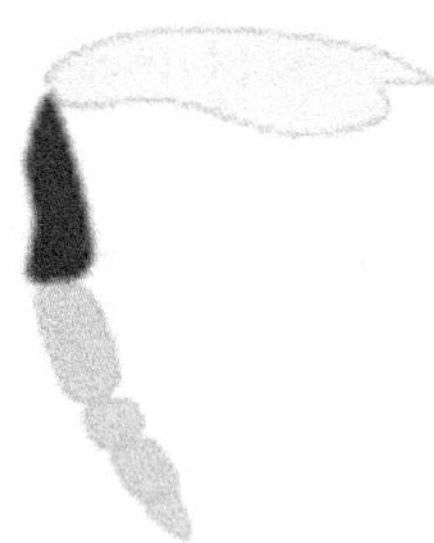 Water boatman flipper

Homologous (common ancestor) vs. analogous structures (similar function but evolved separately)

5. C is correct.

Homology is a relationship between structures or DNA derived from a common ancestor.

Homologous traits descend from a common ancestor (e.g., the forelimb of a human and a dog).

Analogous organs are the opposite of homologous organs and have similar functions in two taxa that were not in the last common ancestor but instead evolved separately.

Homologous sequences are *orthologous* if they descend from the same ancestral sequence separated by a speciation event. When a species diverges into two separate species, the copies of a single gene in the two resulting species are orthologous.

continued...

Orthologs (i.e., orthologous genes) are genes in different species that originated by vertical descent from a single gene of the last common ancestor. Orthology is strictly defined by ancestry.

The strongest evidence that two similar genes are orthologous is usually a phylogenetic analysis of the gene lineage. Orthologous genes often, but not always, have the same function.

D: wings of a pigeon and a bat represent *analogous structures*.

6. B is correct.

Taxonomy of humans: Domain: Eukaryota → Kingdom: Animalia → Phylum: Chordata →
Subphylum: Vertebrata → Class: Mammalia → Order: Primates →
Family: Hominidae → Genus: Homo → Species: *H. sapiens* → Subspecies: *H. s. sapiens*

7. B is correct.

Ants are not chordates.

C: *tunicates* are marine invertebrates and members of the subphylum Tunicata within Chordata (i.e., phylum includes all animals with dorsal nerve cords and notochords). Tunicates live as solitary individuals or replicate by budding and becoming colonies (each unit is a zooid).

Tunicates are marine filter feeders with a water-filled, sac-like body structure and two tubular openings, known as *siphons*, through which they draw in and expel water. They take in water through the *incurrent siphon* during respiration and feeding and expel filtered water through the *excurrent siphon*.

Most adult tunicates are sessile and permanently attached to rocks or other hard surfaces on the ocean floor; others swim in the pelagic zone (i.e., open sea) as adults.

8. B is correct.

Echinoderms are the invertebrate predecessors of the chordates and include starfish, sea urchins, and sea cucumbers. They are characterized by a primitive vascular system known as the water vascular system.

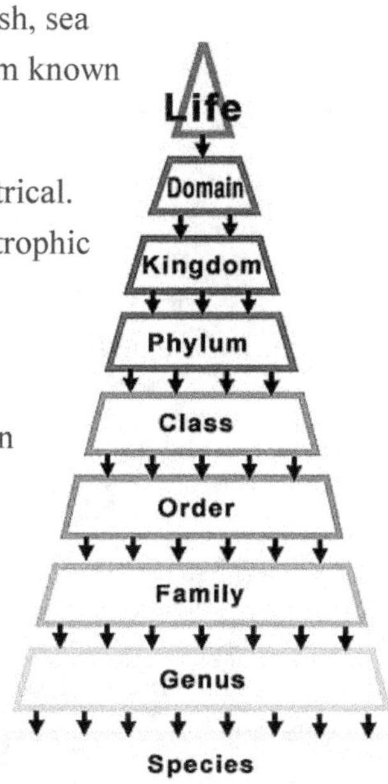

Adult echinoderms have radial symmetry, while the larvae are bilaterally symmetrical. They move with structures known as tube feet, have no backbone, and are heterotrophic (i.e., they do not synthesize their food).

Crayfish are in the phylum Arthropoda, which includes insects.

Arthropods are characterized by segmented bodies covered in a chitin exoskeleton and jointed appendages.

9. B is correct.

Kingdom encompasses many phyla.

Mnemonic: **D**arn **K**ing **P**hillip **C**ame **O**ver **F**or **G**ood **S**oup:

Domain → kingdom → **phylum** → class → order → family → genus → species

10. D is correct.

Homo sapiens belong to the phylum Chordata. *Taxonomy of humans*:

> Domain: Eukaryota → Kingdom: Animalia → Phylum: Chordata → Subphylum: Vertebrata → Class: Mammalia → Order: Primates → Family: Hominidae → Genus: Homo → Species: *H. sapiens* → Subspecies: *H. s. sapiens*

11. B is correct.

Organisms of phylum *Chordata* have a dorsal notochord, while subphylum Vertebrata has a backbone.

Nonvertebrate chordates include amphioxus (lancelet) or unrelated tunicate worm.

Lancelet (or *amphioxus*) is the modern subphylum Cephalochordata and is vital in zoology – it provides indications about the origins of the vertebrates.

Lancelets are comparison points for tracing vertebrates' evolution. Lancets are the archetypal vertebrate form.

Lancelets split from vertebrates more than 520 million years ago; their genomes hold clues about evolution, specifically how vertebrates have adapted old genes for new functions.

Vertebrates include most of the phylum Chordata with about 64,000 species.

Vertebrates include amphibians, reptiles (e.g., lizards), mammals, birds, jawless and bony fish, sharks, and rays.

C and D: shark and lamprey eel are vertebrates with cartilaginous skeletons.

12. A is correct.

Cell, tissue, organ, organism, population, and community represent the correct ordering of the levels of complexity at which life is studied, from simplest to most complex.

13. A is correct.

Notochord is *ventral* to the *neural tube* and forms during gastrulation.

The notochord induces *neural plate formation* (during neurulation) to synchronize neural tube development (precursor to the central nervous system).

The notochord is a flexible rod-shaped structure in embryos of all chordates. It is composed of cells derived from the mesoderm and defines the primitive axis of the embryo.

In lower chordates, this chord remains throughout the life of the animal.

In higher chordates (e.g., humans), the notochord exists during embryonic development and disappears.

If the notochord persists, it functions as the primary axial support. In most vertebrates, it becomes the nucleus pulposus of the intervertebral disc.

B: *notochord remains* in the lower chordates, such as the amphioxus and the tunicate worm. It is not vestigial (not a genetically determined structure that lost its ancestral function) because it disappears.

D: *echinoderms* are the invertebrate predecessors of chordates and include starfish, sea urchins, and cucumber. Echinoderms do *not* possess a notochord.

14. D is correct.

Chordates are animals that, for at least some of their life cycle, possess:

1) a notochord

2) a dorsal neural tube

3) pharyngeal slits

4) an endostyle

5) a post-anal tail

Chordate phylum includes the subphyla Vertebrata (e.g., mammals, fish, amphibians, reptiles, birds), Tunicata (e.g., salps, sea squirts), and Cephalochordata (e.g., lancelets).

Vertebrata subphylums have a backbone, but it is not a shared characteristic of chordates.

Chordates are a phylum that shares a bilateral body plan and have at some stage the following:

Notochord is a relatively stiff rod of cartilage extending inside the body.

Among vertebrate sub-group chordates, the notochord develops into the spine. This helps the animal swim by flexing its tail in aquatic species.

Dorsal neural tube (for vertebrates, including fish) develops into the spinal cord, the leading communications trunk of the nervous system.

Pharyngeal slits comprise the throat immediately behind the mouth (modified fish gills). In some chordates, they are part of a filter-feeding system extracting food from its environment.

Endostyle is a groove in the ventral wall of the pharynx. It stores iodine and may be a precursor of the vertebrate thyroid gland. In filter-feeders, it produces mucus to gather food particles, which help transport food to the esophagus.

Post-anal tail is a muscular tail extending behind the anus.

15. B is correct.

Migratory birds nesting on islands is not behavioral, temporal, or geographical isolation leading to speciation. The birds are migratory and are not geographically isolated.

16. D is correct.

Sponges are Porifera (i.e., pore or Ostia-bearing) phylum animals.

Like other animals, they are multicellular, heterotrophic, and lack cell walls. Sponges lack tissues and organs and have no body symmetry, differentiating them from other animals.

Heterotrophy means obtaining food and energy by consuming other organic substances rather than through sunlight or inorganic compounds.

17. B is correct.

Homology is a relationship between structures or DNA derived from a common ancestor.

Homologous traits are explained by descent from a common ancestor.

Analogous organs have similar functions in two taxa, not in the last common ancestor but instead evolved separately.

Homologous sequences are *orthologous* when descended from ancestral sequences separated by speciation.

When a species diverges into two species, copies of a single gene in the two resulting species are *orthologous*.

Orthologues (i.e., orthologous genes) are genes in different species that originated by vertical descent from a single gene of the last common ancestor.

Orthology is defined by ancestry.

Due to gene duplication and genome rearrangement, the ancestry of genes is difficult to ascertain.

The strongest evidence that two similar genes are orthologous is usually a *phylogenetic analysis* of *gene lineage*.

Orthologous genes often, but not always, have the same function.

	Analogous leg	Analogous flipper
Homologous mammals	 Cat leg	 Whale flipper
Homologous insects	 Praying mantis leg	 Water boatman flipper

Homologous (common ancestor) vs. analogous structures (similar function but evolved separately)

18. C is correct.

Arthropods are invertebrate animals with an exoskeleton, segmented body, and jointed appendages and include insects, arachnids, myriapods, and crustaceans.

19. A is correct.

Animal cells do not have cell walls, distinct from plant cells. Another distinct feature of plant cells is the presence of chloroplasts in the structure.

20. C is correct.

Annelids are a large phylum of segmented worms, with each segment having the same set of organs.

Coelom is a cavity lined by a mesoderm-derived epithelium. Organs formed inside a coelom can grow, freely move, and develop independently of the body wall while protected by fluid cushions.

Organisms are classified into three groups:

1) *Coelomates* are animals with a fluid-filled body cavity (i.e., coelom) and complete lining (peritoneum derived from mesoderm), allowing organs to be attached and suspended while moving freely within the cavity.

Coelomates include most bilateral animals, including all vertebrates.

2) *Pseudocoelomates* are animals with a pseudocoelom fully functional body cavity (e.g., roundworm).

Pseudocoelomate *organ*s are held loosely and not as well organized as coelomates.

Mesoderm-derived tissue partly lines the fluid-filled body cavity.

Pseudocoelomates are *protostomes*, but not all protostomes are pseudocoelomates.

3) *Acoelomates* are animals with no body cavity (i.e., flatworms). Semi-solid mesodermal tissues between the gut and body wall hold their organs in place.

21. B is correct.

From embryological studies, ancient chordates were closely related to the ancestors of echinoderms, which are marine animals such as sea stars and sea urchins.

Sea anemones are a group of water-dwelling, predatory animals.

22. C is correct.

Echinoderms are marine animals such as sea stars, sea urchins, and sand dollars characterized by radial (usually five-point) symmetry.

Most echinoderms can regenerate tissue, organs, and limbs and reproduce asexually.

23. A is correct.

Cnidaria is a phylum of animals with over 10,000 species exclusively in aquatic and marine environments.

Cnidaria bodies consist of a *non-living jelly-like substance* sandwiched between two layers of epithelium that are mostly one cell thick.

Their distinguishing feature is *cnidocytes*, specialized cells mainly for capturing prey.

continued…

Cnidaria has two basic body forms: *swimming medusae* and *sessile polyps*; each is radially symmetrical with mouths surrounded by tentacles that bear cnidocytes.

Both forms have a single orifice and body cavity used for digestion and respiration.

Many cnidarian species form colonies of single organisms with Medusa-like or polyp-like zooids or both.

B: *echinoderms* are a phylum of marine animals, and adults have radial (usually five-point) symmetry (e.g., sea urchins, sand dollars, starfish, and sea cucumbers).

24. C is correct.

Chordates are animals with a *notochord* (stiff rod of cartilage that extends inside the body), *dorsal neural tube*, and *pharyngeal slits*, which are openings of the pharynx to allow aquatic organisms to extract oxygen from water and excrete carbon dioxide (i.e., gills).

25. A is correct.

A hollow *nerve cord* (a single hollow nerve tissue tract) constitutes the central nervous system of chordates.

26. B is correct.

Cephalization is an evolutionary trend when nervous tissue becomes concentrated toward one end of an organism over many generations. This process produces a head region with sensory organs.

The process of cephalization is intrinsically connected with a change in symmetry. Cephalization accompanied the shift to bilateral symmetry in flatworms.

In addition to a concentration of sense organs, annelids place the mouth in the head region. This process is tied to developing an anterior brain in the chordates from the notochord.

An exception to the trend of cephalization throughout evolutionary advancement is the phylum Echinodermata.

Echinodermata, despite having a bilateral ancestor, developed with 5-point symmetry) and no concentrated neural ganglia or sensory head region.

Some echinoderms have developed bilateral symmetry secondarily.

27. B is correct.

Deuterostomes differ from protostomes in embryonic development.

In deuterostomes, the first opening (i.e., blastopore) becomes the anus. In protostomes, it becomes the mouth.

Chordates, echinoderms, and hemichordates are deuterostomes.

28. A is correct.

Pharyngeal pouches are filter-feeding organs in fish that develop into gills used for respiration.

29. D is correct.

Corals are marine invertebrates (phylum Cnidaria) typically in compact colonies with identical individual *polyps*. The group includes the reef builders of tropical oceans secreting calcium carbonate, forming a hard skeleton.

A coral "head" is a colony of myriad genetically identical polyps. Each polyp is a spineless animal, typically a few millimeters in diameter and a few centimeters in length.

Tentacles surround a central mouth opening, and an *exoskeleton* is excreted near the base.

Over many generations, colonies create a large skeleton structure.

Individual heads grow by the asexual reproduction of polyps.

Corals breed *sexually* by *spawning:* polyps of the same species release gametes simultaneously for several nights, cycling around full moons.

30. B is correct.

The first amphibians developed from lobe-finned fish about 370 million years ago during the Devonian period.

Lobe-finned fish had multi-jointed leg-like fins that allowed crawling at the bottom of the sea. Their bony fins evolved into limbs, becoming the ancestors of tetrapods, including amphibians, reptiles, birds, and mammals.

31. D is correct.

Notochord is a flexible rod essential in vertebrate development skeletal element of developing embryos.

The notochord is related to cartilage, serving as the axial skeleton of the embryo until other elements form.

32. D is correct.

Chordates are animals that, for at least some period of their life cycles, possess the following: a notochord, a dorsal neural tube, pharyngeal slits, an endostyle, and a post-anal tail.

Notochord is a flexible rod-shaped structure in embryos of chordates. It is composed of mesoderm-derived cells that define the embryo axis.

Notochord is a relatively stiff rod of cartilage that extends along the inside of the body.

Among the vertebrate sub-group of chordates, the notochord develops into the spine, and in wholly aquatic species, this helps the animal swim by flexing its tail.

In lower chordates, the notochord remains throughout life.

In higher chordates (e.g., humans), the notochord exists only during embryonic development and disappears.

If it persists throughout life, it functions as the primary axial support, while in most vertebrates, it becomes the nucleus pulposus of the intervertebral disc.

Notochord is ventral to the neural tube and forms during gastrulation. The notochord *induces neural plate formation* (during neurulation) to synchronize neural tube development (precursor to the central nervous system).

33. C is correct.

Earthworms are part of the Annelid phylum. *Pseudocoelom* is a fluid-filled cavity between the body wall and intestine of certain invertebrates such as roundworms and hookworms.

All other statements are true.

34. C is correct.

Notochords are similar to cartilage, made of type II collagen. Collagen makes notochords soft and flexible.

The notochord is a flexible rod between the nerve cord and the digestive tract.

35. B is correct.

Cnidaria bodies consist of a non-living jelly-like substance sandwiched between two layers of epithelium that are mostly one cell thick.

Cnidaria has two body forms: swimming medusae and sessile polyps, each *radially symmetrical* with mouths surrounded by tentacles with cnidocytes.

Echinoderms are a phylum of marine animals, and adults have radial (usually five-point) symmetry (e.g., sea urchins, sand dollars, starfish, and sea cucumbers).

36. A is correct.

Chordates share several key features, such as a notochord, a dorsal hollow nerve cord, pharyngeal slits, and a post-anal tail.

Of the characteristics visible on a cat, a tail extending beyond the anus makes it a chordate.

37. B is correct.

Vertebral column is the backbone or spine, a segmented series of bones separated by intervertebral discs.

38. B is correct.

Ovoviviparous fish (e.g., guppies, angel sharks) have eggs that develop inside the mother's body after *internal fertilization*, whereby each embryo develops within its egg.

Yolk nourishes the embryo, which directly receives little or no nourishment from the mother.

Viviparous fish retain the eggs and nourish the embryos. They have a structure analogous to the placenta (mammals), connecting the mother's blood with the embryo.

39. C is correct.

Chordates are defined by the presence of a *notochord.*

Not all chordates are vertebrates, have paired appendages, or have backbones.

Chordates are *deuterostomes* where the first opening (i.e., blastopore) becomes the anus.

Chordates are a phylum that shares a bilateral body plan and have at some stage the following:

Notochord is a relatively stiff rod of cartilage extending inside the body.

Among vertebrate sub-group chordates, the notochord develops into the spine. This helps the animal swim by flexing its tail in aquatic species.

Dorsal neural tube (for vertebrates, including fish) develops into the spinal cord, the leading communications trunk of the nervous system.

Pharyngeal slits comprise the throat immediately behind the mouth (modified fish gills). In some chordates, they are part of a filter-feeding system extracting food from its environment.

Endostyle is a groove in the ventral wall of the pharynx. It stores iodine and may be a precursor of the vertebrate thyroid gland. In filter-feeders, it produces mucus to gather food particles, which help transport food to the esophagus.

Post-anal tail is a muscular tail extending behind the anus.

40. A is correct.

Chordates are animals with a notochord, a dorsal neural tube, pharyngeal slits, endostyle, and a post-anal tail for at least some of their life cycles.

41. B is correct.

Hominoids are called *apes*, but *ape* is used differently.

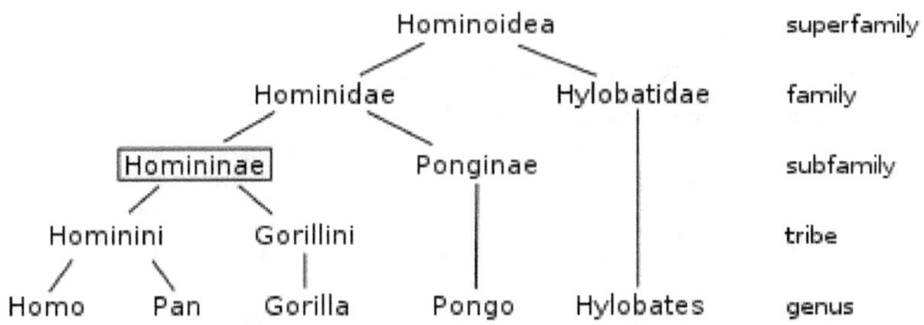

Homininae cladogram divides into genus homo, pan and gorilla

Until 1970, Hominidae meant humans only, with non-human great apes assigned to the family Pongidae. Later discoveries led to revised classifications, with the great apes then united with humans (now in subfamily Homininae) as members of the family Hominidae

By 1990, molecular biology techniques classified gorillas and chimpanzees as more closely related to humans than orangutans. Gorillas and chimpanzees are in the subfamily Homininae.

Hominines are a subfamily of Hominidae that includes humans, gorillas, chimpanzees, bonobos, and hominids which arose after the evolutionary split from orangutans.

Homininae cladogram has three main branches, which lead to gorillas, chimpanzees, bonobos, and humans.

continued...

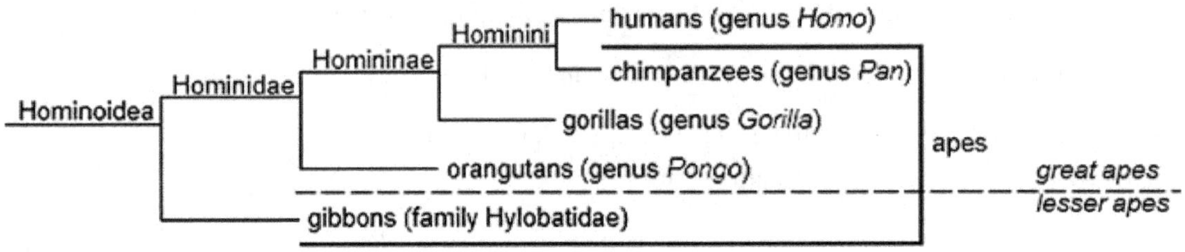

Homininae cladogram includes gorillas, chimpanzees, and humans

42. D is correct.

Fish evolved with *jawless fish*, followed by *bony fish with a jaw.*

Lastly, fish with leg-like fins, which allowed for movement along the seabed, evolved into limbs of *tetrapods.*

43. B is correct.

Jaws and limbs are characteristic of *vertebrates.*

44. A is correct.

Yolk nourishes the embryo, which directly receives little or no nourishment from the mother.

Amniotic fluid, surrounded by the amnion, is a liquid environment around the egg, protecting it from shock.

Embryonic membranes include:

1) *chorion* lines the inside of the shell and permits gas exchange;

2) *allantois* is a saclike structure developed from the digestive tract and functions in respiration, excretion, and gas exchange with the external environment;

3) *amnion* encloses *amniotic fluid* as a watery environment for embryogenesis and shock protection;

4) *yolk sac* encloses the yolk and transfers food to the developing embryo.

45. C is correct.

External fertilization is male sperm fertilizing an egg outside of the female's body, seen in amphibians and fish.

Mammals, which are further in the evolution of vertebrate groups, use internal fertilization for reproduction.

46. A is correct.

Prehensile refers to an appendage or organ adapted for holding or grasping.

47. D is correct.

Mammals are *endothermic* (warm-blooded) vertebrates with *body hair* and the ability to feed babies *milk.*

In the embryonic development of vertebrates (e.g., mammals), *pharyngeal pouches* form that develop into essential structures such as the eardrum and thymus gland.

48. A is correct.

Hominoids are called *apes*, but *ape* is used differently.

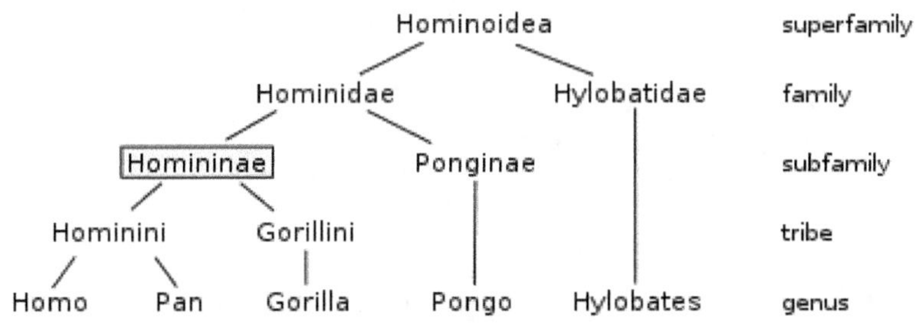

Homininae cladogram divides into genus homo, pan and gorilla

Until 1970, Hominidae meant humans only, with non-human great apes assigned to the family Pongidae. Later discoveries led to revised classifications, with the great apes then united with humans (now in subfamily Homininae) as members of the family Hominidae

By 1990, molecular biology techniques classified gorillas and chimpanzees as more closely related to humans than orangutans. Gorillas and chimpanzees are in the subfamily Homininae.

Hominines are a subfamily of Hominidae that includes humans, gorillas, chimpanzees, bonobos, and hominids which arose after the evolutionary split from orangutans.

Homininae cladogram has three main branches, which lead to gorillas, chimpanzees, bonobos, and humans.

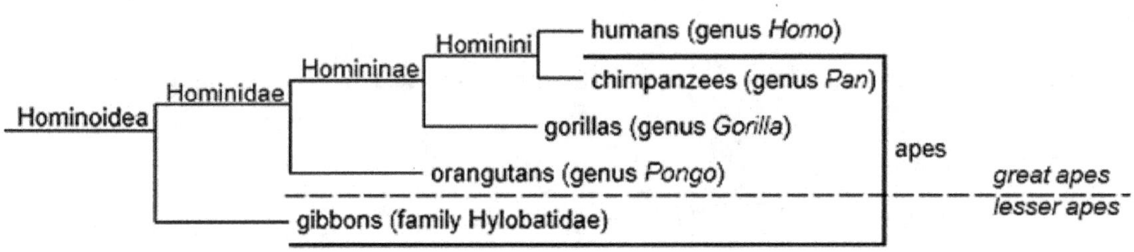

Homininae cladogram includes gorillas, chimpanzees, and humans

49. C is correct.

Cartilaginous fishes have a skeleton made of cartilage rather than bone.

Sharks, skates, rays, and sturgeons have cartilaginous skeletons, while hagfish have cartilaginous skulls.

50. A is correct.

Old World monkeys are primates in the clade of Catarrhini. They are native to Africa and Asia today, inhabiting environments from tropical rainforest to savanna, shrubland, and mountainous terrain.

Old World monkeys include familiar species of nonhuman primates (e.g., baboons, macaques). They are medium to large and range from arboreal (i.e., locomotion in trees) to fully terrestrial (e.g., baboons).

New World monkeys are the five families of primates in Central and South America and portions of Mexico. They are small to mid-sized primates.

continued...

New World monkeys differ from Old World monkeys in several aspects.

The prominent phenotypic distinction is the nose, commonly used to distinguish between groups. New World monkeys have noses flatter than the narrow noses of Old-World monkeys and have side-facing nostrils.

New World monkeys are the only monkeys with prehensile tails.

Old-World monkeys have shorter, non-grasping tails.

51. C is correct.

A reptile's body temperature is determined by the temperature of its surrounding environment, meaning it could fluctuate more often than endothermic (*warm-blooded*) animals.

Birds are *endothermic*, so their body temperature remains constant and is warmer than reptiles.

52. D is correct.

Earthworms are soft-bodied segmented worms with nephridia, paired organs in each worm segment.

Nephridia remove nitrogenous waste.

53. D is correct.

Monotremes are mammals laying eggs instead of birthing live young like marsupials and placental mammals.

The surviving species of monotremes are indigenous to Australia and New Guinea, though evidence suggests they were once widespread.

The existing monotreme species are the *platypus* and four species of echidnas (i.e., spiny anteaters).

54. B is correct.

Legs and limbs indicate that the animal is *segmented.*

55. D is correct.

Lemurs are a clade of *strepsirrhine primates*, characterized by a *moist nose tip.*

Lemurs are primarily *nocturnal animals* and are on the island of Madagascar.

56. A is correct.

Birds have a *complex respiratory system.*

Upon inhalation, 75% of the fresh air bypasses the lungs and flows directly into a posterior air sac extending from the lungs, connects with air spaces in the bones, and fills them with air.

25% of the air goes directly into the lungs. When the bird exhales, the used air flows out of the lung, and the stored fresh air from the posterior air sac is simultaneously forced into the lungs.

The bird's lungs receive fresh air during inhalation and exhalation. Sound production uses the *syrinx*, a muscular chamber incorporating multiple tympanic membranes that diverge from the trachea's lower end. In some species, the *trachea* is elongated to increase *vocalization volume* and *perceived size.*

57. D is correct.

Grasping hands of primates (i.e., having opposable thumbs) are an adaptation to life in the trees.

Thumbs help primates grasp objects (e.g., tree branches) firmly.

58. B is correct.

Chordates are both vertebrates and invertebrates.

Chordates share features, such as a:

notochord,

dorsal hollow nerve cord,

pharyngeal slits, and

post-anal tail.

In vertebrate chordates, the notochord is replaced by a *vertebral column.*

59. B is correct.

Radial symmetry is present in organisms with symmetry around a central axis.

Radial symmetry is in *starfish* or *sea urchins*, whose body parts extend equally outward from a center point.

60. C is correct.

Deuterostome is distinguished by embryonic development, whereby the first opening (i.e., blastopore) becomes the anus.

Protostomes develop with the first opening (i.e., blastopore) as the *mouth.*

Deuterostome mouth develops at the opposite end of the embryo from the blastopore. A digestive tract develops in the middle connecting the two.

Humans are deuterostomes.

Invertebrates are *protostomes.*

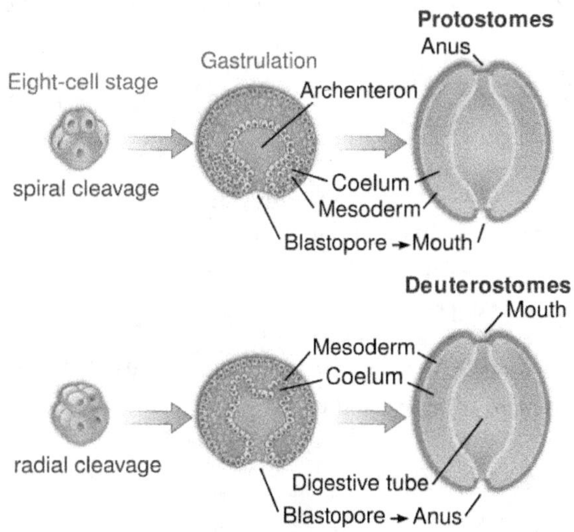

Protostomes (mouth at blastopore) vs. deuterostomes

61. A is correct.

Shapes of forelimbs are an expected variant for vertebrates using these appendages for utility.

62. C is correct.

Homo habilis (or *Australopithecus habilis*) is a species of the *Hominini* tribe, which lived from approximately 2.33 to 1.44 million years ago.

There has been controversy regarding its placement in the genus *Homo* rather than *Australopithecus*; its brain size is 550-687 cm^3, rather than 363-600 cm^3.

Recent findings on brain size support its traditional placement in the genus *Homo* and the theory that *Homo habilis* is the common ancestor.

63. B is correct.

Feathers are the most critical characteristic separating birds from other living animals.

Birds are the only *chordates* with feathers.

Other chordates, such as mammals, have *fur or hair*.

64. B is correct.

Double-loop circulatory system is one in which blood flows through the heart twice.

Pulmonary circulation involves blood flowing between the heart and lungs.

Systemic circulation is blood movement from the heart to the body and back.

Lungs are essential for *double-loop circulatory systems*.

65. A is correct.

Cladograms of animals support segmentation evolving more than once in different branches.

B: *cephalization* is an evolutionary trend whereby nervous tissue (over many generations) becomes concentrated toward one end of an organism to produce a head region with sensory organs.

66. D is correct.

Green algae are classified as *plants* because they have *gene segments* similar to land plants, use chlorophyll *a*, and have cellulose cell walls.

67. C is correct.

Evidence suggests that green algae were the first plants.

Land plants evolved from *filamentous green algae*, which invaded land about 410 million years ago

Land plants and green algae have *cellulose* in their cell walls and share many *photopigments for photosynthesis*.

Notes for active learning

Notes for active learning

Evolution and Natural Selection – Detailed Explanations

1. D is correct.

Charles Darwin (1809-1882), during his trip to the Galapagos Islands in the 1830s, observed several species of finches that varied among islands.

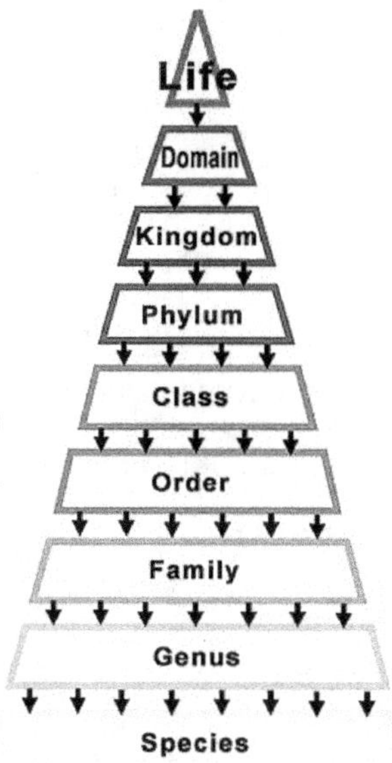

2. A is correct.

Organisms are classified by evolutionary relationships:

Kingdom → Phylum → Class → Order → Family → Genus → Species

The largest group (i.e., kingdom) divides into smaller subdivisions.

Each smaller group has common characteristics. Of the answers, a genus is the smallest subdivision, and organisms in the same genus are more similar than organisms classified as the same family, order, class, or kingdom.

3. D is correct.

Mutations are the source of genetic variation for natural selection.

4. A is correct.

The chloroplast is a plant cell organelle that may have originated from cyanobacteria through *endosymbiosis*.

Endosymbiotic theory proposes that a eukaryotic cell engulfed a photosynthesizing cyanobacterium about one billion years ago. The photosynthesizing cyanobacterium became a permanent resident in the cell because it escaped the phagocytic vacuole it was contained in.

Chloroplasts contain their DNA and ribosomes (similar to prokaryotic ribosomes) and undergo autosomal replication (i.e., replicating independently of the cell cycle), supporting this theory.

Cyanobacteria obtain energy through photosynthesis and have the color of the bacteria (i.e., blue).

Named blue-green algae, this is a misnomer as cyanobacteria are prokaryotes, while algae are eukaryotes.

By producing oxygen as a by-product gas of photosynthesis, cyanobacteria are thought to have converted Earth's early reducing atmosphere into an oxidizing one, dramatically changing the composition of life forms by stimulating biodiversity and leading to the near-extinction of oxygen-intolerant organisms.

Endosymbiotic theory proposes that chloroplasts in plants, mitochondria in eukaryotes, and eukaryotic algae evolved from cyanobacterial ancestors.

C: *mitochondria* evolved from free-living prokaryotic heterotrophs that, similarly to chloroplasts, entered eukaryotic cells and established a symbiotic relationship with the host (the theory of *endosymbiosis*).

Mitochondria contain circular DNA (similar in size and composition to prokaryotes), ribosomes (similar to prokaryotic ribosomes), and unique proteins in the organelle membrane (similar in composition to prokaryotes), and replicate independently, providing further evidence to support this theory.

5. D is correct.

Evolution is the long-term changes in a population gene pool caused by environmental selection pressures.

I: *random mutation* creates new alleles selected for (or against) as a phenotypic variation for natural selection (i.e., survival of the fittest).

II: *reproductive isolation* of a population is a crucial component for speciation.

Geographically isolated populations often diverge to yield reproductive isolation leading to speciation.

III: *speciation* might happen in a population with no specific extrinsic barrier to gene flow.

For example, a population extends over a broad geography, and mating in the population is not random.

Individuals in the far west would have zero probability of mating with individuals in the far east range. This results in reduced *gene flow* but not total isolation. Such a situation may (or not) be sufficient for speciation.

Speciation would be promoted by different selective pressures at opposite ends of the range, which would alter gene frequencies in groups at different ranges so they would not be able to mate if they were reunited.

6. C is correct.

Darwin (1809-1882) hypothesized that the mechanism of evolution involves natural selection.

7. D is correct.

Prokaryotes are primarily unicellular organisms, although a few, such as mycobacterium, have multicellular stages in their life cycles or create large colonies like cyanobacteria.

Eukaryotes are often multicellular and are typically much larger than prokaryotes.

Eukaryotes have *internal membranous structures* (i.e., organelles) and a cytoskeleton composed of microtubules, microfilaments, and intermediate filaments, essential for cellular organization and shape.

8. D is correct.

Archaean Eon contains the oldest known fossil record, dating to about 3.5 billion years.

9. A is correct.

Urey and Miller demonstrated in 1953 that organic molecules might be created from inorganic molecules under the primordial earth conditions. They conducted an experiment that simulated the conditions thought at the time to be on the early Earth and tested for the occurrence of chemical origins of life.

Urey-Miller did not prove the existence of life on earth. However, the original experiments were able to show that there were over 20 different amino acids produced in Miller's original experiments.

10. B is correct.

Nature uses many methods for fertilization, development, and care of offspring.

Internal fertilization involves internal development and much care for offspring. These organisms (e.g., humans and many mammals) produce few offspring, but a large percentage reach adulthood.

External fertilization involves external development and little care for offspring. These organisms (e.g., many species of fish) produce large numbers of sperm and eggs because relatively few sperm and eggs interact to produce a zygote. Few zygotes survive without physical protection from predators, and millions of eggs and sperm must be released to perpetuate the species.

A: *protective coloring* might help for survival against predators, but not related to caring for the young.

C: *laying eggs* is not related to not caring for the young (most birds care for their young).

D: not relevant to the survival rate of the offspring.

11. C is correct.

Mutualism is when two species live in close association, and both benefit.

A: *parasitism* is a relationship between two organisms where one organism benefits while the other is harmed.

B: *commensalism* involves a relationship between organisms where one benefits without affecting another.

D: *mimicry* (in evolutionary biology) is the similarity of one species to another, which protects one or both and occurs when a group (the mimics) evolve to share common characteristics with another group (the models). This similarity can be in appearance, behavior, sound, scent, and location, with the mimics found in similar places to their models.

12. D is correct.

Disruptive selection shifts allele frequencies towards variants of extremes, leading to two subpopulations.

These subpopulations do not favor the intermediate variants of the original population and grow disparate over time, likely leading to speciation.

A: *sexual selection* gives an individual an advantage in finding a mate.

Sexual selection often acts opposite to the effects of natural selection (e.g., a brighter plume may make a bird vulnerable to predators but gives it an advantage during reproduction).

It is hypothesized that sexual selection leads to sexual dimorphism (i.e., phenotypic differences between males and females of the same species).

B: *stabilizing selection* is a shift in phenotypes towards intermediate by disfavoring variants at either extreme.

Stabilizing selection reduces variation within the species and perpetuates similarities between generations.

C: *directional selection* is the population shift in allele frequencies towards variants of one extreme.

13. D is correct.

The similarity among the embryos of fish, amphibians, reptiles, and humans is evidence of common ancestry.

14. A is correct.

Mutualism (i.e., symbiotic) is a relationship between two organisms where both benefit. For example, soybeans depend on humans to survive while humans are provided with food. *Both species* benefit from the relationship.

Commensalism is a class of relationships where one organism benefits without affecting the other.

Mutualism (i.e., symbiotic) is a relationship between two organisms where both benefit.

Parasitism is a relationship between two organisms where one benefits while the other is harmed.

15. B is correct.

Natural selection leads to a population most adapted (*fittest*) to its current environment.

16. A is correct.

From an evolutionary perspective, *passing genetic information* to progeny is paramount.

17. D is correct.

Inherited traits determined by heredity are genes.

18. D is correct.

The evolutionary process of natural selection *works on existing genetic variation* within a population.

19. C is correct.

Genetic drift changes allele frequency within a population due to random sampling.

Allele frequency of a population is the fraction of the same alleles.

Offspring have the same alleles as the parents, and probability determines if an individual survives and reproduces.

Genetic drift may cause allele variants to disappear and thereby reduce genetic variation.

20. A is correct.

The characteristic of human height likely results from polygenic (many genes contribute) inheritance.

21. C is correct.

The oldest known fossilized prokaryotes were laid down approximately 3.6 billion years ago, about 1 billion years after the formation of the Earth's crust.

Eukaryotes appear in the fossil record later and may have formed from the aggregation of multiple prokaryotes.

The oldest known fossilized eukaryotes are about 1.7 billion years old.

However, some genetic evidence suggests eukaryotes appeared as early as 3 billion years ago.

D: *protists* are a large and diverse group of eukaryotic microorganisms which belong to the kingdom Protista. There have been attempts to remove the kingdom from the taxonomy, but it is commonly used. Some professional organizations and institutions prefer the name Protista.

Protists are unicellular, or they are multicellular without specialized tissues. Besides their relatively simple levels of organization, protists do not have much in common.

The straightforward cellular organization distinguishes protists from other eukaryotes (e.g., fungi, animals).

Protists live in almost any environment with liquid water.

Many protists, such as algae, are photosynthetic and are vital primary producers in ecosystems, particularly in the ocean as part of the plankton.

Protists include pathogenic species, such as the kinetoplastid *Trypanosoma brucei,* causing sleeping sickness, and species of the apicomplexan *Plasmodium,* causing malaria.

22. D is correct.

Morphological or physical similarity between organisms is insufficient for classifying the same species.

Species are organisms that mate and produce *viable* and *fertile* offspring (give rise to additional offspring).

Organisms appearing different (e.g., dog breeds) can be the same species if they mate and produce fertile offspring.

A: organisms can be the same species yet have different gene varieties (i.e., alleles) at the same gene locus.

For example, blue-eyed and green-eyed individuals can mate and produce fertile offspring.

B: *no relationship* between producing viable, fertile offspring and blood type.

C: *somatic cells* are body cells and not reproductive cells (i.e., gametes) or haploid (i.e., 1N or monoploid).

Somatic cells are diploid (2N) and undergo *mitosis* for cell division.

Gametes are haploid (1N) and undergo *meiosis* for cell division.

23. C is correct.

Gene pool is the population's total collection of alleles.

24. C is correct.

A pivotal point in Darwin's explanation of evolution is that any trait that confers an increase in the *probability* of its possessor surviving and reproducing is favored and spread through the population.

25. C is correct.

If an allele's frequency in a population is 0.7, the alternate allele is 0.3.

26. B is correct.

Natural selection is how mutations are selected for (or against) in the environment.

If the resulting phenotype offers some degree of fitness, the genes are passed to the next generation.

A: *natural selection* includes selection pressures and needs a population with genetic variation (random mutations) to select the fittest organisms.

C: *Darwin's theory* depends on more than mere mutations. It is based on over-reproduction. Offspring are selected for (or against) when their genetic makeup is fit for the environment where the organism is located.

Organisms most fit (for their environment) and pass their genes (i.e., gametes) to the next generation. The result is survival (greatest reproduction) of the fittest organisms within the environment.

D: *Lamarck* proposed that if traits were used (e.g., stretching of a giraffe's neck), these acquired traits are passed to the next generation.

Acquired characteristics (phenotypic changes) do not affect the genes and are not passed to the next generation via the gametes (e.g., sperm and egg).

27. B is correct.

Fewer copies of an allele magnify the effect of *genetic drift*.

Genetic drift is negligible when there are many copies of an allele.

28. D is correct.

A population that survived a bottleneck and recovered to its original size has less genetic variation than before.

29. B is correct.

Founder effect and *population bottleneck* differ because the founder effect requires isolating a small group from a larger population.

30. B is correct.

Life originated in an atmosphere with *little or no oxygen.*

Until the Great Oxygenation Event (GOE) about 2.4 billion years ago, there was no atmospheric free oxygen.

Cyanobacteria appeared about 200 million years before the GOE began producing oxygen by photosynthesis.

Before the GOE, free oxygen produced was chemically captured by dissolved iron or organic matter.

The GOE was when oxygen sinks became saturated with oxygen produced by cyanobacterial photosynthesis. After GOE, excess free oxygen accumulated in the atmosphere.

Free oxygen is toxic to obligate anaerobic organisms. The rising concentrations eradicated most of Earth's anaerobic organisms. Cyanobacteria were responsible for one of Earth's most significant extinction events.

Free oxygen reacts with atmospheric methane (a greenhouse gas), which is oxidized to CO_2 and H_2O.

Free oxygen has been an essential constituent of the atmosphere ever since. Periods with much oxygen in the atmosphere are associated with the rapid development of animals.

Today's atmosphere contains about 21% oxygen, which is enough for the rapid development of animals.

31. B is correct.

Founder effect is the likely explanation that a population of humans has a higher rate of polydactyly (extra fingers or toes) than the human population.

32. C is correct.

Gene flow occurs through migration.

33. C is correct.

D: first organisms had nothing else to eat.

Even with millions of organisms, one organism has to eat many, which would have exhausted the supply.

34. B is correct.

Evolutionary fitness measures *reproductive success* (i.e., the hereditary succession of gene pools).

35. A is correct.

Lamarck's theory of evolution proposed that enhanced structures (e.g., a giraffe's neck) arise based on use.

Lamarck (1744-1829) had a false premise that favorable characteristics obtained by use were inheritable.

For example, giraffes stretch their necks to reach leaves on branches. He proposed that offspring inherit the trait (longer necks) based on use. However, only changes in the DNA of gametes (e.g., egg or sperm) are inherited.

B: de Vries (1848-1935) confirmed Mendel's observations with different plant species.

C: Mendel (1822-1884) defined classical genetics by experiments with inheritable traits (e.g., seed color, wrinkled vs. round seeds) in pea plants.

Mendel described the principles of dominance, segregation, and independent assortment. The unit of inheritance (i.e., genes) was unknown.

D: *Darwin's Theory of Natural Selection* states that pressures in the *environment select the fittest organism* to survive and reproduce.

Natural Selection by Mendel is based on six principles.

1) **Overpopulation**: more offspring are produced than can survive from insufficient food, air, light, and space to support the population;

2) **Variations**: offspring have differences (i.e., variations) in characteristics compared to the population. Darwin did not know the reason, but de Vries later suggested that *genetic mutations* cause variations. Some mutations are *beneficial* fittest), but most are *detrimental*;

continued...

3) **Competition**: developing populations compete for necessities (e.g., food, air, light). Many young die, while the number of adults remains about constant for generations;

4) **Natural selection**: some organisms have phenotypes (i.e., variations) conferring an advantage over others;

5) **Inheritance of variations**: individuals reproduce and transmit favorable phenotypes to offspring. Favored alleles (i.e., variations of genes) eventually dominate the gene pool;

6) **Evolution of a new species**: natural selection (fittest) propagates favorable genes and phenotypes over generations.

Favorable changes result in significant changes in the gene pool for the evolution of a new species (i.e., organisms that reproduce and yield fertile offspring).

36. D is correct.

Non-poisonous butterflies evolve color changes to look like poisonous butterflies in *directional selection.*

37. A is correct.

Genetic variation would likely be decreased by stabilizing selection.

38. C is correct.

Asexual reproduction is more efficient than sexual reproduction in the number of offspring produced per reproduction, in the amount of energy invested in this process, and in the amount of time invested in the development of the young (before and after birth).

Asexual reproduction relies on genetic mutation for phenotypic variability to be passed to future generations since it produces genetic clones of the parent.

Sexual reproduction involves the process of meiosis – two rounds of cell division and the likelihood of cross-over (during prophase I).

A: new phenotype may be disadvantageous *or* advantageous.

B: a much greater probability of a mutation occurring with sexual reproduction (i.e., at the chromosomal level), known as chromosomal aberrations.

D: sexual reproduction requires more energy and time per progeny (i.e., offspring).

Species reproducing sexually have a selective advantage because of *gene recombination* during fertilization.

The fusion of two genetically unique nuclei (haploid sperm and haploid egg nucleus) yields a unique (2N) zygote.

Fertilizing two unique gametes introduces phenotypic variability into a population.

This genetic and phenotypic variability may benefit or harm individuals in their environment.

Advantageous phenotypes will likely survive and pass genes to future generations consistent with natural selection (i.e., survival of the fittest).

39. D is correct.

Birds laying an intermediate number of eggs have the highest reproductive success, likely resulting from stabilizing selection.

40. D is correct.

Nucleotides are sugar (ribose for RNA or deoxyribose for DNA), a phosphate group, and a base.

Nitrogenous bases are guanine, adenine, cytosine, thymine (for DNA), or uracil (for RNA).

Adenine and guanine are purines, while thymine, cytosine, and uracil are pyrimidines.

A: l*ipids* (i.e., fats) are composed of glycerol (i.e., 3-carbon chain) and (up to) three fatty acids.

B: *monosaccharides* are the monomers (e.g., glucose, fructose) of carbohydrates (e.g., glycogen, cellulose).

C: *nucleosides* have a nitrogenous base and sugar (without phosphate). Nucleotides have a phosphate group, nitrogenous base, and sugar.

41. A is correct.

Sexual selection likely results from a female deer choosing to mate with males with the biggest antlers.

42. A is correct.

Embryos indicate that fossilized organisms reproduce sexually.

The development of an embryo involves two distinct reproductive cells fusing, characteristic of sexual reproduction. This single-cell zygote results from fertilizing the female egg cell with a male sperm cell.

43. D is correct.

Cambrian Explosion occurred around 542 million years ago with the rapid appearance of most major animal phyla. This was accompanied by significant diversification of other organisms, as shown by the fossil record.

Before about 580 million years ago, most organisms were simple, composed of individual cells occasionally organized into colonies.

Over the following 70-80 million years, the rate of evolution accelerated by order of magnitude, and the diversity of life became similar to today.

44. C is correct.

During evolution, primates needed to walk on two legs (i.e., *bipedal*) to free their hands for other tasks.

Hands were used in creating weapons and tools to gather food and protect the young.

45. B is correct.

Eyes facing forward allow two fields of vision to overlap, enabling the perception of depth.

Animals without facing forward eyes could not likely judge the distance to objects (e.g., tree branches) due to flawed depth perception. As primates moved into trees to escape predators, they needed to navigate tree branches, meaning evolution favored good depth perception.

46. D is correct.

Sexual reproduction increases fitness and allows for modifications in traits, which can help adapt to new living conditions in a changing environment. Sexual reproduction adds to the diversity of the offspring. All other answer choices characterize asexual reproduction.

47. D is correct.

Arboreal locomotion is the movement of animals in *trees*.

Some animals have evolved to move in habitats where trees are present.

Animals may only scale trees occasionally or become exclusively arboreal.

48. A is correct.

Australopithecus afarensis is an extinct species with hip, knee, and foot morphology distinctive to bipedalism.

Footprints show *Australopithecus afarensis* walked with an upright posture and a strong heel strike, early evidence of bipedalism.

49. D is correct.

Amniotic eggs are laid on land by certain reptiles, birds, and mammals.

Amniotic eggs are different from *anamniotic eggs*, typically laid in water (by fishes and amphibians).

50. A is correct.

Gluteus muscles of the pelvis are important for propulsion and stability while walking.

Bipedal humans are different from quadrupedal apes in the lateral orientation of the ilium, which constitutes the bowl-shaped pelvis.

51. C is correct.

Homo sapiens evolved after the *Homo erectus* and are *modern man*.

52. B is correct.

Neanderthals are closely related to modern humans, with differences based on DNA by 0.3%, but twice the greatest DNA difference among contemporary humans.

Genetic evidence suggests Neanderthals contributed to the DNA of anatomically modern humans, probably through interbreeding between 80,000 and 30,000 years ago.

Recent evidence suggests that Neanderthals practiced burial behavior and buried their dead.

Archaeological remains by Neanderthals include bones and stone tools.

53. D is correct.

The decimating fire was a random event unrelated to the apparent fitness of the fly in its typical environment.

Genetic drift is the random change over time in the allele frequency within a population, such as the one caused by the decimating fire in the loss of allele(s) for the altered structure.

A: reproduction is not involved in the loss of the advantageous modification.

B: natural selection is not the cause since the death of the flies was unrelated to their fitness, and a decimating fire would likely have killed all flies, regardless of their advantageous modification.

C: Hardy-Weinberg describes ideal circumstances which do not apply to this situation.

Notes for active learning

Animal Behavior and Evolution – Detailed Explanations

1. A is correct.

Potato washing requires learning.

Learning behavior includes habituation, sensitization, classical conditioning, operant conditioning, observational learning, play, and insight learning.

Innate behaviors do not need prior experience or learning.

2. A is correct.

Behavior is the *internally coordinated responses* (i.e., actions or inactions) of organisms (individuals or groups) to internal or external stimuli. Behaviors can be *innate* (instinct) or *learned.*

Behavior is an act of an organism that changes its relationship to its environment; it provides outputs from the organism to the environment.

3. B is correct.

Evolution is the long-term changes in a population *gene pool* caused by environmental selection pressures.

Mutations are the source of genetic variation for natural selection.

Random mutations create new alleles selected for (or against) as a phenotypic variation for natural selection (i.e., survival of the fittest).

4. C is correct.

Operant conditioning is behavior modified by *antecedents* (i.e., before trained behavior) and consequences.

Operant conditioning is distinguished from *classical conditioning* (or *respondent conditioning*) because operant conditioning addresses reinforcement and punishment to change behavior.

Imprinting is *phase-sensitive* learning (learning during a life stage) rapid and independent of the consequences of behavior. Imprinting has a critical period.

Insight learning is the faculty of reason or rationality and involves a sudden realization distinct from cause-and-effect problem-solving. It manifests *spontaneously* and is a noteworthy event in the learning process.

Insight learning is *problem-solving* occurring suddenly by understanding relationships between parts of a problem rather than through test and error.

Classical conditioning differs from *operant conditioning* because the behavior is strengthened or weakened, depending on its consequences (i.e., reward or punishment).

Classical conditioning involves reflexive (reflex) behaviors elicited by prior exposure. It occurs when a conditioned stimulus is paired with an unconditioned stimulus.

Consequences do not maintain behaviors conditioned through classical conditioning.

5. A is correct.

Innate (or *instinct*) behavior is the inherent inclination towards a particular complex behavior. A truncated sequence of actions, without variation, is carried out in response to a clearly defined stimulus.

Instinctive behavior is *without prior experience* (i.e., absence of learning) and use innate biological factors.

6. D is correct.

Innate (or *instinct*) behavior is the inherent inclination towards a particular complex behavior; a truncated sequence of actions, without variation, is carried out in response to a clearly defined stimulus.

Instinctive behavior is performed *without prior experience* (i.e., absence of learning) and expresses innate biological factors.

7. B is correct.

Moths communicate with preditors in this example by sight. Animals generally *communicate* using four methods: visual, auditory, tactile, and chemical.

8. D is correct.

Classical conditioning differs from *operant conditioning* because the behavior is strengthened or weakened, depending on its consequences (i.e., reward or punishment).

Classical conditioning involves reflexive (reflex) behaviors elicited by prior exposure. It occurs when a conditioned stimulus is paired with an unconditioned stimulus.

Consequences do not maintain behaviors conditioned through classical conditioning.

9. C is correct.

Imprinting is *phase-sensitive* learning (learning during a life stage) rapid and independent of the consequences of behavior. Imprinting has a critical period.

10. A is correct.

Imprinting is *phase-sensitive* learning (learning during a life stage) rapid and independent of the consequences of behavior. Imprinting has a critical period.

11. A is correct.

Habituation diminishes a physiological or emotional response to a frequently repeated stimulus.

12. D is correct.

Classical conditioning differs from *operant conditioning* because the behavior is strengthened or weakened, depending on its consequences (i.e., reward or punishment).

Classical conditioning involves reflexive (reflex) behaviors elicited by prior exposure. It occurs when a conditioned stimulus is paired with an unconditioned stimulus. Consequences do not maintain behaviors conditioned through classical conditioning.

13. C is correct.

Animals secrete *pheromones* (e.g., sex attractants) to influence the behavior of species members.

Animals generally *communicate* using four methods: visual, auditory, tactile, and chemical.

14. B is correct.

Imprinting is *phase-sensitive* learning (learning during a life stage) rapid and independent of the consequences of behavior.

Imprinting has a critical period.

15. C is correct.

Learning occurs when an external stimulus causes an animal to change behavior (e,g, response to stimuli).

16. A is correct.

Circadian rhythms are biological processes reoccurring every 24 hours.

Circadian clock drives these rhythms widely observed in plants, animals, fungi, and cyanobacteria.

17. D is correct.

Migrating birds that nest in the Northern Hemisphere tend to fly northward in the spring for the burgeoning insect populations, budding plants, and abundance of nesting locations. As winter approaches and the availability of insects and other food drops, the birds move south again.

Migration is based on seasons.

18. C is correct.

Kin selection is the evolutionary strategy favoring the reproductive success of an organism's relatives, even at a cost to the organism's survival and reproduction.

Kin altruism is unselfish behavior driven by *kin selection.*

Kin selection is *inclusive fitness*, combining the number of offspring produced with the number an individual can produce by supporting others (e.g., siblings).

Examples of kin selection include honeybees that do not reproduce by deferring to relatives, adopting orphans in animal populations, and familial caring for the young.

19. C is correct.

Hibernation is how animals conserve energy to survive adverse weather conditions or food scarcity.

Hibernation involves *physiological changes* such as decreased body temperature, heart rate, respiration rate, and slowed metabolism. It is a form of torpor (lethargy); metabolism is less than five percent.

20. B is correct.

Operant conditioning is behavior modified by *antecedents* (i.e., before trained behavior) and consequences. It is distinguished from *classical conditioning* (or *respondent conditioning*) because operant conditioning addresses reinforcement and punishment to change behavior.

Insight learning is problem-solving occurring suddenly by understanding relationships between parts of a problem rather than through test and error. It is the faculty of reason or rationality and involves a sudden realization distinct from cause-and-effect problem-solving. Insight learning manifests spontaneously and is a noteworthy event in the learning process.

Classical conditioning differs from *operant conditioning* because the behavior is strengthened or weakened, depending on its consequences (i.e., reward or punishment).

Classical conditioning involves reflexive (reflex) behaviors elicited by prior exposure. It occurs when a conditioned stimulus is paired with an unconditioned stimulus.

Consequences do not maintain behaviors conditioned through classical conditioning.

Habituation diminishes a physiological or emotional response to a frequently repeated stimulus.

21. A is correct.

Fertility and reproduction in animals follow circadian and seasonal rhythms.

Reproductive rhythm aims to maximize the mating opportunities and the birth of offspring during favorable seasons for climate and food availability. Their purpose is to increase offspring survival.

Chronobiology studies the rhythms of organisms and how external conditions influence them.

22. A is correct.

Courtship is mate-selection behavior (i.e., *rituals*).

Animal courtship occurs out of sight of human observation, so it is the least documented animal behavior.

Animals generally *communicate* using four methods: visual, auditory, tactile, and chemical.

Courtship may include complicated protocols, vocalizations, beauty, or prowess.

23. D is correct.

Members of a society belong to the same species and interact closely.

Social animals (e.g., ants, termites) live together in large groups and often cooperate in conducting many tasks.

24. B is correct.

Insight learning is the faculty of reason or rationality and involves a sudden realization distinct from cause-and-effect problem-solving. It manifests *spontaneously* and is a noteworthy event in the learning process.

Insight learning is *problem-solving* occurring suddenly by understanding relationships between parts of a problem rather than through test and error.

25. C is correct.

Animals require resources (e.g., food, water, biotic factors) for survival. An animal may benefit most by defending a territory if that area has more resources than the surrounding areas.

Competitive exclusion principle (*Gause's law of competitive exclusion* or *Gause's law*) states that two species competing for the same resources cannot coexist when ecological factors are constant.

26. D is correct.

Innate (or *instinct*) behavior is the inherent inclination towards a particular complex behavior.

A truncated sequence of actions, without variation, is carried out in response to a clearly defined stimulus.

Instinctive behavior is performed *without prior experience* (i.e., absence of learning) and expresses innate biological factors.

27. A is correct.

Social animals (e.g., ants, termites) live together in large groups and often cooperate in conducting many tasks.

Groups offer advantages such as greater protection from predation.

28. C is correct.

Nocturnality is animal behavior characterized by *activity during the night* and *sleep during the day.*

Nocturnal creatures generally have highly developed senses of hearing to avoid predators and communicate.

Diurnal animals are awake during the day and sleeping at night.

29. A is correct.

Animals generally *communicate* using four methods: visual, auditory, tactile, and chemical.

Some species rely more on one form to communicate; however, all use methods to show affection, ward off threats or attract mates.

Animals often use chemical communication (e.g., *pheromones*) to mark territory. Some animals (e.g., dogs, lions) make *scent posts* to mark their territory and set territory boundaries.

30. B is correct.

Operant conditioning is behavior modified by *antecedents* (i.e., before trained behavior) and consequences.

Operant conditioning is distinguished from *classical conditioning* (or *respondent conditioning*) because operant conditioning addresses reinforcement and punishment to change behavior.

Imprinting is *phase-sensitive* learning (learning during a life stage) rapid and independent of the consequences of behavior. Imprinting has a critical period.

Habituation diminishes a physiological or emotional response to a frequently repeated stimulus.

Classical conditioning differs from *operant conditioning* because the behavior is strengthened or weakened, depending on its consequences (i.e., reward or punishment).

Classical conditioning involves reflexive (reflex) behaviors elicited by prior exposure. It occurs when a conditioned stimulus is paired with an unconditioned stimulus.

Consequences do not maintain behaviors conditioned through classical conditioning.

31. A is correct.

Animals generally *communicate* using four methods: visual, auditory, tactile, and chemical.

For example, changes in skin color and patterns are visual signals.

Animals secrete *pheromones* (e.g., sex attractants) to influence the behavior of species members.

32. C is correct.

Animals generally *communicate* using four methods: visual, auditory, tactile, and chemical.

For example, changes in skin color and patterns are visual signals.

Pheromones are *chemical signals* released by animals to influence other animals.

33. A is correct.

Imprinting is when environmental patterns (or objects) presented to a developing organism during a *critical period* in early life become accepted as permanent aspects of their behavior.

For example, a duckling passes through a critical period when its mother is the first large moving object it sees.

B: *instrumental conditioning* involves reward or reinforcement to stimuli to establish a conditioning response.

C: *discrimination* involves learning organisms that respond differently to slightly different stimuli.

D: *pheromones* (e.g., sex attractants) are secreted by animals to influence the behavior of species members.

Notes for active learning

Notes for active learning

Cell – the Basic Unit of Life – Detailed Explanations

1. D is correct.

Facilitated transport (or *passive transport*) uses a protein pore/channel but is differentiated from active transport that requires energy.

2. A is correct.

Cell is the unit of function and reproduction because subcellular components cannot regenerate whole cells.

3. C is correct.

I: water relies on *aquaporins* to readily diffuse across a plasma membrane.

Without aquaporins, only a small fraction of water molecules can diffuse through the cell membrane per unit of time because of the polarity of water molecules.

II: *small hydrophobic molecules* readily diffuse through the hydrophobic tails of the plasma membrane.

III: *small ions* rely on ion channel transport proteins to diffuse across the membrane.

IV: *neutral gas molecules* (e.g., O_2, CO_2) readily diffuse through the hydrophobic tails of the plasma membrane.

4. B is correct.

Lysosome is the digestive region of the cell and is a membrane-bound organelle with a low pH (around 5) that stores hydrolytic enzymes.

A: *vacuoles* and *vesicles* are membrane-bound sacs involved in the transport and storage of materials ingested, secreted, processed, or digested by cells.

Vacuoles are larger than vesicles and are likely in plant cells (e.g., central vacuole).

C: *chloroplasts* are the site of photosynthesis and are only in algae and plant cells. Chloroplasts contain their DNA and ribosomes and may have similarly evolved via endosymbiosis like mitochondria.

D: *phagosomes* are vesicles for transporting and storing materials ingested by the cell through phagocytosis. Vesicles form by fusion of the cell membrane around the particle. Phagosome is a cellular compartment in which pathogenic microorganisms are digested. Phagosomes fuse with lysosomes in their maturation process to form *phagolysosomes.*

5. C is correct.

DNA damage checkpoints are *signal transduction pathways* that block cell cycle progression in G1, G2, and metaphase and slow the S phase progression rate when DNA is damaged.

It pauses the cell cycle, allowing the cell to repair the damage before dividing.

G1 → S → G2 → prophase → metaphase → anaphase → telophase
Interphase (G1, S, G2) — Mitosis (PMAT) (prophase, metaphase, anaphase, telophase)

Cell cycle divides into interphase (G1. S, G2) and mitosis (prophase, metaphase, anaphase, and telophase)

6. B is correct.

Cytoskeleton determines the overall shape of a cell.

7. C is correct.

The two ribosomal subunits are synthesized in the nucleolus, a region within the nucleus. The ribosomes are the sites of protein production.

Prokaryotic ribosomes (30S small + 50S large subunit = 70S complete ribosome) are smaller than *eukaryotic ribosomes* (40S small + 60S large subunit = 80S complete ribosome).

A: *Golgi apparatus* is a membrane-bound organelle that modifies (e.g., glycosylation), sorts, and packages proteins synthesized by the ribosomes.

B: *lysosomes* have a low pH of about 5 and contain hydrolytic enzymes involved in digestion.

D: *rough endoplasmic reticulum* (RER) is part of the endomembrane extending from the nuclear envelope.

The RER has ribosomes associated with its membrane and is the site of the production and folding of proteins.

Misfolded proteins exit the rough ER and are sent to the *proteasome* for degradation.

8. B is correct.

Codon is a three-nucleotide segment of an mRNA that hybridizes (via complementary base pairing) with the appropriate anticodon on the tRNA to encode one amino acid in a polypeptide chain during protein synthesis.

The tRNA molecule interacts with the mRNA codon after the ribosomal complex binds the mRNA.

A: tRNA molecules interact with the mRNA codon after (not before) the ribosomal complex binds mRNA

C: *translation* involves the conversion of mRNA into protein.

D: *operons* regulate the transcription of genes into mRNA and are not involved in translating mRNA into proteins.

9. C is correct.

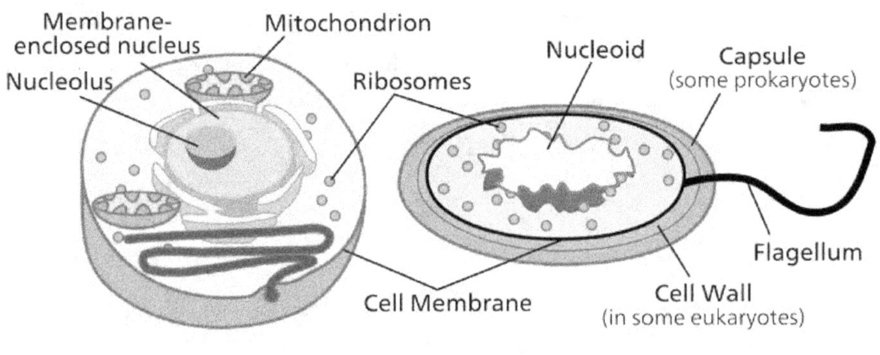

Structures within eukaryotic (left) and prokaryotic (right) cells

Peroxisomes are organelles in most eukaryotic cells; a major function is the breakdown of very-long-chain fatty acids through *beta-oxidation*. Peroxisome convert the very long fatty acids to medium-chain fatty acids in animal cells, subsequently shuttled to the mitochondria. Medium-chain fatty acids are degraded, via oxidation, into CO_2 and H_2O.

10. C is correct.

Vacuole is a membrane-bound organelle in plant and fungal cells and some protist, animal, and bacterial cells.

Vacuoles are enclosed compartments filled with water. They contain inorganic and organic molecules (including enzymes in solution) and may contain solids that have been engulfed.

The function and significance of vacuoles vary by cell type, with much greater prominence in plants, fungi, and certain protists than in animals or bacteria.

11. D is correct.

B: *cytochrome c oxidase* is a large transmembrane protein complex in bacteria and the mitochondrion of eukaryotes. It receives an electron from each of four cytochrome c molecules and transfers the electrons to an O_2 molecule, converting molecular oxygen to two molecules of H_2O.

Cytochrome c oxidase is the last enzyme in the electron transport chain of mitochondria (or bacteria) in the mitochondrial (or bacterial) inner membrane.

12. A is correct.

Cytoskeleton is integral to proper cell division because it forms the mitotic spindle and separates sister chromatids during cell division.

The cytoskeleton is composed of microtubules and microfilaments, provides mechanical cell support to maintain shape, and functions in cell motility.

13. B is correct.

Biochemical events during the cell cycle include DNA damage repair and replication completion, centrosome duplication, assembly of the spindle, and attachment of the kinetochores to the spindle.

14. A is correct.

Mitochondria divide autonomously to produce daughter mitochondria that incorporate some new nonradioactive phosphatidylcholine and inherit radioactive phosphatidylcholine from the parent via semiconservative replication.

Therefore, the daughter mitochondria have equal radioactivity.

B: mitochondria divide autonomously, and daughter mitochondria retain parental (original) radioactive label.

C: original sample was 100% radiolabeled.

DNA replication is *semiconservative*, whereby one strand of parental DNA (radiolabeled) and one strand of newly replicated (non-radiolabeled) are in each daughter cell after the first round of division.

D: requires the daughter mitochondria to be synthesized *de novo* (i.e., new) with newly-synthesized, nonradioactive phosphatidylcholine.

If the mitochondria divide autonomously, the daughter mitochondria retain the radioactive label evenly via *semiconservative replication.*

15. C is correct.

Golgi apparatus (i.e., Golgi complex) is a eukaryotic cell organelle. It processes proteins via post-translational modifications. These proteins have three destinations: 1) *secreted* from the cell, 2) transported into *organelles,* or 3) targeted to the *plasma membrane*.

16. B is correct.

Most animal and plant cells range between 1 and 100 micrometers and are visible with a microscope.

Light microscopes visualize objects from 1 millimeter (10^{-3} m) to 0.2 micrometers (2×10^{-7} m).

Electron microscopes visualize objects as small as an atom (1 angstrom or 10^{-10} m).

The microscopic scale is from 1 millimeter (10^{-3} m) to a ten-millionth of a millimeter (10^{-10} m).

There are immense variations in objects' sizes even within the microscopic scale.

10^{-3} m is 10 million times larger than 10^{-10} m, equivalent to the Earth's size *vs.* a beach ball.

Use the magnification to determine the size of the objects observed.

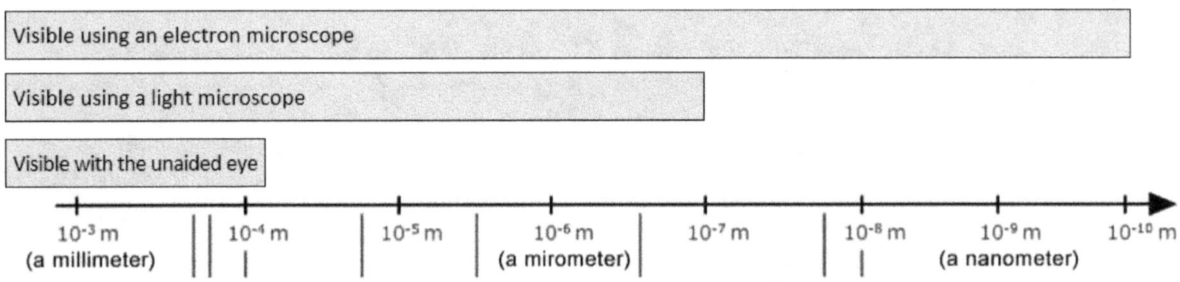

Comparison of resolution by unaided eyes, light, and electron microscopes

17. B is correct.

Ribosomes are composed of specific rRNA molecules and associated proteins.

Ribosomes are identified by the sedimentation coefficients (i.e., S units for Svedberg units) for density.

Prokaryotes have a 30S small and a 50S large subunit (i.e., complete ribosome = 70S; based on density).

Eukaryotes have a 40S small and 60S large subunit (i.e., complete ribosome = 80S).

A: *peroxisomes* are organelles involved in hydrogen peroxide (H_2O_2) synthesis and degradation. They function in cell detoxification and contain the catalase that decomposes H_2O_2 into H_2O and O_2.

C: *mitochondria* are organelles as the site of cellular respiration (i.e., oxidation of glucose to yield ATP) and plentiful in cells with high demands for ATP (e.g., muscle cells).

The number of mitochondria within a cell varies widely by organism and tissue type. Many cells have a single mitochondrion, whereas others contain several thousand mitochondria.

D: *nucleus* is the largest membrane-bound organelle in eukaryotes, containing the genetic code (i.e., DNA). The *nucleus* directs the cell by storing and transmitting genetic information. Cells can contain multiple nuclei (e.g., skeletal muscle cells), one nucleus, or none (e.g., red blood cells).

18. A is correct.

Osmosis is a type of diffusion involving water and is a form of passive transport.

Hypertonic means a solution of high solute and low solvent concentrations.

Hypotonic means a solution of high solvent and low solute concentrations.

Solvents flow spontaneously from an area of high solvent to a low solvent concentration.

During *osmosis*, water flows from a hypotonic to a hypertonic environment.

19. C is correct.

Smooth endoplasmic reticulum (smooth ER) is involved in synthesizing phosphatidylcholine.

20. D is correct.

Plastids (e.g., chloroplast and chromoplast) are major organelles in plants and algae. They are the site of manufacturing and storing critical chemical compounds used by cells. They often contain pigments used in photosynthesis. Pigments change to determine cell color.

Plastids, like prokaryotes, contain a circular double-stranded DNA molecule.

21. B is correct.

Rough *endoplasmic reticulum* (rough ER) is involved in synthesizing plasma membrane proteins.

22. B is correct.

Active transport uses a carrier protein and energy to move a substance across a membrane against (i.e., up) a concentration gradient: from low solute to a region of high solute concentration.

Donnan equilibrium refers to some ionic species passing through the barrier while others cannot.

Charged substances that cannot pass through the membrane create an uneven electrical charge. The electric potential arising between the solutions is the Donnan potential.

23. B is correct.

Most proteins that are secretory, membrane-bound, or targeted to an organelle use the N-terminal signal sequence (i.e., 5 to 30 amino acids) to target the protein. The signal sequence of the polypeptide is recognized by a signal recognition particle (SRP) while the protein is synthesized on the ribosome.

Synthesis pauses while the ribosome-protein complex transfers to an SRP receptor on the ER (in eukaryotes) or the plasma membrane (in prokaryotes) before polypeptide translation resumes.

24. C is correct.

Difference between "*free*" and "*attached*" ribosomes:

1) *free ribosomes* are in the cytoplasm, while *attached ribosomes* are anchored to the endoplasmic reticulum, and

2) *free ribosomes* produce proteins in the cytosol, while *attached ribosomes* produce proteins inserted into the ER lumen.

25. C is correct.

Urea, a byproduct of amino acid metabolism, is a small uncharged molecule that crosses cell membranes by simple diffusion – a passive process that does not require energy.

A: export of Na^+ from a neuron is coupled with the import of K^+; the sodium-potassium-ATPase pump – an ATP-dependent process necessary to maintain a voltage potential across the neuron membrane.

B: movement of Ca^{2+} into a muscle cell occurs against the concentration gradient, and Ca^{2+} enters cells by active transport, which requires ATP.

D: *synaptic vesicles* contain neurotransmitters, and their exocytosis at a nerve terminus is an ATP-dependent process triggered by an action potential propagating along the neuron.

Vesicle fusion requires Ca^{2+} to enter the cell upon axon depolarization reaching the terminus.

26. D is correct.

Newly synthesized secretory protein pathway:

rough ER → Golgi → secretory vesicles → exterior of the cell (via exocytosis)

A and C: peroxisomes and lysosomes are destinations for proteins but do not involve secretory path exocytosis.

B: ribosomes synthesize proteins, but the Golgi is the final organelle before exocytosis.

27. B is correct.

Plant cell membranes have higher amounts of unsaturated fatty acids.

The ratio of saturated and unsaturated fatty acids determines membrane fluidity.

Unsaturated fatty acids have kinks in their tails (due to double bonds) that push the phospholipids apart, so the membrane retains its fluidity.

A: most textbooks state that plant membranes have no cholesterol, but a small amount is present, which is negligible compared to animal cells.

Cholesterol is usually dispersed in varying amounts throughout animal cell membranes in the irregular spaces between the hydrophobic lipid tails of the membrane.

Cholesterol functions as a bidirectional buffer; at high temperatures, it decreases the membrane's fluidity because it confers stiffening and strengthening effects on the membrane; at low temperatures, cholesterol intercalates between the phospholipids, preventing clustering and stiffening of the membrane.

28. D is correct.

Nucleus, chloroplast, and mitochondria are organelles enclosed in a double membrane.

29. C is correct.

Golgi apparatus processes *secretory proteins* via post-translational modifications.

Proteins targeted to the Golgi (from the rough ER) have three destinations: secreted out of the cell, transported into organelles, or targeted to the plasma membrane (as a receptor, channel, or pore).

30. D is correct.

Glycolysis occurs in the *cytoplasm*.

Krebs (TCA) cycle and *pyruvate oxidation* to acetyl-CoA occurs in the mitochondrial *matrix*.

Electron transport chain (ETC) uses cytochromes in the inner mitochondrial membrane and a proton (H^+) gradient in the intermembrane space).

A: *pyruvate* is oxidized to acetyl-CoA preliminarily and transported from the cytoplasm into the matrix before joining oxaloacetate in the Krebs cycle.

B: Krebs cycle is the second stage of cellular respiration and occurs in the matrix of the mitochondrion.

C: *electron transport chain* is the final stage of cellular respiration occurring in the inner membrane (i.e., cytochromes) / intermembrane space (i.e., H^+ proton gradient) of the mitochondrion.

31. D is correct.

Secretory sequence in the flow of newly synthesized protein for export from the cell is:

rough ER → Golgi → plasma membrane

32. D is correct.

Recycling of organelles within the cell is accomplished through autophagy by *lysosomes*.

33. B is correct.

Human gametes, formed during meiosis, are cells with a single copy (1N) of the genome.

After the second meiotic division, cells have a single unreplicated copy (i.e., devoid of a sister chromatid).

34. D is correct.

Smooth endoplasmic reticulum is the organelle for fatty acid, phospholipid, and steroid synthesis.

35. C is correct.

Centrioles are cylindrical structures mainly of tubulin in eukaryotic cells (except flowering plants and fungi).

Centrioles are involved in the organization of the mitotic spindle and the completion of cytokinesis.

Centrioles contribute to the structure of centrosomes and to organizing microtubules in the cytoplasm.

The position of the centriole determines the position of the nucleus and plays a crucial role in the spatial arrangement of cells.

36. A is correct.

Centrioles are the organizational sites for microtubules (i.e., spindle fibers) that assemble during cell division (e.g., mitosis and meiosis).

Four phases of mitosis are prophase, metaphase, anaphase, and telophase. These are followed by cytokinesis, which physically divides cells into two identical daughter cells.

Condensed chromosomes align along the *equatorial plane* in metaphase before the centromere (i.e., heterochromatin region of DNA) splits. The two sister chromosomes begin their journey to the respective poles of the cell.

37. A is correct.

cAMP is a second messenger triggered when a ligand (e.g., peptide hormone or neurotransmitter) binds a membrane-bound receptor.

Adenylate cyclase (enzyme) is activated through a G-protein intermediate and converts ATP into cAMP.

Adenylate cyclase is attached to the inner layer of the phospholipid bilayer and is not in the cytoplasm.

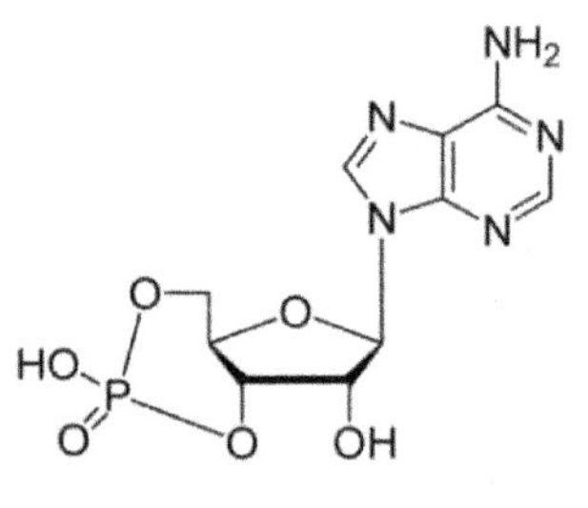

cAMP

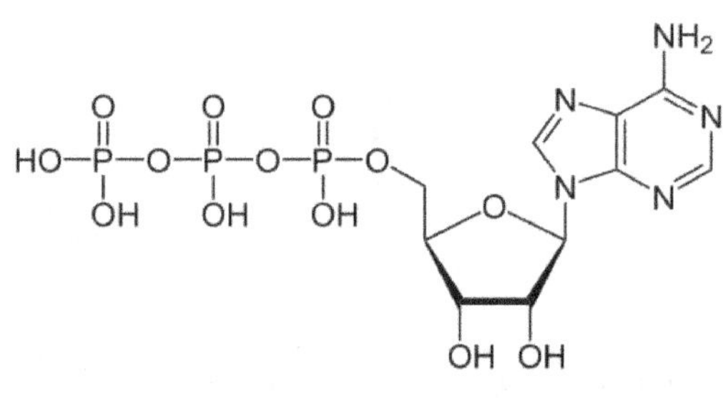

ATP

38. C is correct.

Cyclin and its associated *protein kinase* ensure a proper progression of cell division.

Cyclins are phosphorylated proteins responsible for specific events during cycle division (e.g., microtubule formation, chromatin remodeling).

Cyclins are divided into four classes based on their behavior in the cell cycle: G1/S, S, M, and G1 cyclins.

39. A is correct.

Hypertonic solution is when there is a higher concentration of solutes outside the cell than inside.

When a cell is in a *hypertonic solution*, the tendency is for *water to flow out* of the cell to balance the concentration of the solutes.

Osmotic pressure draws water out of the cell, and the cell shrivels (i.e., crenation).

40. A is correct.

Peroxisomes are tiny organelles abundant in the liver containing oxidases and detoxifying substances (i.e., alcohol and hydrogen peroxide).

41. B is correct.

Microtubules are hollow proteins composed of *tubulin* monomers necessary for:

1) formation of the *spindle apparatus* that separates chromosomes during cell division,

2) synthesis of *cilia* and *flagella*, and

3) formation of the *cytoskeleton* (within the cytoplasm).

II: actin and myosin are contractile fibers in muscle cells composed of microfilament (not microtubules).

42. D is correct.

If a cell (e.g., pancreatic exocrine cell) is producing large amounts of proteins (i.e., enzymes) for export, this involves the *rough endoplasmic reticulum* (RER).

Protein synthesis begins in the nucleus with mRNA transcription, translated into polypeptides with the RER.

Exported proteins are packaged and modified in the Golgi; this cell would have a prominent nucleolus for rRNA (i.e., ribosomal components) synthesis.

43. A is correct.

Mitosis is cell division when new somatic (i.e., body) cells are added to multicellular organisms as they grow and tissues are repaired or replaced.

Mitosis does not produce genetic variations. A daughter cell is identical in chromosome number and genetic makeup to the parental cell.

Eukaryotes divide by mitosis. Mitosis distributes identical genetic material to two daughter cells. The fidelity of DNA transmission between generations without dilution is remarkable.

44. B is correct.

Apoptosis is programmed cell death that occurs during fetal development and aging.

Synaptic cleft development, the formation of separate digits in the hand of a fetus, and tadpole tail reabsorption are examples of apoptosis during development.

The synthesis of the uterine lining is an *anabolic* process that involves mitosis (i.e., cell division).

45. A is correct.

A cell involved in active transport (e.g., intestinal epithelial cells) requires much ATP.

Therefore, many mitochondria are needed to meet cellular respiration needs (i.e., glucose → ATP).

B: high levels of *DNA synthesis* occur in cells using mitosis for rapid reproduction (i.e., skin cells).

C: high levels of *adenylate cyclase* (i.e., cAMP second messenger) are in the target cells of peptide hormones.

D: *polyribosomes* are in cells with a high protein synthesis level.

46. D is correct.

Cells respond to peptide hormones through biochemical reactions involving membrane receptors and kinase activation. The first and second messengers communicate for cells to respond because the G protein links the first and second messengers.

cAMP is a second messenger triggered when a ligand (e.g., peptide hormone or neurotransmitter) binds a membrane-bound receptor.

The adenylate cyclase enzyme is activated through a G-protein intermediate and converts ATP into cAMP.

Adenylate cyclase is attached to the inner layer of the phospholipid bilayer and is not in the cytoplasm.

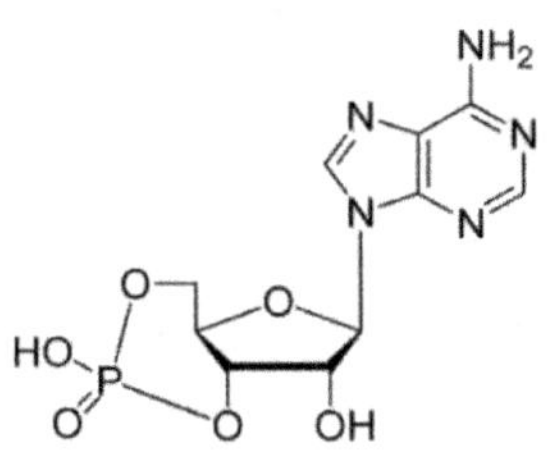

cAMP

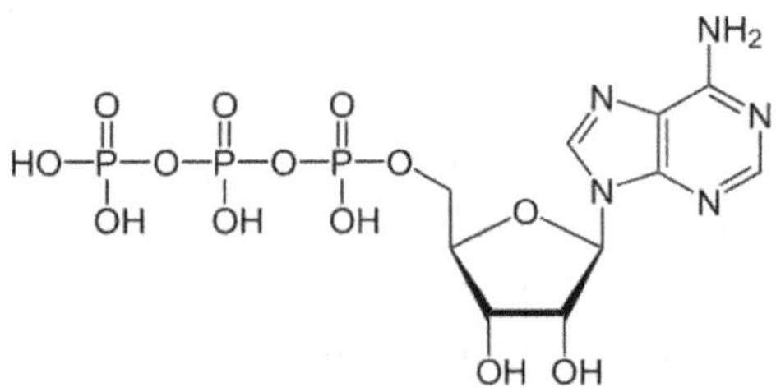

ATP

Notes for active learning

Notes for active learning

Enzymes and Cellular Metabolism – Detailed Explanations

1. B is correct.

Cellular respiration uses enzyme-catalyzed reactions and is a catabolic pathway to use the potential energy stored in glucose.

While glycolysis yields two ATP per glucose, cellular respiration (i.e., glycolysis, Krebs/TCA cycle, and electron transport chain) produces about 30–32 ATP. Around the year 2004, research results reduced the original estimate of 36-38 ATP due to inefficiencies of the ATP synthase.

Cellular respiration is an aerobic process whereby O_2 is the final acceptor of electrons passed from cytochrome electron carriers during the final stage of the electron transport chain.

Glycolysis (glucose to pyruvate) occurs in the cytoplasm.

Krebs cycle (TCA) occurs in the matrix (i.e., the cytoplasm of mitochondria), while the electron transport chain occurs in the intermembrane space of the mitochondria for eukaryotes.

For prokaryotes, both processes occur in the cytoplasm because prokaryotes lack mitochondria.

Some forms of anaerobic respiration use an electron acceptor other than O_2 (e.g., iron, cobalt, or manganese reduction) at the end of the electron transport chain. Glucose molecules are entirely oxidized in these cases because glucose products enter the Krebs cycle.

The highly reduced chemical compounds (e.g., NADH or $FADH_2$) establish an electrochemical gradient across a membrane in the Krebs cycle. The reduced chemical compounds are oxidized by integral membrane proteins that transfer the electron to the final electron acceptor.

2. C is correct.

Metabolism means change and refers to life-sustaining chemical transformations within cells.

Enzyme-catalyzed reactions allow organisms to maintain structures, grow, reproduce, and respond.

Metabolism is divided into anabolism (buildup) and catabolism (break down).

Catabolism is the breakdown of organic matter and harvests energy from cellular respiration (e.g., glycolysis, Krebs cycle, and the electron transport chain).

Anabolism utilizes energy (e.g., ATP) to *synthesize biomolecules* (e.g., lipids, nucleic acids, proteins).

3. D is correct.

Fermentation is an anaerobic process and occurs in the absence of oxygen.

The purpose of fermentation is to regenerate the high-energy nucleotide intermediate of NAD^+.

Oxidative phosphorylation occurs during the electron transport chain as the final stage of cellular respiration.

4. B is correct.

When plasma glucose levels are low, the body utilizes other energy sources for cellular metabolism.

These energy sources are used in preferential order:

glucose → other carbohydrates → fats → proteins

These molecules are converted to glucose or glucose intermediates degraded in the glycolytic pathway and the Krebs cycle (i.e., citric acid cycle).

Proteins are used last for energy because there is no protein storage in the body.

Catabolism of protein results in muscle wasting and connective tissue breakdown, which is harmful in the long term.

5. D is correct.

Activation energy is the minimum amount of energy needed for a reaction. It is thought of as an "energy barrier" because energy must be added to the system.

Enzymes act by decreasing the activation energy and lowering the energy barrier, allowing the reaction and increasing the reaction rate.

6. A is correct.

A phosphate group from ATP is transferred to glucose, which creates a phosphodiester bond.

7. A is correct.

The greatest direct source of ATP synthesis involves the electron transport chain. Glycolysis occurs in the cytoplasm, the Krebs cycle (TCA) in the matrix of mitochondria, and oxidative phosphorylation (i.e., the electron transport chain) in the intermembrane space and inner membrane of mitochondria.

Glycolysis produces pyruvate, converted into acetyl CoA, and joined to oxaloacetate in the Krebs cycle, producing citrate as the first intermediate.

The Krebs cycle produces two GTP (i.e., ATP) per glucose molecule (or 1 GTP per pyruvate). Most ATP is formed by *oxidative phosphorylation* when NADH and $FADH_2$ nucleotides are oxidized and donate their electrons to the cytochromes in the electron transport chain.

8. C is correct.

Cofactors are organic or inorganic molecules classified depending on how tightly they bind to an enzyme.

Coenzymes are loosely bound organic cofactors released from the enzyme's active site during the reaction (e.g., ATP, NADH).

Prosthetic groups are tightly bound organic cofactors.

9. A is correct.

Mitochondrion is the organelle bound by a double membrane, where the Krebs (TCA) cycle, electron transport, and oxidative phosphorylation occur.

In eukaryotes, the Krebs cycle (i.e., TCA) occurs in the matrix of the mitochondria.

Matrix is an interior space in the mitochondria analogous to the cell cytoplasm.

B: *smooth ER* (endoplasmic reticulum) is connected to the nuclear envelope and synthesizes lipids, phospholipids, and steroids.

Cells secreting lipids, phospholipids, and steroids (e.g., testes, ovaries, skin oil glands) have prominent smooth endoplasmic reticulums.

The smoother ER carries out the metabolism of carbohydrates, drug detoxification, attachment of receptors on cell membrane proteins, and steroid metabolism.

C: *cytosol* is common to eukaryotes and prokaryotes and equivalent to the cytoplasm.

D: *nucleolus* is the organelle within the nucleus responsible for synthesizing ribosomal RNA (rRNA).

10. B is correct.

Pyruvate is the product of glycolysis and is converted into acetyl Co-A as the starting reactant for the Krebs cycle. Pyruvate is *not* a waste product of cellular respiration. Metabolic waste products are produced during enzymatic processes.

Pyruvate, made during glycolysis (as an intermediate of cellular respiration), is converted to acetyl CoA and enters the Krebs cycle to be oxidized into CO_2 and H_2O or converted into the waste product of lactate under anaerobic conditions. Waste products can damage cells and must be removed.

A: *lactate* is produced from pyruvate as the waste product of anaerobic respiration. Lactate is converted to pyruvate in the liver when O_2 becomes available.

If lactate is not metabolized, it can lead to lactic acidosis (acidification of the blood) and death.

C and D: CO_2 and H_2O are waste products of aerobic respiration.

CO_2 and H_2O are removed from the lungs via expiration during regular breathing.

CO_2 increases lead to acidosis (lowering blood pH) because CO_2 reacts with H_2O to form carbonic acid (H_2CO_3).

11. C is correct.

Apoenzyme, with its cofactors, is a *holoenzyme* (an active form). Enzymes that require a cofactor, but do not have one bound, are *apoenzymes* (or apoproteins).

Cofactors can be inorganic (e.g., metal ions) or organic (e.g., vitamins). Organic cofactors are coenzymes or prosthetic groups. Most cofactors are not covalently attached to an enzyme but are tightly bound. Organic prosthetic groups can be covalently bound.

Holoenzyme refers to enzymes containing multiple protein subunits (e.g., DNA polymerases), whereby the holoenzyme is the complete complex with all subunits needed for the activity.

12. C is correct.

Electron transport chain (ETC) uses cytochromes in the inner mitochondrial membrane and is a complex carrier mechanism of electrons that produces ATP through *oxidative phosphorylation.*

A: Krebs (i.e., TCA) cycle occurs in the *matrix* of the mitochondria. It begins when acetyl CoA (i.e., 2-carbon chain) combines with oxaloacetic acid (i.e., 4-carbon chain) to form citrate (i.e., 6-carbon chain).

A series of additional enzyme-catalyzed reactions result in the oxidation of citrate and the release of two CO_2, and oxaloacetic acid is regenerated to begin the cycle again once it is joined to an incoming acetyl CoA.

B: *long-chain fatty acid degradation* occurs in peroxisomes that hydrolyze fat into smaller molecules fed into the Krebs cycle and used for cellular energy.

D: *glycolysis* occurs in the cytoplasm as the oxidative breakdown of glucose (i.e., 6-carbon chain) into two pyruvates (i.e., 3-carbon chain).

13. A is correct.

Apoenzyme is an inactive enzyme without its cofactor. An apoenzyme, with its cofactor(s), is a *holoenzyme* (an active form).

Enzymes that require a cofactor, but do not have one bound, are *apoenzymes* (or apoproteins).

Cofactors are inorganic (e.g., metal ions) or organic (e.g., vitamins).

Organic cofactors are *coenzymes* or *prosthetic groups.*

14. D is correct.

Cofactors are organic or inorganic molecules classified depending on how tightly they bind to an enzyme.

Coenzymes are loosely bound organic cofactors released from the enzyme's active site during the reaction (e.g., ATP, NADH).

Prosthetic groups are tightly bound organic cofactors.

15. B is correct.

$FADH_2$ is produced in the Krebs cycle (TCA) in cellular respiration.

A: final product of glycolysis is pyruvate (i.e., pyruvic acid), which is converted to lactic acid (in humans) or ethanol & CO_2 (in yeast) *via* anaerobic conditions.

Yeast is used for alcoholic production and baking for the dough to rise.

C: in glycolysis, the first series of aerobic (or anaerobic) reactions occur in the cytoplasm as the breakdown of the 6-carbon glucose into two 3-carbon pyruvate molecules.

D: two ATP are required, and four ATP are produced = net of two ATP at the end of glycolysis.

16. A is correct.

During glycolysis, two net (four gross = total) molecules of ATP are produced by *substrate-level phosphorylation*, and two molecules of NAD^+ are reduced (i.e., gain electrons) to form NADH.

B: during glycolysis, *two pyruvates* form from each starting molecule of glucose.

C: *glycolysis* is an anaerobic process and does not require O_2.

D: *glucose* is partially *oxidized* (not reduced) into *two pyruvate* molecules during glycolysis.

17. B is correct.

The Krebs (TCA) cycle and the *electron transport chain* are metabolic processes occurring in the mitochondria.

18. C is correct.

Feedback inhibition occurs when a product binds an enzyme to prevent it from catalyzing further reactions.

Since it is *allosteric inhibition,* this product binds a site other than the active site and changes the conformation of the active site to inhibit the enzyme.

Although the allosteric inhibitor can be the same product of the enzyme-catalyzed reaction *or* the product of the metabolic pathway, it is usually the final product that inhibits the entire metabolic pathway, thus saving the energy necessary to produce the intermediates as well as the final product.

When levels of the final product are low, the enzyme resumes its catalytic activity.

19. A is correct.

Oxidative phosphorylation (O_2 as substrate/reactant) occurs only in the electron transport chain (the last step for cellular respiration) in the inner membrane (i.e., cristae) of the mitochondria.

NADH and $FADH_2$ donate electrons to a series of cytochrome (i.e., protein) molecules embedded in the inner membrane and create an electron gradient, which establishes a proton (H^+) gradient within the intermembrane space of the mitochondria.

H^+ gradient drives a proton pump coupled with an enzyme, producing ATP via *oxidative phosphorylation.*

Electrons from NADH and $FADH_2$ are transferred to ½ O_2 (i.e., oxidative phosphorylation) to generate ATP and form H_2O as a metabolic waste (along with CO_2) from cellular respiration.

H_2O and CO_2 (as metabolic wastes) are expired from the lungs during breathing.

Each Krebs cycle $FADH_2$ yields two ATP, while each NADH yields three ATP in the electron transport chain.

NADH from glycolysis only produces two ATP because energy is expended to shuttle the NADH from glycolysis (in the cytoplasm) through the double membrane of the mitochondria.

20. C is correct.

Enzymes do not affect free energy (ΔG); they decrease the activation energy required, increasing the reaction rate.

Enzymes do not affect the position or direction of equilibrium; they only affect the speed of equilibrium.

21. C is correct.

Zymogen (or proenzyme) is an inactive enzyme precursor. Proteolysis (i.e., cleavage) of the zymogen makes it active to catalyze reactions. Changes in substrate concentration regulate enzyme activity; an increase in substrate concentration increases enzyme activity (at saturation, V_{max} is achieved).

Post-translational modifications occur during protein biosynthesis (after translation), and they may involve cleaving proteins or introducing new functional groups to arrive at the mature protein product.

The enzyme may not be functional until post-translational modifications, a crucial regulatory aspect.

22. C is correct.

Allosteric regulators bind enzymes at allosteric sites (i.e., other than active sites) with *non-competitive inhibition.*.

Once bound to the allosteric site, the enzyme changes its conformational shape. If the shape change causes the ligand (i.e., molecule destined to bind to the active site) to bind to the active site with less efficiency, the modulator is inhibitory. The modulator is excitatory if the shape change causes the ligand to bind to the active site with greater efficiency (i.e., the active site becomes open and accessible).

23. B is correct.

Biological reactions can be *exergonic* (releasing energy) or *endergonic* (consuming energy).

Biological reactions have *activation energy*, the minimum amount of energy required for the reaction. These reactions use *enzymes* (i.e., *biological catalysts*) to lower activation energy and increase the reaction rate.

24. C is correct.

Oxidative phosphorylation (in the electron transport chain) uses ½ O_2 as the ultimate electron acceptor.

Electron transport chain (ETC) is on the inner mitochondrial membrane. It uses cytochromes (i.e., proteins) in the inner membranes of the mitochondria to pass electrons released from the oxidation of NADH & $FADH_2$.

Mitochondrial *matrix* (i.e., the cytoplasm of mitochondria) is the site for the Krebs cycle (TCA).

Outer mitochondrial membrane does not directly participate in oxidative phosphorylation or the Krebs cycle.

Glycolysis and the Krebs cycle produce ATP via *substrate-level phosphorylation*, while *oxidative phosphorylation* requires O_2 and occurs during the ETC.

25. D is correct.

Enzymes are often proteins and function as biological catalysts at an optimal temperature (physiological temperature of 36 °C) and pH (7.35 is the pH of blood).

At higher temperatures, proteins *denature* (i.e., unfold by disrupting hydrogen and hydrophobic bonds and change shape) and lose their function. Proteins often interact with inorganic minerals (i.e., cofactors) or organic molecules (i.e., coenzymes or tightly bound prosthetic groups) for optimal activity.

Mutations affect DNA sequences that encode for proteins, resulting in a change in the amino acid sequence of the polypeptide and a change in conformation within the enzyme.

A change in the conformation of enzymes often results in a change in function.

26. D is correct.

Allosteric enzymes have allosteric sites that bind metabolites.

Allosteric sites are distinct from the active sites that bind substrate.

An enzyme must have a *quaternary structure* (multiple polypeptides) to bind more than one molecule.

Allosteric enzymes show *cooperative* substrate binding with the binding of a molecule at one site affecting the binding at another. If an enzyme binds an inhibitor at an allosteric site, this *decreases the affinity* of the active site for the substrate.

27. C is correct.

I: *prosthetic groups* are tightly bound chemical compounds required for the enzyme's catalytic activity, and it is often involved at the active site.

II: *zymogens* are inactive enzyme precursors requiring biochemical changes to become active enzymes. This activation is usually a proteolysis (protein cleaving) reaction that reveals the active site.

28. B is correct.

Hydrolysis of ATP → ADP + phosphate forms glucose-6-phosphate from glucose and phosphate.

Enzymatic coupling of reactions allows ATP hydrolysis energy to drive the synthesis of glucose-6-phosphate.

Notes for active learning

Microbiology – Detailed Explanations

1. B is correct.

The *major* distinction between prokaryotic and eukaryotic cells is that prokaryotic cells do not have a nucleus, but eukaryotic cells do.

2. B is correct.

Penicillin prevents the formation of the bacterial *peptidoglycan cell wall* by covalently binding to a serine residue at the transpeptidase active site.

Binding the enzyme's active site is characteristic of an *irreversible competitive* inhibitor.

Irreversible binding with a covalent bond creates a stable and permanent attachment.

Competitive because it competes with the substrate for the active site.

Penicillin (β-lactam antibiotic) inhibits the formation of peptidoglycan cross-links in the bacterial cell wall. The enzymes that hydrolyze the peptidoglycan cross-links continue to function, but those that form crosslinks do not. This weakens the bacterium cell wall, and osmotic pressure rises, causing cell death (cytolysis).

A: *reversible competitive inhibitors* utilize weak molecular attachment (i.e., van der Waals or hydrogen bonds) when attaching to the enzyme active site.

C: penicillin does not digest the cell wall but prevents the crosslinking of the peptidoglycan of the bacterial cell wall. The effect inhibits the bacteria's ability to remodel their cell wall or prevent progeny from forming a necessary cell wall to counter osmotic pressure.

3. D is correct.

Viruses are simple, non-living organisms that take on living characteristics when they infect a host cell. Their genetic material is RNA or DNA; not arranged into chromosomes but associated with a complex of nucleic acids and histone proteins.

Viruses contain DNA or RNA and a protein coat (i.e., capsule).

Cellular machinery and biomolecules in a host cell (prokaryotic or eukaryotic) are required for viral replication.

4. D is correct.

Replica plating technique of Joshua and Esther Lederberg demonstrated in 1952 that streptomycin revealed the presence of streptomycin-resistant bacteria.

5. B is correct.

Operon is a functional unit of genomic DNA with a cluster of genes under the control of a single regulatory signal or promoter. Genes (i.e., nucleotide sequences) are transcribed together into mRNA strands.

The genes contained in the operon are expressed together or not at all.

Several genes must be *co-transcribed* and *co-regulated* to be defined as an operon.

6. C is correct.

Neurospora is a fungus and haploid for most of its life cycle.

For fungus, there is a brief diploid (2N) stage after fertilization that transitions via meiosis to produce haploid (1N) cells, which repeatedly divide via mitosis before entering another sexual cycle.

A: most fungi undergo meiosis and a sexual cycle.

B: separation of fertilization and meiosis is characteristic of the life cycle of plants.

D: *fertilization* immediately following meiosis is characteristic of diploid organisms.

7. D is correct.

Enzymes are biomolecules in prokaryotic cells.

8. A is correct.

Endomembrane system extends the nuclear envelope and includes the endoplasmic reticulum and Golgi apparatus.

The Golgi receives proteins within vesicles from the RER and modifies, sorts, and packages proteins destined for the secretory pathway (i.e., plasma membrane, exocytosis from the cell, or organelles within the cell).

Golgi is essential for synthesizing *proteoglycans* (i.e., components of connective tissue) in the extracellular matrix of animal cells.

Golgi primarily modifies proteins delivered from the rough endoplasmic reticulum. It is involved in lipid transport, the synthesis of lysosomes, and is the site of carbohydrate synthesis.

B: *lysosomes* are organelles with low pH that function to digest intracellular molecules.

C: *peroxisomes* are organelles in most eukaryotic cells. A major function of the peroxisomes is the breakdown of very-long-chain fatty acids through *beta-oxidation.*

In animal cells, the very-long-chain fatty acids are converted to medium-chain fatty acids and subsequently shuttled to mitochondria. They are degraded, via oxidation, into *carbon dioxide* and *water.*

D: *smooth endoplasmic reticulum* connects to the nuclear envelope and functions in metabolic processes. It synthesizes lipids, phospholipids, and steroids. It carries out the metabolism of carbohydrates, drug detoxification, attachment of receptors on cell membrane proteins, and steroid metabolism.

Cells secreting lipids, phospholipids, and steroids (e.g., testes, ovaries, skin oil glands) have a robust, smooth endoplasmic reticulum.

The smooth endoplasmic reticulum contains glucose-6-phosphatase (enzyme), which converts glucose-6-phosphate to glucose during gluconeogenesis.

9. B is correct.

Retrotransposons (or transposons via RNA intermediates) are a subclass of transposons. They are endogenous genetic elements that amplify in a genome as ubiquitous components of DNA in many eukaryotic organisms.

Around 42% of the human genome is made up of retrotransposons.

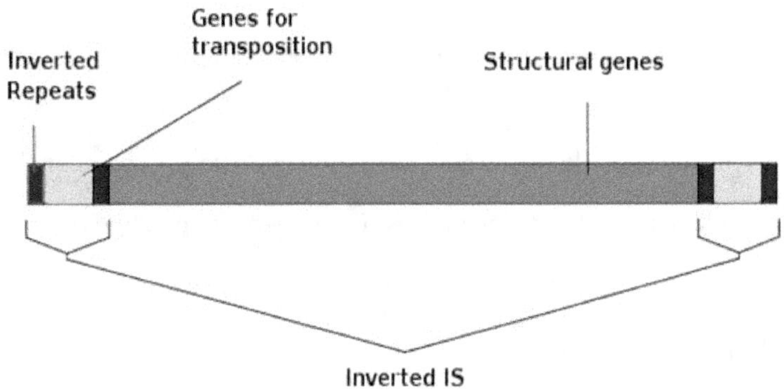

Bacterial DNA transposon with flanking inverted sequences encompassing structural genes

Retrotransposons are copied into RNA and back into DNA that may integrate into the genome.

The second step of forming DNA may be carried out by a reverse transcriptase encoded by the retrotransposon.

10. B is correct.

Plasmids are small circular double-stranded DNA in bacteria carrying extrachromosomal genetic information. Plasmids (with an origin of replication) are replicated by bacterial proteins and inherited by progeny.

The proteins encoded by genes on plasmids often provide resistance to antibiotics by degrading the antibiotic.

Through recombinant DNA technology, plasmids are engineered to carry genes not typically in bacteria.

Plasmids are introduced into cells by transforming DNA across the cell wall and plasma membrane without killing bacterial cells.

After transformation, exposure to the specific antibiotic allows the selection of bacteria that received plasmid.

Bacteriophages such as bacteriophages λ are viruses that infect bacterial cells.

Bacteriophage λ are engineered to carry novel genes, not in bacteria or bacteriophages.

I: *plasmids* are extra-chromosomal circular DNA molecules and not organelles because organelles are membrane-bound cellular components present only in eukaryotes.

II: like the bacterial genome, plasmids are in the cytoplasm.

Without a nucleus, prokaryotic ribosomes translate plasmid mRNA into proteins while transcribed into mRNA from DNA.

III: plasmids rely on bacterial machinery for metabolic processes (e.g., replication, transcription, translation).

11. A is correct.

Plants are generally autotrophs; animals are heterotrophs, and bacteria can be either.

Auxotrophs are unable to synthesize an organic compound required for growth. In genetics, a strain is auxotrophic if it has a mutation that renders it unable to synthesize an essential compound (i.e., *essential* is needed for growth but cannot be synthesized).

For example, a uracil auxotroph is a yeast mutant with an inactivated uracil synthesis pathway gene. Such a strain cannot synthesize uracil and will grow if uracil can be taken up from the environment.

B: *chemotrophs* obtain energy by the oxidation of electron donors in their environments.

These molecules can be organic (e.g., chemoorganotrophs) or inorganic (e.g., chemolithotrophs).

The chemotroph designation contrasts with phototrophs, which utilize solar energy.

Chemotrophs can be autotrophic or heterotrophic.

C: *heterotrophs* use energy derived from another organism's metabolism (*hetero* means other).

D: *prototrophs* are characterized by synthesizing the compounds needed for growth.

12. B is correct.

Endospore is a dormant, tough, non-reproductive structure produced by certain bacteria.

The endospore consists of bacterium DNA and cytoplasm, surrounded by a tough outer coating.

Endospores can survive without nutrients. A lack of nutrients usually triggers endospore formation.

13. A is correct.

Fungi are a large group of eukaryotes, including microorganisms such as mold, yeasts, and mushrooms.

Fungi are a kingdom (Fungi) separate from plants, animals, protists, and bacteria.

One significant difference is that fungal cells have cell walls with chitin.

The cell walls of plants and some protists contain cellulose, while bacterial cell walls contain peptidoglycan.

Fungi reproduce by sexual and asexual means and produce spores like basal plant groups (e.g., ferns, mosses).

Like algae and mosses, fungi typically have haploid (1N) nuclei with a small percentage of their life cycle in a diploid (2N) phase.

14. C is correct.

Cyanide is a poison interfering with the electron transport chain (ETC) of the inner plasma membrane (for prokaryotes) or the inner mitochondrial membrane (for eukaryotes) by binding to cytochromes (mainly cytochrome c oxidase) of the electron transfer complexes.

Cyanide inhibits the flow of electrons down their reduction potential and effectively inhibits the transport chain.

continued...

Consequently, the proton pump stops, and ATP cannot be generated aerobically.

Electron carriers such as NADH and $FADH_2$ cannot be oxidized and yield high-energy electrons to the electron transport chain because they are blocked. NAD^+ and FAD are not regenerated, and aerobic *respiration ceases*.

DNA replication, RNA transcription, and protein synthesis require energy (ATP or GTP) and are processes necessary for host cell functions and viral replication.

If aerobic ATP formation is inhibited, there will be a deficiency of ATP for cell functions and viral replication.

Energy is required for a bacteriophage (a virus that infects bacteria) to replicate.

15. A is correct.

Transformation is the genetic alteration of a cell resulting from the direct uptake, incorporation, and expression of exogenous genetic material (exogenous DNA) from its surroundings and passes through the cell membrane.

Transformation introduces genetic material into nonbacterial cells, including animal and plant cells.

B: *conjugation* transfers genetic material (plasmid) between bacterial cells by direct cell-to-cell contact or a bridge-like connection between two cells (i.e., pili).

Conjugation is a mechanism of *horizontal gene transfer*, as are transformation and transduction.

Other gene transfer mechanisms do not use cell-to-cell contact (conjugation uses a pilus).

C: *transduction* is how a virus transfers DNA from one bacterium to another.

In molecular biology, foreign DNA is introduced via a viral vector into another cell.

D: *recombination* is when DNA molecules exchange genetic information, forming new allele combinations.

In eukaryotes, genetic recombination between homologous chromosomes during prophase I of meiosis leads to a novel set of genetic information passed on to progeny.

Most recombination occurs spontaneously to increase genetic variation.

16. D is correct.

Viruses can have a genome consisting of double-stranded DNA (dsDNA).

When the genome is RNA (single-stranded or double-stranded), the virus is a *retrovirus*. After infecting the host cell, the retrovirus uses *reverse transcriptase* to convert its RNA into DNA.

17. B is correct.

Fungi are members of a large group of eukaryotic organisms that includes microorganisms such as yeasts, molds, and mushrooms.

These organisms are classified as a Fungi kingdom (separate from plants, animals, protists, and bacteria).

One major difference is that fungal cells have cell walls with *chitin.*

The cell walls of plants contain *cellulose*), some protists contain *cellulose,* and bacteria contain *peptidoglycan.*

A: *protoplasts* are plant, bacterial or fungal cells that lose their cell wall by mechanical or enzymatic means.

C: L forms are strains of bacteria that lack a cell wall.

D: viruses do not have cell walls.

A virus particle (i.e., virion) has nucleic acid surrounded by a protective protein coat as the *capsid.*

18. C is correct.

Prokaryotic ribosomes (30S small & 50S large subunit = 70S complete ribosome).

Eukaryotic ribosomes (40S small & 60S large subunit = 80S complete ribosome).

19. A is correct.

Retroviruses are a family of enveloped viruses that use reverse transcription to replicate within a host cell.

Retrovirus is a single-stranded RNA virus with nucleic acid as a single-stranded mRNA genome (including 5' cap and 3' poly-A tail).

Retrovirus is an obligate parasite within the host cell.

Once inside the host cell cytoplasm, the virus uses its reverse transcriptase (enzyme) to produce DNA from its RNA genome.

Retro describes this backward flow of genetic information.

Integrase (enzyme) incorporates this new reverse-transcribed DNA into the host cell genome.

Provirus is the integrated retroviral DNA.

The host cell treats the viral DNA as part of its genome, translating and transcribing the viral genes and the cell's genes, producing the proteins required to assemble new copies of the virus.

20. C is correct.

Virus is a simple non-living organism that takes on living characteristics when it enters host cells.

Since viruses are not free-living organisms, they must replicate within a host cell.

Viruses contain DNA (single-stranded or double-stranded) or RNA (single-stranded or double-stranded) and have a protein coat.

continued...

Bacteriophage (i.e., a virus that infects bacteria) injects its genome into the bacterium while leaving the protein coat on the cell surface.

However, in eukaryotes, the entire virus (including the protein capsid) may enter the host cell. After entering the cytoplasm, the protein coat is removed (i.e., a virus is unencapsulated).

21. A is correct.

Bacterial conjugation transfers genetic material (plasmid) between bacterial cells by direct cell-to-cell contact or a bridge-like connection between two cells (i.e., pili).

Conjugation is *horizontal gene transfer*, as are transformation and transduction.

The other mechanisms do not involve cell-to-cell contact (conjugation requires a pilus).

B: *transformation* is the genetic alteration resulting from the direct uptake, incorporation, and expression of exogenous genetic material (exogenous DNA) from its surroundings and through the cell membranes.

In molecular biology, transformation introduces genetic material into nonbacterial cells, including animal and plant cells.

C: *transduction* is how a virus transfers DNA from one bacterium to another.

In molecular biology, *transduction* is when foreign DNA is introduced into another cell via a viral vector.

D: *recombination* is when DNA molecules exchange genetic information, resulting in a new combination of alleles.

Most recombination occurs spontaneously to increase genetic variation.

In eukaryotes, *genetic recombination* between *homologous chromosomes* during prophase I of meiosis leads to a novel set of genetic information passed on to progeny.

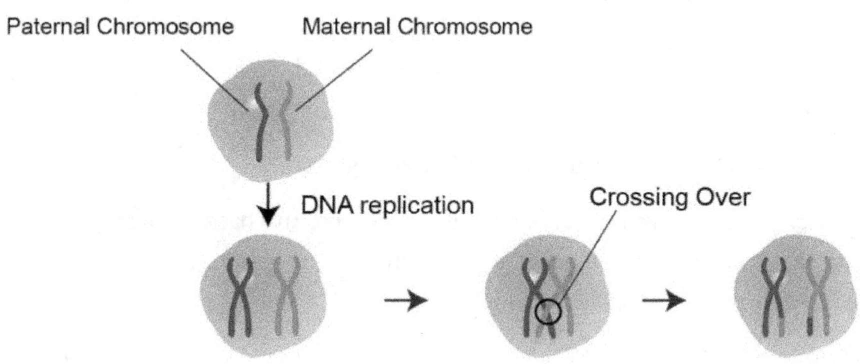

Crossing over occurs during prophase I at the tetrad during prophase I of meiosis as the product of recombination

continued…

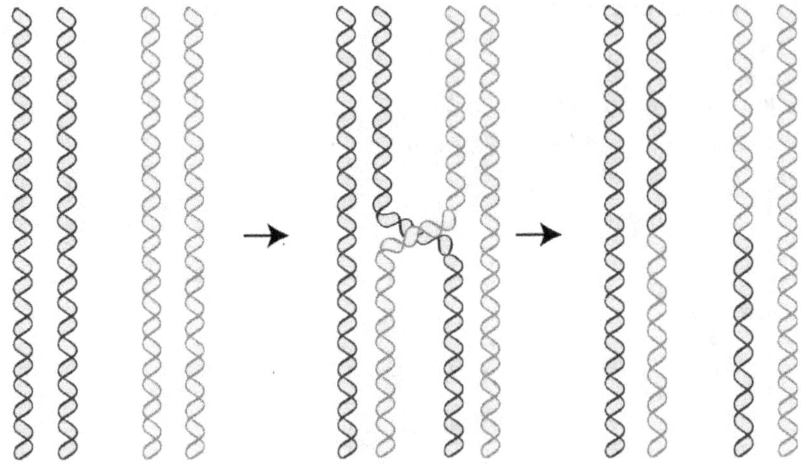

Recombination between homologous chromosomes during prophase I of meiosis

22. D is correct.

E. coli on the soft agar produces a solid growth (i.e., lawn) covering the agar.

Phages lyse cells, releasing phages (i.e., viruses). *Clear spots on the agar plate* correspond to where the bacteria were lysed and are *plaques* (i.e., clear spots).

A: *viral replication* produces clear spots on the lawn of *E. coli* growth.

B: *colonies* on the agar surface indicate *E. coli* growth, not locations where the virus lysed the bacteria cells.

C: *smooth layer* of bacterial growth on the agar plate indicates the absence of the virus.

23. D is correct.

Gram staining differentiates bacterial species into Gram-positive and Gram-negative by the chemical and physical properties of peptidoglycan cell walls.

Gram stain is usually the first assay in identifying a bacterial organism.

Gram-negative bacteria are more resistant to antibiotics than Gram-positive bacteria, despite their thinner peptidoglycan layer. The pathogenic capability of Gram-negative bacteria is often associated with specific components of their membrane, in particular, the lipopolysaccharide layer (LPS).

Penicillins are β-lactam antibiotics to treat bacterial infections by, usually, Gram-positive organisms.

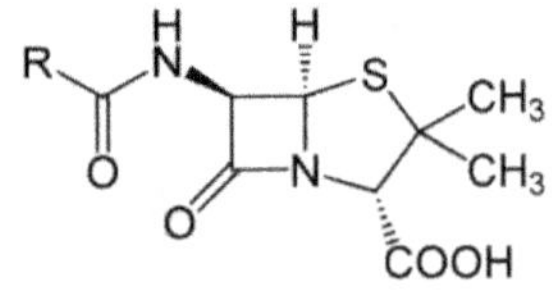

Penicillin core structure, where "R" is the variable group

continued...

Gram-positive bacteria take up the violet stain used in Gram staining. This distinguishes them from the other large group of bacteria, the Gram-negative bacteria that cannot retain the crystal violet stain.

Gram-negative bacteria take up the counterstain (e.g., safranin or fuchsine) and appear red or pink.

Gram-positive bacteria retain crystal violet stain by a thick *peptidoglycan layer* superficial to cell membranes.

In Gram-negative bacteria, this peptidoglycan layer is thinner and between two cell membranes.

The cell membrane, peptidoglycan layer, and cell wall are three distinct structures.

Cell walls provide structural support, protection, and rigidity to the cell.

Gram-positive bacteria have a thicker peptidoglycan cell wall, giving them a purple-blue color from staining.

Gram-negative bacteria stain a pink-red color.

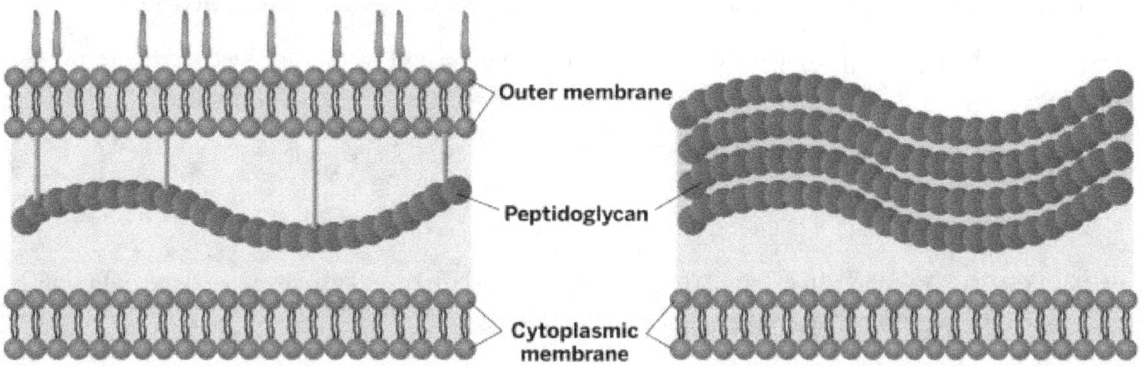

Gram-negative (left) comparison with gram-positive (right) cell walls

24. B is correct.

Viral genomes are single-stranded DNA, double-stranded DNA, single-stranded RNA, or double-stranded RNA.

Retroviruses contain a single-stranded RNA (i.e., mRNA as its genome).

Retroviruses have *reverse transcriptase* and *integrase* to facilitate their infective ability within the host cell.

25. C is correct.

Operons are mainly in prokaryotes and encode for clusters of related (similar function) genes.

Lac operon of *E. coli* allows the digestion of lactose and consists of a set of control and structural genes.

Three structural genes are controlled by an operator on another part of the genome.

Lac genes encoding enzymes are *lacZ*, *lacY*, and *lacA*.

The fourth *lac* gene is *lacI*, encoding the lactose repressor—"I" stands for *inducibility*.

Without lactose, a repressor protein (*lacI*) is bound to the operator, preventing RNA polymerase from binding.

The binding of the repressor protein prevents the translation of the structural genes (*lacZ*, *lacY*, and *lacA*).

When lactose is present (i.e., glucose is absent), *lacI* binds the repressor that dissociates from the operator region. RNA polymerase (i.e., *inducible system*) attaches to the promoter, and translation occurs.

26. D is correct.

Prophage is a viral genome inserted and integrated into the circular bacterial DNA chromosome or existing as an extrachromosomal plasmid. *Lysogenic phase* (i.e., latent form) of a bacteriophage is when the viral genes are in the bacterium without disrupting the cell.

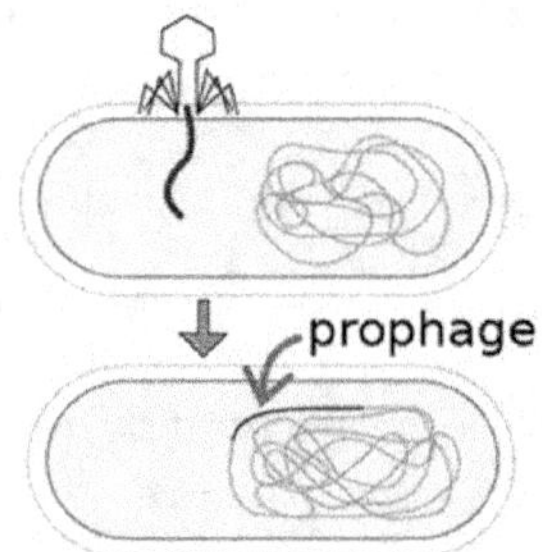

Prophase formation with a virus infecting a bacterial cell

Lytic cycle is one of the two cycles of viral reproduction, the other being the lysogenic cycle.

The lytic cycle destroys the infected cell and its membrane.

A critical difference between the lytic and lysogenic phage cycles is that in the lytic phage, the viral DNA exists as a separate molecule within the bacterial cell and replicates separately from the host bacterial DNA.

The location of viral DNA in the lysogenic phage cycle is within the host DNA (i.e., integrated); the virus (phage) replicates using the host DNA machinery.

Phage is a free-floating separate molecule to the host DNA in the lytic phage cycle.

27. C is correct.

Retrovirus has an RNA genome transcribed (by the virus-encoded reverse transcriptase enzyme) into single- and then double-stranded DNA incorporated into the host cell genome.

B: *reverse transcriptase* (enzyme) converts the RNA of the virus into DNA when infecting host cells.

28. B is correct.

E. coli are normal flora of the human gut and are adapted to live in the human intestine. Therefore, the optimal temperature for growth is 37 °C, whereby enzyme activity is optimal.

A: temperature of the human intestine is 37 °C. Bacteria growing optimally at this temperature exhibit the typical phenotype.

C: *conjugation* by bacteria is not necessary for growth. Conjugation is mediated by a small genetic element (i.e., fertility or F factor) as independent or integrated into the bacterial chromosome. The F factor encodes for the F pili, which forms a conjugation bridge and allows for the transfer of genetic material between mating cells.

Cells carrying the F factor are F^+ and transfer it to an F^- cell. Part of the bacterial chromosome can be transferred during conjugation, but the point of origin and the gene order are the same.

D: oxygen utilization is irrelevant to the growth temperature of bacteria.

29. A is correct.

Some bacteria can propel themselves through liquid using a flagellum.

30. C is correct.

When a virus infects a cell (eukaryotic or prokaryotic), it must attach to a host cell via specific proteins on the viral capsid or envelope. These proteins bind to receptor proteins on the membrane of the target cell.

Following this process, the virus enters cells via one of the three mechanisms: membrane fusion, endocytosis, or viral penetration. The later mechanism is specific to bacteriophages.

Membrane fusion is the most well-known mechanism for enveloped viruses to enter eukaryotic cells.

The viral envelope fuses with the host cell's membrane, and the contents of the virus are released into the cell.

Endocytosis is utilized by enveloped and non-enveloped viruses that infect eukaryotic cells.

In endocytosis, the virus is engulfed by the cell and shuttled by cellular vesicles to the cytoplasm.

31. C is correct.

Bacteria are prokaryotes, and they have no membrane-bound organelles such as peroxisomes or nucleolus.

32. A is correct.

Lysogen is a bacterial cell where a phage (i.e., virus) exists as DNA in its dormant state (prophage).

Prophage is integrated into the host bacterial chromosome (i.e., lysogenic cycle) or (rarely) exists as a stable plasmid in the host cell.

The prophage expresses genes that repress the phage's lytic action until this repression is disrupted.

The virus enters the *lytic life cycle* that ultimately ruptures the cell and releases virions.

B: *temperate* refers to the ability of some bacteriophages to enter the lysogenic lifecycle.

33. D is correct.

Protein synthesis (like for eukaryotes) occurs in three phases: initiation, elongation, and termination.

For translation, hydrogen bonds join codons on mRNA and anticodons on tRNA, not amino acid and mRNA.

The prokaryotic cellular translation machinery is the 70S ribosome (30S small and 50S large subunits).

F-met is used only in prokaryotic translation initiation. First, an initiation complex forms with the 30S subunit, mRNA, initiation factors, and a special initiator formyl methionine F-met tRNA.

50S subunit binds the initiation complex to form the 70S ribosome.

The complete large subunit of the ribosome complex has two binding sites: the P (peptidyl transferase) site and the A (aminoacyl) site.

Initiator formyl methionine (F-met) tRNA is in the P site.

Elongation begins with binding a second tRNA (charged with its corresponding amino acid) to the vacant A site of the large ribosomal subunit.

continued...

The appropriate amino acid is determined by the complementary hydrogen bonding between the anticodon of the tRNA and the next codon of mRNA.

The orientation for the mRNA (from 5' to 3') is E-P-A (ribosomes move towards the 3' end of mRNA).

Peptide bonds are catalyzed by peptidyl transferase (at the P site) with a nucleophile attack from the lone pair of electrons on the N of the amino terminus on the carbonyl of the C terminus (at the A-site).

tRNA in the P site, now uncharged (without amino acid), moves to the E site and dissociates.

tRNA in the A site (after peptide bond formation and the ribosome translocates) is now in the P site, the A site is vacant, and this is where the incoming tRNA (carrying the next amino acid to join the polypeptide) binds.

This binding cycle to the A site, peptide bond formation, and translocation create a growing polypeptide chain.

Termination occurs when one of the three tRNA stop codons (i.e., not charged with an amino acid) hybridizes to the mRNA in the A site.

Termination factors catalyze the hydrolysis of the polypeptide chain from the tRNA, and the ribosomal complex dissociates.

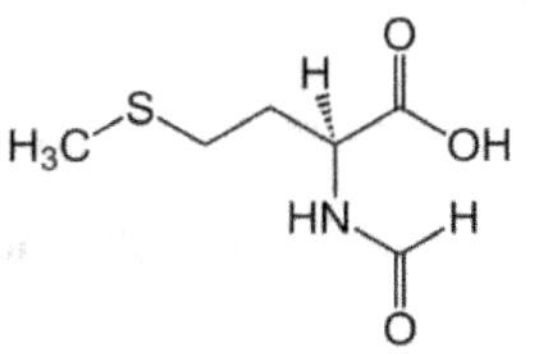

N-Formylmethionine (fMet) is a derivative of methionine whereby a formyl group adds to the amino group

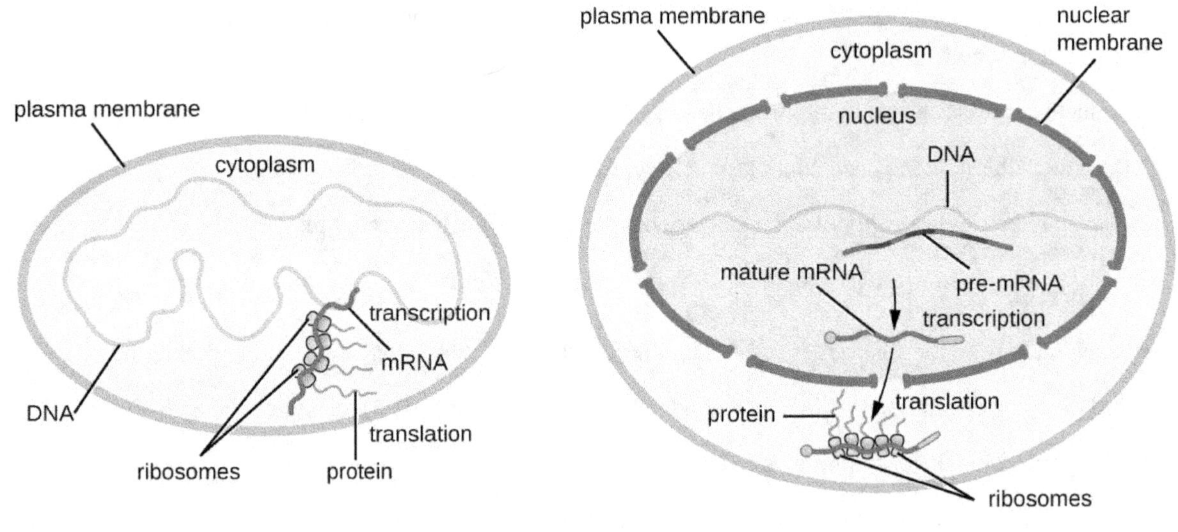

Prokaryotic cell (left) and eukaryotic cell (right). Prokaryotes have concurrent transcription and translation, while eukaryotes separate the processes by the presence of a nucleus.

A: prokaryote mRNA does not undergo splicing to remove introns.

B: *translation* in prokaryotes uses N-terminal formyl methionine (F-met) as the initiator amino acid.

C: *prokaryotes* have no nucleus, the mRNA has no nuclear membrane to cross, and translation begins before its synthesis is complete.

Eukaryotic mRNA must be processed (5' G-cap, 3' poly-A tail, and splicing of exons with the removal of introns) and transported across the nuclear membrane before translation begins in the cytoplasm.

34. B is correct.

Surface-to-volume ratio limits the size of biological cells because when the volume increases, so does the surface area, but the surface area increases at a slower rate.

Eukaryotes grow beyond the apparent limitation of surface-to-volume ratios because of organelles (specialized environment) and extensive cytoplasmic matrix, utilizing microfilaments and motor proteins for cargo transport.

35. D is correct.

Prokaryotes have *peptidoglycan cell walls*, 30S and 50S ribosomes, and plasma membranes without cholesterol.

36. B is correct.

Viruses consist of genetic material with single-stranded DNA, double-stranded DNA, single-stranded RNA, or double-stranded RNA packaged within a protein coat.

A: viruses do not have membrane-bound organelles but consist entirely of nucleic acid (RNA or DNA) and, for retroviruses, reverse transcriptase (enzyme).

C: bacteria have *peptidoglycan* (i.e., N-linked glucose polymers) cell walls.

D: viruses do not have a phospholipid bilayer membrane.

37. B is correct.

Prokaryotes do *not* have membrane-bound organelles (e.g., nucleus, mitochondria, lysosomes).

C: prokaryotes have *peptidoglycan* cell walls, while fungi have cell walls with chitin.

38. B is correct.

Reverse transcriptase (enzyme) converts RNA of the virus into ds-DNA when infected into the host cell.

The host cell treats the viral DNA as part of its genome, translating and transcribing the viral genes and the cell's genes, producing the proteins required to assemble new copies of the virus.

This new DNA is incorporated into the host cell genome by integrase (enzyme), at which point the retroviral DNA is a *provirus*.

Retrovirus flow of genetic information:

RNA (virus) → DNA (from virus template) → RNA (host cell machinery) → polypeptide

39. C is correct.

Bacteriophages are viruses that infect bacteria and typically consist of a protein coat (i.e., head) and a core containing nucleic acid. Like viruses, bacteriophages contain host-specific protein tail fibers for attachment.

Upon infection, bacteriophages can enter one of two life cycles (image below).

continued...

In the *lytic cycle*, the viral nucleic acid enters the bacterial cell. It begins using host machinery to produce new virions (i.e., virus particles), lysing the host cell and infecting other cells.

Lysogenic cycle is when viral DNA integrates into the bacterial chromosome, replicating with it and being passed to daughter cells (integrated into their genome) in this inactive form.

The lytic cycle destroys the infected cell and its membrane. A critical difference between the lytic and lysogenic phage cycles is that in the lytic phage, the viral DNA exists as a separate molecule within the bacterial cell and replicates separately from the host bacterial DNA.

The location of viral DNA in the lysogenic phage cycle is within the host DNA (i.e., integrated); the virus (phage) replicates using the host DNA machinery.

Phage is a free-floating separate molecule to the host DNA in the lytic phage cycle.

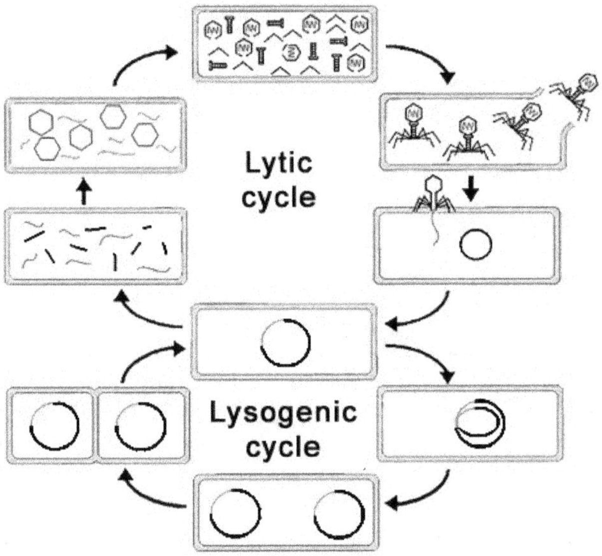

Lytic phase lysis the cell and releases virons while lysogenic cycles involve the integration of the viral genome into the host cell

When an integrated virus (i.e., prophage) becomes activated, it exits the lysogenic and enters the lytic cycle. Activation triggers for the virus to enter the lytic cycle include *cellular or metabolic stresses* such as UV light, depleted nutrients, or cellular abnormalities.

Retroviruses are RNA-containing viruses that replicate by reverse transcriptase (enzyme) through a DNA intermediate by a viral-coded RNA-dependent DNA polymerase. Human immunodeficiency virus (HIV), the causative agent of AIDS, is an example of a retrovirus.

Retroviral life cycle consists of four main events:

1) virus binds its host and injects its RNA and a few viral enzymes;

2) RNA is converted to DNA by reverse transcriptase;

3) DNA integrates into the host cell's genome; and

4) viral genes are expressed, and virions are assembled and released from the cell by budding.

Bacteriophage and retrovirus integrate into the host genome (i.e., lysogenic) or lyse cells (i.e., lytic cycle).

A: *bacteriophages* have host-specific protein tail fibers and infect bacteria.

B: bacteria (the target of bacteriophage) have no immune system. Bacteria have a defense against infection by nucleotide-specific *restriction enzymes* used for molecular biology and biotechnology.

D: *retroviruses* contain an RNA genome (ss-RNA or ds-RNA) and reverse transcriptase, not in bacteriophages.

40. C is correct.

Gram-positive bacteria take up the violet stain used in the Gram staining. This distinguishes them from the other large group of bacteria, the Gram-negative bacteria that cannot retain the crystal violet stain.

Gram-negative bacteria take up the counterstain (e.g., safranin, fuchsine) and appear red or pink.

Gram-positive bacteria retain the crystal violet stain due to a thick peptidoglycan layer superficial to the cell membrane. In Gram-negative bacteria, this peptidoglycan layer is much thinner between cell membranes.

The cell membrane, peptidoglycan layer, and cell wall are three distinct structures.

Cell walls provide structural support, protection, and rigidity to the cell.

D: *teichoic acids* are in cell walls of Gram-positive bacteria (e.g., *Staphylococcus*, *Streptococcus*, *Bacillus*, *Clostridium, Listeria*) and extend to the surface of the peptidoglycan layer.

Teichoic acids are not in Gram-negative bacteria.

41. C is correct.

A bacterium with an outer lipopolysaccharide layer is Gram-negative and protects against penicillin.

A: *fimbriae* (i.e., a proteinaceous appendage in many Gram-negative bacteria) is thinner and shorter than a flagellum and allows a bacterium to attach to solid objects.

B: *bacterial cell membranes* have a phospholipid bilayer like eukaryotes, except it lacks cholesterol.

D: *Gram-negative bacteria* have a thinner peptidoglycan cell wall that does not retain Gram stain.

42. B is correct.

DNase is an enzyme that degrades DNA by hydrolysis of the DNA molecule.

Bacterial cells treated with DNase die because bacteria have DNA genomes, while RNA viruses are unaffected by DNase and continue to synthesize proteins following treatment with DNase.

A: having multiple copies of a gene is not enough to prevent DNase from degrading the DNA.

C: *viral genomes* typically contain multiple reading frames to efficiently use their limited nucleic acid, but multiple reading frames do not prevent DNA hydrolysis.

D: viral protein coat is not able to denature DNase. Denaturation involves breaking weak (e.g., hydrogen, dipole, and hydrophobic) bonds from heat or chemical treatment.

43. C is correct.

Integration of the phage DNA into bacterial chromosomes uses site-specific recombinase encoded in the phage.

44. A is correct.

The sequences recognized by most restriction enzymes are *inverted repeats* (i.e., palindromes) that read the same if inverted (i.e., rotated by 180°).

The *sticky ends* for DNA fragments are the same in either orientation, so ligation occurs from either orientation.

B: DNA strands do serve as primers for DNA polymerase

C: DNA *ligase* creates a phosphodiester bond that covalently links sticky ends of DNA.

D: *plasmid DNA*, like bacterial DNA, is double-stranded.

45. C is correct.

As a large polysaccharide, dextran does *not pass* through the cell membrane.

Osmotic pressure of the solution increases and H_2O moves out of the cell to reduce the osmotic pressure difference, and the cell undergoes crenation (i.e., shrinking).

46. A is correct.

Gram staining differentes bacterial species into two large groups: Gram-positive and Gram-negative. Gram staining differentiates the chemical and physical properties of *peptidoglycan cell walls.*

Gram-positive bacteria have a thicker peptidoglycan cell wall which gives Gram-positive bacteria a purple-blue color from staining, while Gram-negative results in a pink-red color. The Gram stain is usually the first assay in identifying a bacterial organism.

47. D is correct.

Virulence describes disease severity or a pathogen's infectivity.

The virulence factors of bacteria are typically proteins or other molecules synthesized by enzymes.

Virulent viruses use the *lytic* lifecycle.

These proteins are coded for by genes in chromosomal DNA, bacteriophage DNA or plasmids.

48. C is correct.

Promoter is a region on DNA recognized and bound by RNA polymerase as the initiation site for transcription.

B: *inducer* is the molecule that inactivates the repressor.

When the inducer binds the *repressor*, the bound repressor dissociates from the DNA and allows the polymerase to bind and activate the operon.

D: *repressor* binds the *operator* (i.e., a segment of DNA within the operon).

Notes for active learning

Notes for active learning

Plants: Structure, Function and Reproduction – Detailed Explanations

1. D is correct.

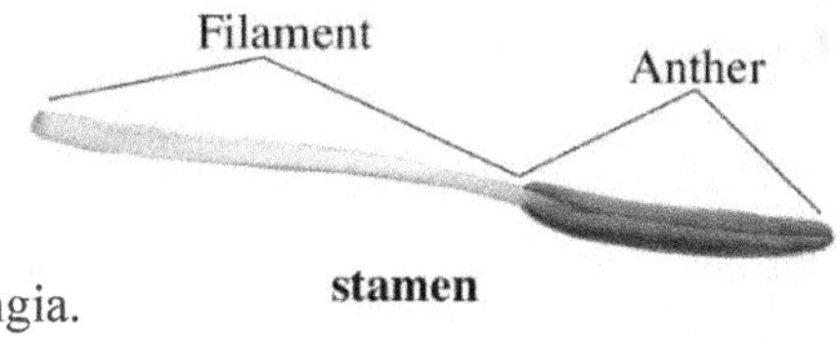

Angiosperms produce pollen grains in the *anther*.

Stamen is a flower pollen-producing reproductive organ.

Stamen typically has a *filament* stalk and an *anther* with microsporangia.

Anthers are part of the stamen, often two-lobed, and attached to the filament at the base or in the middle portion.

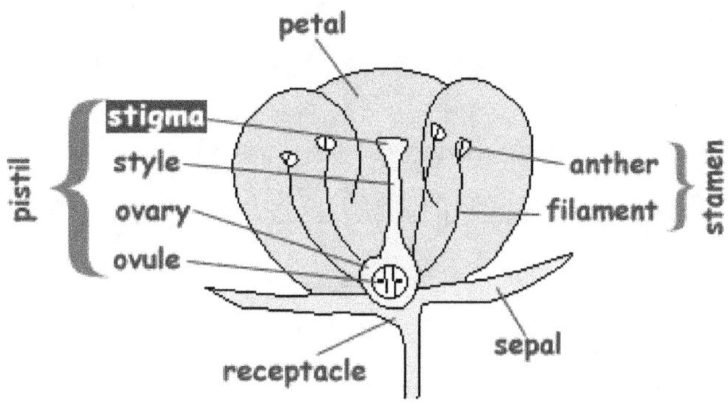

Anatomy of an angiosperm flower with female pistil and male stamen

A: *ovule* is the structure that gives rise to and contains female reproductive cells with the megaspore-derived female *gametophyte* (*megagametophyte*) in the center.

Megagametophyte (i.e., embryo sac in flowering plants) produces egg (or several) cells for fertilization.

Ovule develops into a seed after fertilization.

Ovary is the enlarged basal portion of the pistil where *ovules* are produced.

B: *sepals* are a part of the flower of *angiosperms* (i.e., flowering plants), and morphologically, sepals and petals are modified leaves.

C: *stigma* is part of the pistil where pollen germinates.

2. A is correct.

Leaves are plant organs with a primary function of *photosynthesis*.

B: *flowers* have the sole function of sexual reproduction. They are often the most prominent part of a plant.

Flowers' beauty and fragrance evolved pollinators (e.g., insects or birds), which are essential for reproduction. Flowers also help classify plants.

C: *root* is part of a vascular plant, usually underground. Its functions are anchorage of the plant, absorption of water and dissolved minerals and conduction of these nutrients to the stem, and storage of reserve foods.

continued...

D: *stems* primary functions are to support the leaves; conduct water and minerals to the leaves, where they can be converted into usable products by photosynthesis; and transport these products from the leaves to other parts of the plant, including the roots.

3. B is correct.

Most plants get the water they need from the soil. *Roots* absorb the bulk of water uptake by most plants.

Plants can absorb water through leaves, but it is not efficient. Plants can take in some of that surface water if water condenses on the leaf during high humidity (e.g., fog).

4. A is correct.

In angiosperms, *xylem* consists of *tracheids* and *vessel elements.*

Tracheids and *vessel elements* are elongated cells in vascular plant xylem that transport water and mineral salts.

Tracheids are elongated cells transporting water and mineral salts through the xylem of vascular plants.

Tracheids provide *physical support* and *conduct water and dissolved minerals upwards* and are the only such elements in conifers and ferns.

Tracheids, unlike vessel elements, do not have perforations.

B: *parenchyma* is a type of simple permanent tissue and a significant part of ground tissues in plants, where other tissues, like vascular tissues, are embedded.

Parenchyma is *non-vascular*, with *undifferentiated cells* modified to perform various functions.

C: *sieve tube elements* are specialized cells essential for the phloem's function, a highly organized tissue that transports organic compounds from photosynthesis.

D: *albuminous cells* have a similar role to companion cells but are associated with sieve cells only and are only in seedless vascular plants and gymnosperms.

5. D is correct.

Flowers in angiosperms have the sole function of sexual reproduction and are often the most prominent part of a plant. Flowers that animals pollinate do not exist as small structures.

Flowers' beauty and fragrance evolved pollinators (e.g., insects or birds), which are central to the reproductive process. Flowers also help classify plants.

6. C is correct.

Without gas exchange, a plant could not make food (i.e., carbohydrates such as glucose).

Stomata (stoma, singular) and *lenticels* regulate gas exchange in plants.

Plant epidermis has stomata, tiny pores that control transpiration and gas exchange with air. During the day, photosynthesis occurs, carbon dioxide (CO_2) is consumed, and oxygen (O_2) is released into the atmosphere.

continued...

Without gas exchange, plants cannot complete photosynthesis without carbon dioxide from the atmosphere.

Photosynthesis

$6CO_2 + 6H_2O \longrightarrow C_6H_{12}O_6 + 6O_2$

carbon dioxide + water ⟶ glucose + oxygen

Photosynthesis reaction:

$$6\ CO_2 + 6\ H_2O \rightarrow C_6H_{12}O_6 + 6\ O_2$$

Six carbon dioxide (CO_2) and six water (H_2O) molecules are converted by light energy (photons from the sun) captured by the photopigment chlorophyll into a sugar molecule (e.g., $C_6H_{12}O_6$ as glucose), and six expired oxygen (O_2) molecules.

7. D is correct.

Filament and *anther* comprise a stamen as a male reproductive structure of flowering plants.

Stamen (male organ) consists of an *anther* (site of pollen development) and, in most species, a stalk-like *filament*, which transmits water and nutrients to the anther and positions it for pollen dispersal.

Pistil (female organ) is the *ovule-producing* organ of a flower with a stigma, style, and ovule.

Ovary often supports a long *style*, topped by a *stigma.*

Fruit is a mature ovary, and the *mature ovule* is a *seed.*

Stigma is part of the pistil where *pollen germinates.*

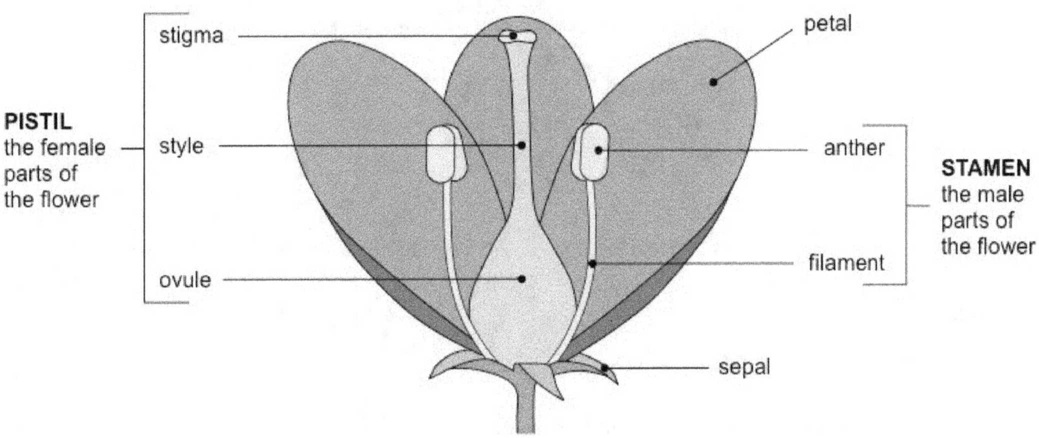

Pistil (stigma, style, and ovule) is female, and stamen (anther and filament) is the male part of the flower

8. C is correct.

Plants living on land must *conserve water*.

Water loss was the primary challenge for plants moving from aquatic to terrestrial environments.

Cutin, stomata, roots, and root hairs help terrestrial plants absorb and conserve water.

Roots and root hairs allow plants to absorb nutrients and water from the soil.

Plants have adapted to environments to sustain water balance (e.g., closing of stomata), as, without it, all processes stop, and the plant would not be able to survive.

For example, when the humidity is low, water is more likely to evaporate quickly from the leaf surface, and plants often close or partially *close their stomata* to maintain a stable water balance in the leaf.

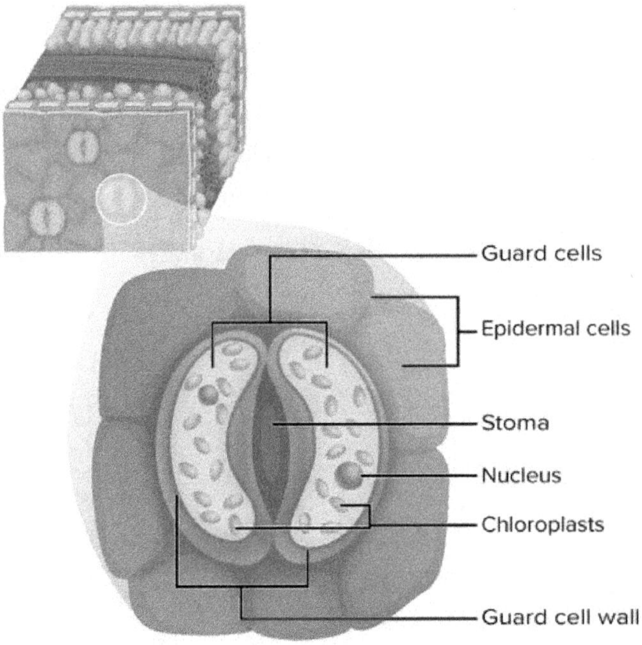

Leaf underside cross-section with an open stoma, guard cells, and support structures

9. B is correct.

Vascular tissue in plants consists of *xylem* and *phloem.*

Xylem is in vascular bundles (non-woody plants and non-woody parts of woody plants), secondary xylem from vascular cambium meristem in woody plants, and non-bundles (e.g., many ferns).

Xylem transports water, soluble minerals, and nutrients from the roots throughout the plant and replaces water lost during transpiration (water movement upwards) and photosynthesis (sugars formed in leaves).

Phloem is living tissue in vascular plants that transports (i.e., translocation) organic nutrients (e.g., fructose, sucrose) synthesized during photosynthesis to vascular plant parts.

Phloem tissue consists of *conducting cells* (i.e., sieve elements), *parenchyma cells* (specialized companion and albuminous cells and unspecialized cells), and *supportive cells* (i.e., fibers and sclereids).

10. D is correct.

Vascular tissue in plants consists of *xylem* and *phloem.*

Xylem is in vascular bundles (non-woody plants and non-woody parts of woody plants), secondary xylem from vascular cambium meristem in woody plants, and non-bundles (e.g., many ferns).

Xylem transports water, soluble minerals, and nutrients from roots and replaces water lost during photosynthesis.

Phloem is living tissue in vascular plants that transports downward (*translocation*) organic nutrients (e.g., sucrose) synthesized during photosynthesis to vascular plant parts.

Phloem tissue consists of *conducting cells* (i.e., sieve elements), *parenchyma cells* (specialized companion and albuminous cells and unspecialized cells), and *supportive cells* (i.e., fibers and sclereids).

For example, the phloem is the innermost layer of bark in trees.

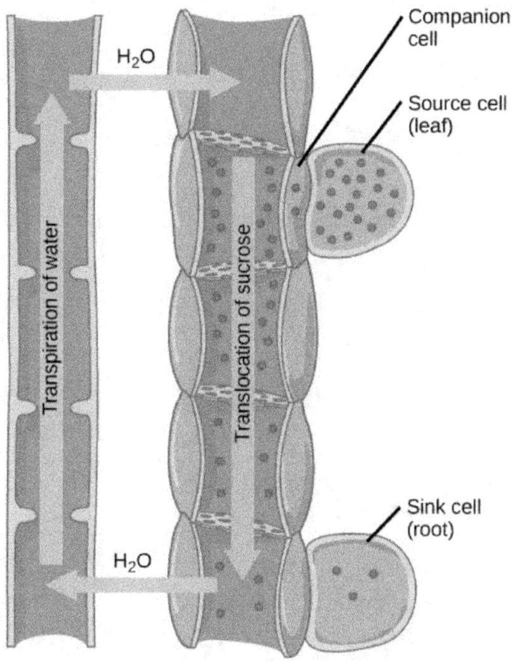

Xylem (on the left) functions for the transpiration of H_2O from roots, with phloem (on right) for the translocation of sugars from leaves during photosynthesis

A: *epidermis* (or d*ermal tissue*) is the outer covering of a plant (roots, stems, leaves, flowers, fruits, and seeds); usually, one cell layer thick, and its cells lack chloroplasts.

B: *parenchyma* is the most versatile and abundant cell of the three types of ground tissue.

Sclerenchyma is dead ground tissue that supports thick (primary and secondary) cell walls.

Collenchyma is living ground tissue that supports living tissue with irregular walls. They have thick deposits of *cellulose* in cell walls and appear *polygonal* in cross-section.

C: *meristem* is tissue in most plants containing undifferentiated cells in growth zones that give rise to plant organs and keep the plant growing.

11. C is correct.

Ancestors of land plants likely evolved from *streptophyte algae* (freshwater green algae) 500 million years ago.

Likely, plants evolved from living in water to inhabiting land because of genes they took up from bacteria.

12. B is correct.

Plants reproduce vegetatively by mitosis, whereby a single parent produces cloned offspring; no seeds or spores are generated, and does not involve the fusion of gametes.

Vegetative reproduction is *asexual* reproduction in plants when a new daughter plant grows from a fragment of the parent plant or grows from a *specialized reproductive structure* (e.g., stolon, rhizome, tuber, corm, or bulb).

Vegetative part (i.e., root, stem, leaf) of the parent is preserved and propagated to generate offspring.

13. D is correct.

Plants relying on animals for dispersion have *seeds enclosed in fleshy fruits*, encouraging animals to eat them.

Animal digests fruit flesh, but the seed enclosed in a thick indigestible coating passes through an animal's digestive tract without being digested.

When an animal excretes undigested seeds, they often land in a place distant from their parent and can germinate when conditions are right.

14. D is correct.

Plant shoots consist of any plant stem with appendages, leaves and lateral buds, flowering stems, and flower buds that can regenerate the plant.

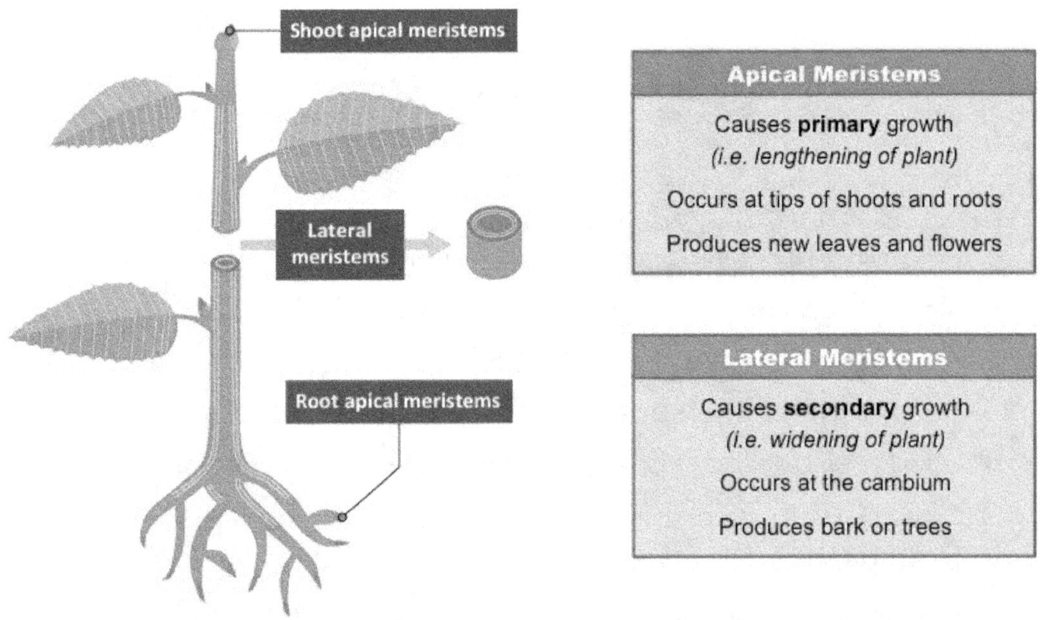

Apical (shoots and roots) and lateral (cambium) meristems are regions of mitotic cell growth

continued…

Meristematic cells (i.e., centers of mitotic cell division) at the tip of roots render desirable and suitable raw material to study the stages of mitosis.

Meristems mitotic cells are undifferentiated *self-renewing stem cells* from which most structures arise.

Apical meristems cause primary growth to elongate the plant (e.g., shoot growth)

Lateral meristems cause secondary growth to widen the plant (e.g., growth rings of trees)

Meristematic cells are also responsible for keeping the plant growing.

15. C is correct.

Animals would not likely live on land if plants did not migrate to land.

Land could not be colonized by other organisms until land plants became established because plants became a food source for many land animals.

16. A is correct.

Xylem and *phloem* are the two types of transport tissue in vascular plants.

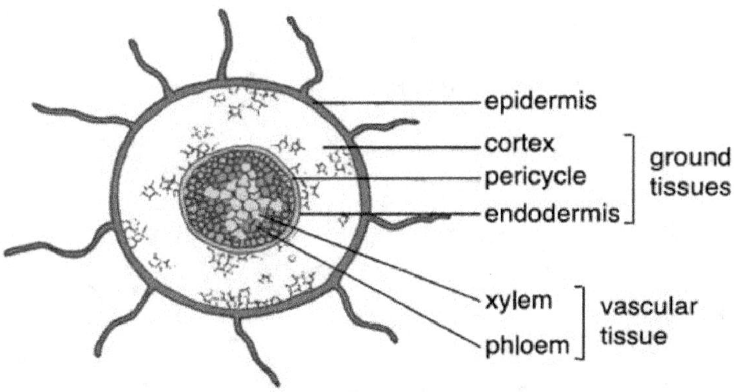

Root cross-section with xylem and phloem vascular tissue

Xylem is in vascular bundles (non-woody plants and non-woody parts of woody plants), secondary xylem laid down by vascular cambium (meristem) in woody plants, and not divided into bundles (e.g., many ferns).

Xylem transports water, soluble minerals, and nutrients from roots and replaces water lost during photosynthesis.

Phloem is the tissue that transports (i.e., translocation) soluble organic nutrients synthesized during photosynthesis (e.g., glucose) to all vascular plant parts where needed; it is the innermost layer of tree bark.

Phloem tissue is *conducting cells* (i.e., sieve elements), *parenchyma cells* (specialized companion and albuminous cells and unspecialized cells), and *supportive cells* (i.e., fibers and sclereids).

D: *meristem* is the tissue in most plants that contains undifferentiated cells in the growth zones of plants. Meristematic cells give rise to various plant organs and keep the plant growing.

17. D is correct.

Vegetative reproduction is asexual reproduction in plants when a daughter plant grows from a fragment of the parent or grows from a specialized reproductive structure (e.g., stolon, rhizome, tuber, corm, or bulb).

Plants reproduce vegetatively by mitosis, whereby a single parent produces *identical cloned offspring*; no seeds or spores are generated, and do not involve genetic diversity with the fusion of gametes.

Offspring are produced by mitosis and lack genetic recombination generated by meiosis.

Progeny are identical clones of the parent and lack genetic diversity to survive a shift in the physical environment.

18. D is correct.

Endodermis is the layers of cells enclosing the vascular tissue (xylem and phloem) in a root's central region.

A: *apical meristem* is meristematic tissue at the tips of stems and roots, enabling an extension in length.

B: *epidermis* (or d*ermal tissue*) is the outer covering of a plant (roots, stems, leaves, flowers, fruits, and seeds); usually, one cell layer thick, and its cells lack chloroplasts.

C: *cortex* (woody plants) is between the *periderm* (bark) and *vascular tissue* (phloem) that transports materials into the central cylinder of the root through *diffusion.*

Cortex may be used to store food in the form of *starch.*

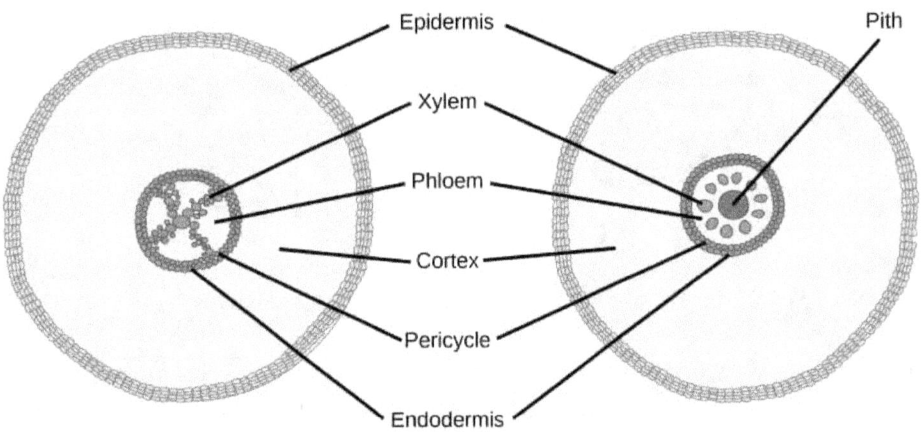

Epidermis, vascular tissue, cortex, and endodermis for dicot root (on left) and monocot root (on right)

19. A is correct.

Ploidy number is the fundamental difference between a sporophyte and a gametophyte.

Gametophyte is haploid (1N)

Gametophyte is the sexual phase with haploid (1N) male and female organs (*gametangia*) developing and producing eggs and sperm gametes through mitosis as sexual reproduction.

Egg and sperm unite in fertilization, forming a 2N zygote (fertilized egg) that develops into a diploid sporophyte, which produces unicellular spores through meiosis.

These develop a new gametophyte phase.

Gametophytes mature within spores, not free-living, as gametophytes of other seedless vascular plants.

Sporophyte is diploid (2N)

Sporophyte is the nonsexual phase when a diploid (2N) plant body grows and produces spores through meiosis.

These spores divide mitotically to produce haploid gametophytes.

Spore release in a suitable environment leads to germination and a new generation of gametophytes.

Completing the life cycle requires water, as male gametes, propelled by flagella, swim to the female gametes.

After fertilization, the zygote matures and grows into a *sporophyte*, forming sporangia (or *spore vessels*) in which mother cells undergo meiosis and produce haploid spores.

Heterosporous seedless plants are seen as the evolutionary forerunners of seed plants.

Seed plants (or *spermatophytes*) form two types of spores:

Megaspores (female) develop into gametophytes that produce eggs.

Microspores (male) mature into gametophytes that generate sperm.

Alternations of generations are distinct phases in the life history of organisms, each phase producing the other.

Gametophytes (haploid) represent the sexual phase in the alteration of generations.

Sporophyte (diploid) is the dominant and nonsexual phase in the life cycle.

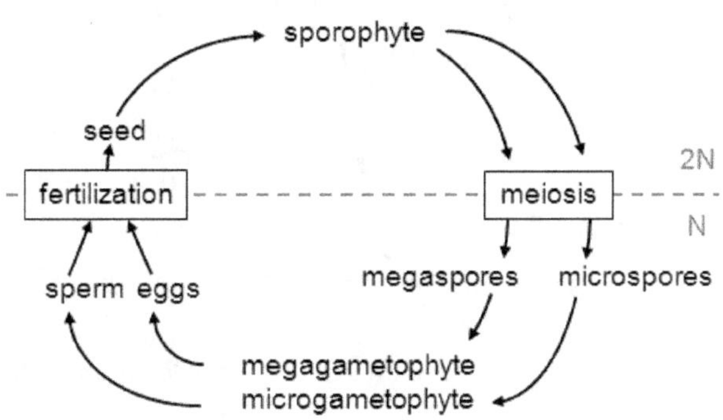

Alternation of generations with sporophyte is diploid (2N), while a gametophyte is haploid (1N) in the life cycle

continued...

Two gametes join during fertilization as a diploid zygote, dividing mitotically to form a new sporophyte.

Sporophytes	Gametophytes
Sporophytes undergo *meiosis*	Gametophytes undergo *mitosis*
Form *spores*	Form *gametes*
Diploid (2N) plants	Haploid (1N) plants
Two sets of chromosomes	Single set of chromosomes
Reproduce *asexually*	Produce *sexually*
Diploid spore cells undergo meiosis, forming haploid (1N) meiospores	Gametes fuse (or undergo fertilization), forming diploid (2N) zygotes

Gametophytes and sporophytes with spore versus gamete distinctions

Bryophytes and pterophytes characterize the alternation of generations.

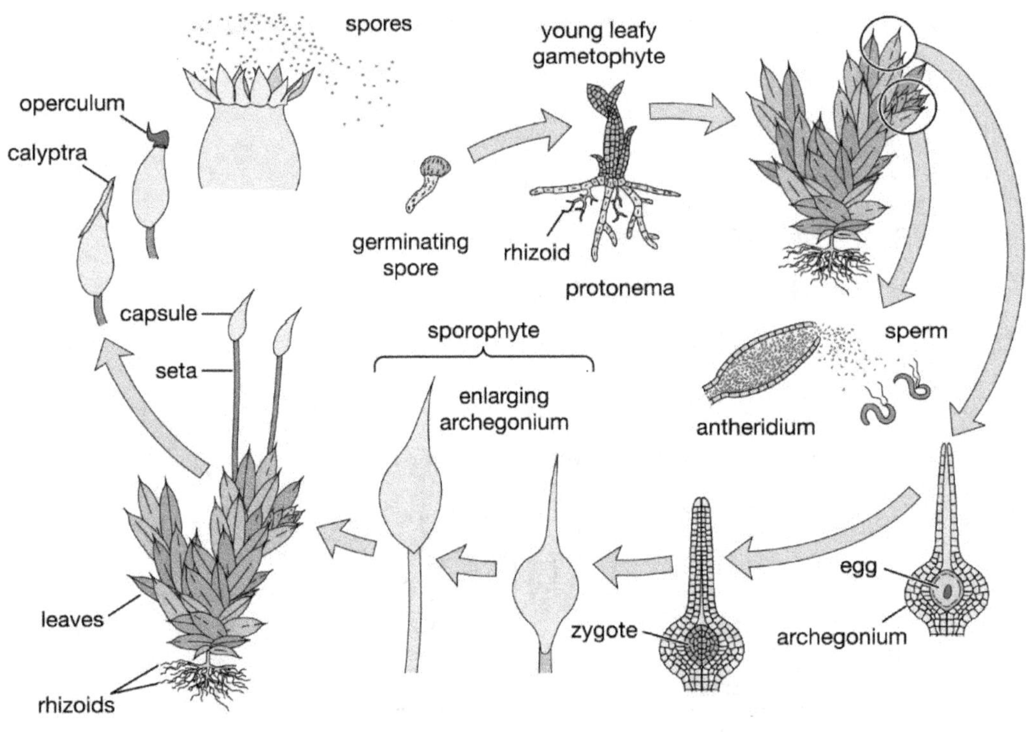

Alternations of generations for gametophytes and sporophytes in mosses

Pollen and seed allowed seed plants to break their dependence on water for reproduction and development of the embryo and thrive on dry land.

Pollen grains carry male gametes with small haploid (1N) cells encased in a protective coat to prevent desiccation (i.e., drying) and mechânical damage.

20. B is correct.

Dermal tissue (epidermis) is the outer covering of a plant (e.g., roots, stems, leaves, flowers, fruits, and seeds), usually one cell layer thick, with cells lacking chloroplasts.

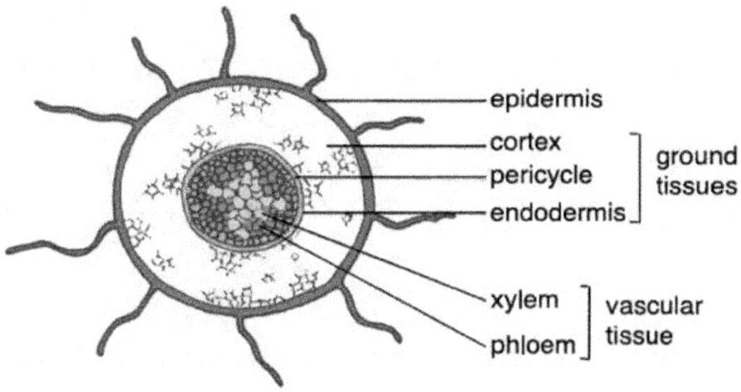

Root cross-section with xylem and phloem vascular tissue

21. A is correct.

Vascular plant gametophytes tend to be much smaller than sporophytes, sometimes only a few cells in size.

Gametophytes have become smaller, while sporophytes have become larger as plants evolved.

For example, a pollen grain is a male gametophyte in vascular plants.

In seed plants, evolutionary trends led to a dominant sporophyte generation with a reduction in gametophyte size from a conspicuous structure to a microscopic cluster of cells enclosed with sporophyte tissue.

> *Gametophyte* is the sexual haploid phase in the life cycle when a haploid (1N) organism develops from a haploid (1N) spore with one set of chromosomes.
>
> *Sporophyte* is the diploid stage in the life cycle and develops from a diploid (2N) zygote produced when a haploid (1N egg) cell is fertilized by a 1N sperm.
>
> Sporophyte cells have a set of chromosomes, one set from each parent.

Diploid plants with alternation of generations begin from a diploid (2N) zygote and produce haploid (1N) spores by meiotic division.

Seeds and pollen (adaptations to drought) distinguish seed plants from seedless vascular plants.

Both adaptations were critical to the colonization of land.

Pollen can travel far from the sporophyte that bore it, spreading the plant's genes and reducing competition with genetically similar plants.

Seed offers embryo protection, nourishment, and a mechanism to maintain dormancy for thousands of years, surviving harsh environments and germinating when growth conditions are optimal.

Seeds disperse through space and time; evolution favored seed plants as successful and familiar plants.

22. B is correct.

Dormant seeds are ripe seeds that do not germinate because they are subject to external environmental conditions that prevent the initiation of metabolic processes and cell growth.

Germination is when an organism grows from a seed or spore. It applies to the sprouting of a seedling from a seed of an angiosperm (flowering plant) or gymnosperm (seed-producing plant) and the growth of the pollen tube from the pollen grain of a seed plant.

When conditions (e.g., water, light) become favorable, the dormant seed germinates, and the embryonic tissues resume growth, developing into a seedling.

23. D is correct.

Seeds and fruits result from fertilization or sexual reproduction in plants.

Seeds are formed in gymnosperms and angiosperms. In angiosperms (flowering plants), fruits are matured ovaries that contain seeds.

Ovaries in angiosperms develop into fruit, whereas *ovules* become *seeds* enclosed within the fruit.

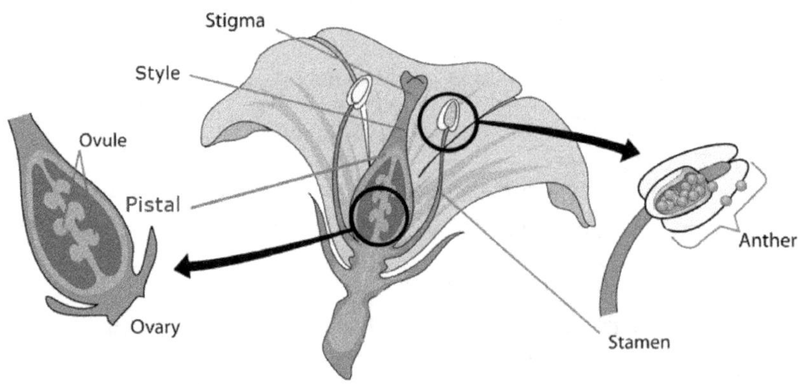

Angiosperm sexual reproduction anatomy with female ovary and male stamen (anther and filament)

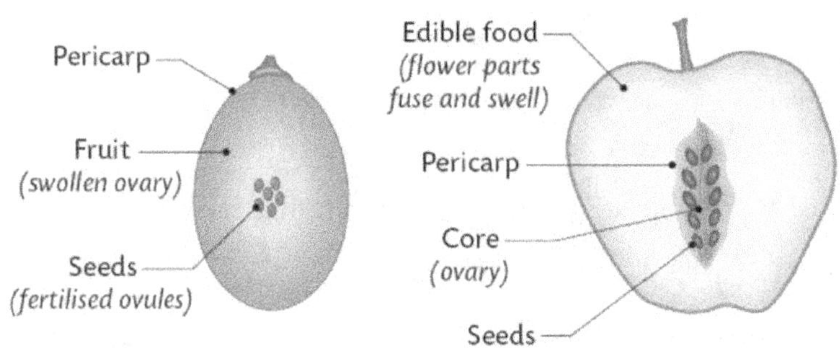

Two types of fruits with grapes as true fruit and apples as false fruit (on right)

True fruit forms from all tissues derived from a ripened ovary and its contents.

False fruit forms from other parts of the plant and the ovary, especially the receptacle.

Receptacle is the part of a flower stalk where parts of the flower are attached.

Sepal is the outer parts of the flower (often green and leaf-like) that enclose a developing bud.

24. D is correct.

Green algae (unicellular or multicellular) are classified as plants because they contain *chloroplasts* and produce food (e.g., carbohydrates such as sucrose) through *photosynthesis*.

Green algae lack many structures of plants.

For example, algae do not have roots, stems, or leaves.

25. C is correct.

Rhizoid is like a root in bryophytes (i.e., mosses, liverworts, and hornworts).

Rhizoids are simple hair-like protuberances extending from bryophyte lower epidermal cells similar in structure and function to *root hairs of vascular land plants*.

Rhizoids form from single cells, whereas roots are multicellular organs composed of multiple tissues that collectively perform a common function.

26. D is correct.

Seeds dispersed to an area far away from the parent plant have less competition with parent plants.

Seeds can be transported (e.g., via animals) to a new location or survive adverse climate conditions.

Seed offers embryo protection, nourishment, and a mechanism to maintain dormancy for thousands of years, surviving harsh environments and germinating when growth conditions are optimal.

Seeds disperse through space and time; evolution favored seed plants as successful and familiar plants.

27. A is correct.

Root pressure is produced in the vascular cylinder by *osmotic pressure* within cells of a root system that causes sap to rise through a plant stem to the leaves.

Osmotic pressure is the minimum pressure that must be applied to a solution to stop the flow of solvent molecules through a semipermeable membrane (osmosis).

Osmotic pressure is a colligative property and depends on the solution's concentration of solute particles.

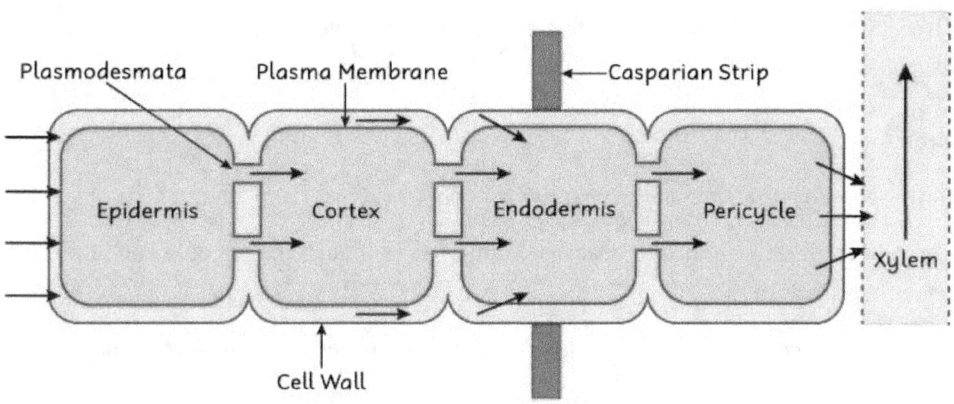

Root pressure in vascular plants is produced by osmotic pressure

continued...

Root pressure occurs in the *xylem* of some vascular plants when the soil moisture level is high (e.g., at night or when transpiration is low during the day).

Diffusion and *osmosis* (movement of water) use passive transport and do *not require energy* input.

In diffusion and osmosis, particles move from an area of *higher concentration* to one of *lower concentration.*

Casparian strips form a barrier to apoplastic flux (i.e., through cell walls), forcing ions through the selectively permeable plasma membrane into the cytoplasm rather than moving along cell walls.

28. C is correct.

Evidence suggests that green-blue algae (protists) were the first plants to evolve from a group of green algae 850 mya, and algae-like plants might have evolved as early as 1 billion years ago.

Green algae are classified as plants because they contain *chloroplasts* and produce food through *photosynthesis*.

Photosynthesis

$$6CO_2 + 6H_2O \longrightarrow C_6H_{12}O_6 + 6O_2$$

carbon dioxide + water → glucose + oxygen

Green algae and plants have cellulose in their cell walls and share many photopigments for color.

However, they lack many other structures of true plants.

For example, algae do *not* have roots, stems, or leaves.

29. B is correct.

Germination is the growth of an embryonic plant within a seed, resulting in the formation of a seedling.

Seeds of vascular plants are small packages with fruits (or cones) from the union of male and female gametes.

Fully developed seeds contain an embryo (most plants also store nutrient reserves) encased in a seed coat.

Some plants produce varying numbers of seeds that lack embryos; these are empty seeds that do not germinate.

Most seeds undergo a period of dormancy when there is no active growth.

Seeds can be transported (e.g., via animals) to a new location or survive adverse climate conditions until favorable circumstances develop.

30. D is correct.

Dormancy allows seeds to *germinate* under ideal living conditions (e.g., temperature, humidity).

Dormancy is when there is no germination in seeds or tubers even though the required conditions (temperature, humidity, oxygen, and light) exist.

Dormancy uses *hard seed coat impermeability* or *lack of enzymes* (internal dormancy) needed for germination.

31. D is correct.

Primary functions of stems:

1) support the leaves,

2) conduct water and minerals to the leaves for photosynthesis, and

3) transport carbohydrate products from leaves to other parts of the plant, including the roots.

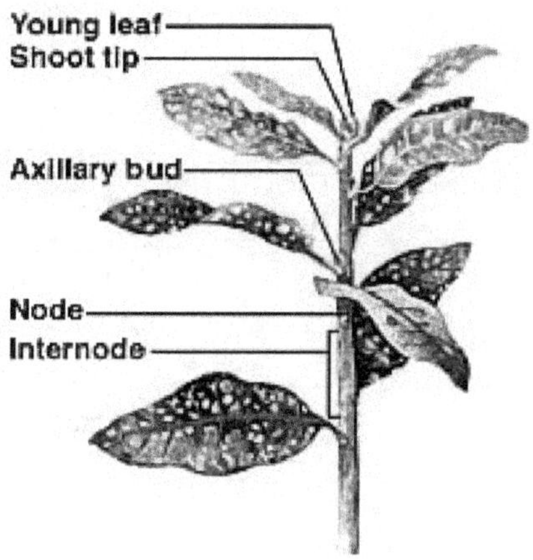

Stems (with meristem growth regions shown) function for support, conduction, and growth

32. D is correct.

Gibberellins (GAs) are a group of plant hormones that stimulate shoot elongation, seed germination, and fruit and flower maturation.

For example, maturing grapes are treated with gibberellins to promote fruit size.

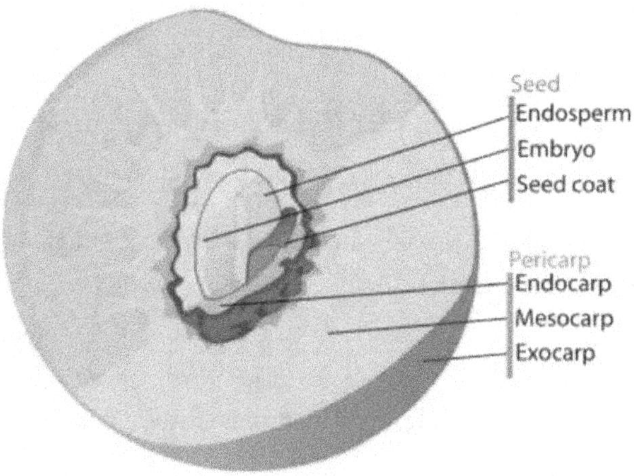

Flowering plant fruit anatomy with seed and pericarp components

Fruit growth is mainly dependent on *auxin* and *cytokinin.*

Seeds are a rich source of hormones, particularly auxins, GA, and cytokinin, which stimulate the growth of surrounding tissues and determine fruit size.

continued…

Auxins (the first plant hormones discovered) function primarily in stem elongation by promoting cell growth and are a major growth factor in plants.

Auxins' roles in plants depend on where they are produced.

Fruit ripening uses *ethylene* for texture changes, softening, color, and other ripening processes.

Ethylene level in under-ripe fruit is very low.

For example, as the fruit develops, it produces larger amounts which speed up ripening or stage of ripening.

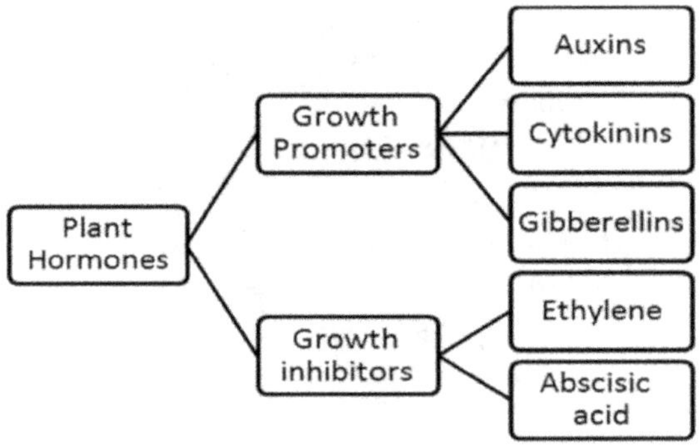

Growth promoters (auxins, cytokinin, gibberellins) and inhibitors (ethylene, abscisic acid)

33. B is correct.

Mosses likely grow in the shaded ground near a small pond.

Mosses are non-vascular plants in the land plant division Bryophyta (mosses, liverworts, and hornworts).

Mosses are a phylum of small soft plants typically 0.4–4 inches (some species grow larger) that commonly grow close together in clumps in damp or shady locations.

Mosses produce spore capsules at certain times.

Mosses differ from vascular plants because they *lack* water-bearing xylem tracheids or vessels.

Mosses do not have flowers or seeds; their simple leaves cover the thin, wiry stems.

Small, non-woody plants (e.g., moss) absorb water and nutrients mainly through their *leaves* and harvest sunlight to create food through photosynthesis.

34. A is correct.

When environmental conditions are unfavorable, is it advantageous for green algae to reproduce sexually.

Sexual reproduction produces offspring with

1) *greater genetic variety* (i.e., recombination during prophase I of meiosis) for survival, and

2) *durable zygotes* help green algae survive unfavorable conditions.

35. D is correct.

Cacti often have large stems that function to produce *food via photosynthesis*.

Cactuses survive in hot, dry climates. Most have thick, fleshy stems with waxy skin that store water and keep the water from evaporating.

Plants undergo photosynthesis by collecting carbon dioxide through stomata (regulated openings) in leaves and converting it into sugar and oxygen with the input of energy (i.e., photons) from the sun.

Cacti utilize CAM photosynthesis, a process unique to succulents whereby stomata open only at night when the plant is relatively cool, so less moisture is lost through transpiration.

Photosynthesis requires sunlight, so water intake and light-stimulated photosynthesis are separated in time.

Cactus leaves are reduced to thin spines to prevent water loss through transpiration.

Cacti spines cannot perform photosynthesis carried out by the thick green stem.

36. C is correct.

Phloem is living tissue in vascular plants that transports (i.e., translocation) organic nutrients (e.g., sucrose) synthesized during photosynthesis to vascular plant parts.

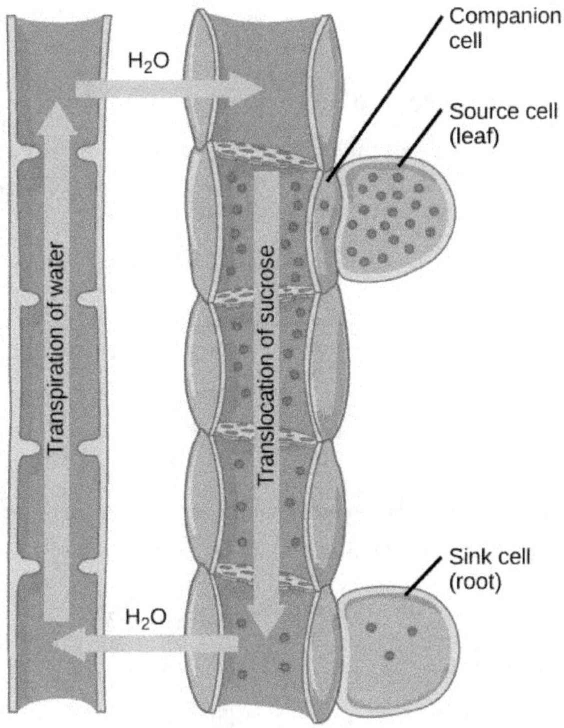

Xylem (on the left) functions for the transpiration of H_2O from roots, with phloem (on right) for the translocation of sugars from leaves during photosynthesis.

Phloem transports (or translocation) *soluble organic material* made during photosynthesis.

continued...

Phloem tissue consists of:

conducting cells (i.e., sieve elements),

parenchyma cells (specialized companion and albuminous cells and unspecialized cells), and

supportive cells (i.e., fibers and sclereids).

For example, the phloem is the innermost layer of bark in trees.

A: *lignin* is an alkyl-aromatic polymer in the cell walls of terrestrial plants.

Lignin provides structure and rigidity, is an effective barrier against microbes, and enables water and nutrient transport through plant tissues.

B: *tracheids* are elongated cells transporting water and mineral salts through the xylem of vascular plants.

Tracheids provide physical support and conduct water and dissolved minerals upwards and are the only such elements in conifers and ferns.

D: *xylem* is in *vascular bundles* in non-woody plants and *non-woody parts* of woody plants.

Xylem transports water and soluble mineral nutrients *from the roots* and replaces water lost during transpiration and photosynthesis.

37. A is correct.

Unlike roots, stems may be involved in *photosynthesis.*

For example, many cacti have large stems that function to produce *food via photosynthesis.*

Photosynthesis

$6CO_2 + 6H_2O \longrightarrow C_6H_{12}O_6 + 6O_2$

carbon dioxide + water ⟶ glucose + oxygen

Photosynthesis reaction:

$$6\ CO_2 + 6\ H_2O \rightarrow C_6H_{12}O_6 + 6\ O_2$$

Six carbon dioxide (CO_2) and six water (H_2O) molecules are converted by light energy (photons) captured by the photopigment chlorophyll into a sugar molecule (e.g., $C_6H_{12}O_6$ as glucose), and six expired oxygen (O_2).

Stems are parts of plants that bear branches, leaves, and fruits.

Roots are underground structures with root hairs.

Roots supply minerals, water, and minerals from the soil and provide *anchorage* for the plant.

38. D is correct.

During primary growth, a stem *increases in length.*

Growth occurs as the stems and roots lengthen.

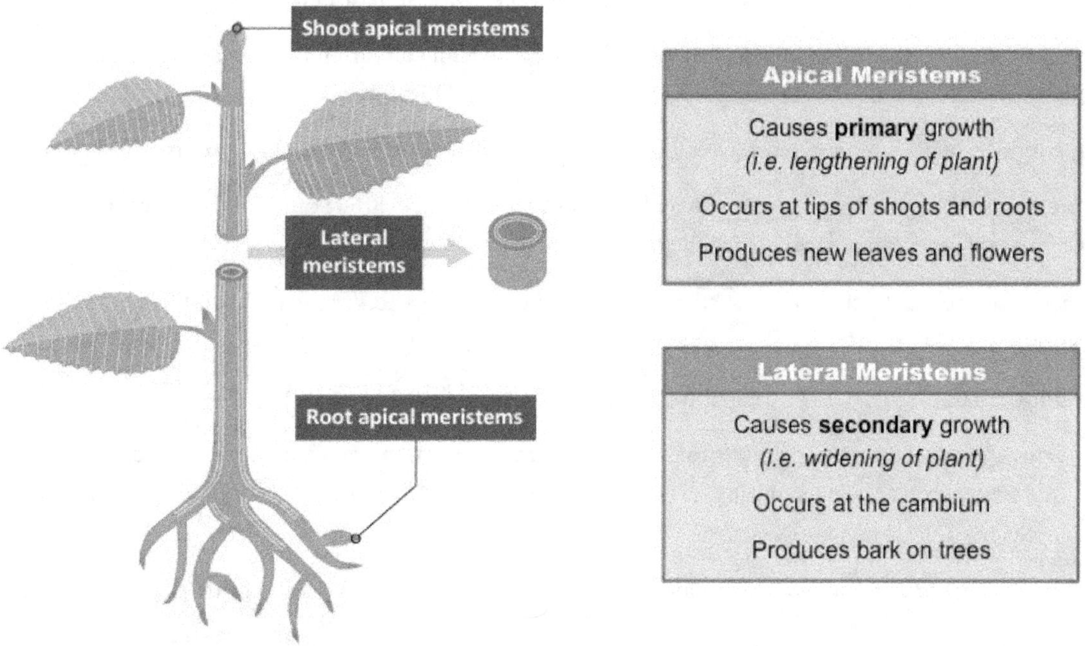

Apical (shoots and roots) and lateral (cambium) meristems are regions of mitotic cell growth

Some plants, especially woody plants, also increase *thickness* during their life span.

Primary growth is the increase in the length of shoots and roots resulting from cell division in the shoot *apical meristem.*

Meristematic tissues consist of three types, based on location, and contribute to *primary* (taller/longer) and *secondary* (wider) growth.

> ***Intercalary*** *meristems* (or *basal meristems*) occur only in some monocots, at the bases of leaf blades, and at nodes (the areas where leaves attach to a stem).
>
> This tissue enables the monocot leaf blade to increase in length from the leaf base.
>
> For example, grass leaves can elongate even after repeated grazing or mowing.
>
> ***Apical meristem*** is the tissue at the tips of stems and roots, enabling increased length.
>
> ***Lateral meristem*** facilitates growth in thickness or girth in a maturing plant.

Primary growth is controlled by *root apical* or *shoot apical meristems.*

Secondary growth is controlled by two lateral meristems, *vascular cambium* and *cork cambium.*

Not all plants exhibit secondary growth.

39. C is correct.

Auxins are hormones for cell elongation and are produced in rapidly growing regions near the tip of roots or stems. They coordinate plant development at the cellular level, through organs, and throughout the plant.

Auxin molecules may trigger responses directly through *gene expression* or inhibition.

Auxin concentration level (along with other local factors) results in cell differentiation and specification. They are essential for cell growth on the cellular level, affecting both cell division and cellular expansion.

A: *cytokinins* are a class of plant hormones promoting cell division (or cytokinesis) in roots and shoots.

They are involved primarily in cell growth and differentiation but also affect apical dominance, axillary bud growth, and leaf senescence.

B: *gibberellins* are plant hormones that stimulate shoot elongation, seed germination, and fruit and flower maturation.

For example, maturing grapes are treated with gibberellins to promote fruit size.

D: *ethylene* is produced from most parts (e.g., leaves, stems, roots, flowers, fruits, tubers, seeds) of higher plants, and production is induced during growth (e.g., germination), fruit ripening, cutting off leaves, and senescence of flowers.

For example, fruit ripening uses *ethylene* for texture changes, softening, color, and other ripening processes.

Ethylene levels are low in under-ripe fruit.

40. B is correct.

Thin tree ring indicates a year of drought (e.g., adverse environmental factor).

Thicker rings indicate a more prolonged or faster growing season when the temperature was high longer and favored growth.

Thinner rings show shorter or slower growing seasons; either the temperature was low, or the tree lacked water.

41. D is correct.

Phototropism is the orientation of an organism (e.g., plant) in response to light.

Phototropism is observed in plants and other organisms, such as fungi.

positive phototropism is toward the light source.

negative phototropism is away from the light source.

Cells farthest from the light contain auxin for phototropism.

The plant elongates cells on the furthest side from the light.

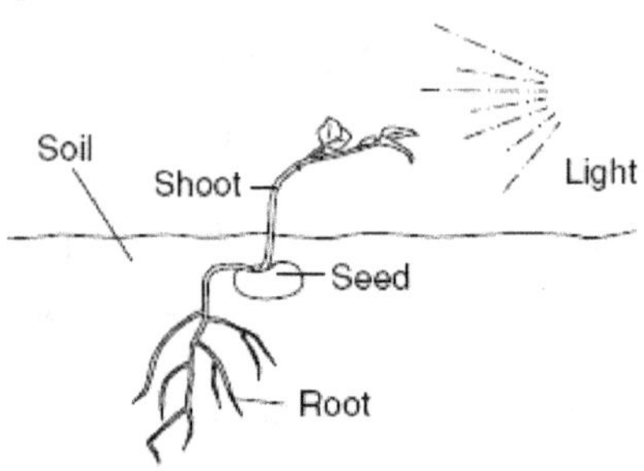

continued...

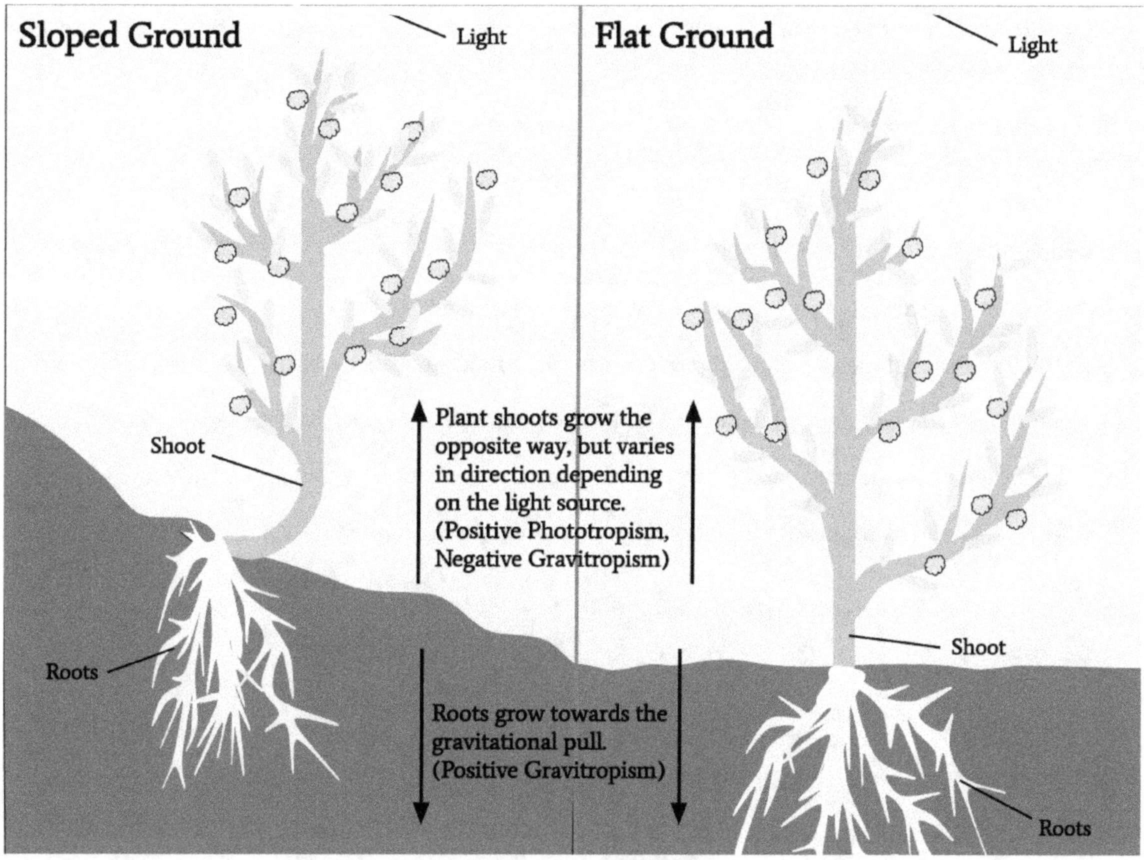

. Roots and shoots grow relative to soil pitch for sloped and flat ground. Shoot grows towards the light (most likely up), and roots grow down. Plants on sloped ground (on left) have shoots that bend opposite the gravitational pull, and roots grow down towards the gravitational pull.

Phototropism is a *tropism* (or movements) that responds to external stimuli.

A: *thigmotropism* is a directional growth occurring as a mechanosensory response to a touch stimulus.

Thigmotropism is typical in twining plants and tendrils; also observed in flowering plants and fungi.

B: *rapid movement* encompasses movement in plant structures over a very short period, usually under 1 sec.

For example, the Venus flytrap closes within 100 milliseconds while Utricularia closes in about 0.5 msec.

C: *gravitropism* is a growth response to the environment that directs shoots upward and roots downward, allowing each organ to reach environments adequate for their primary functions.

42. D is correct.

Bryophytes must live in moist areas because they lack vascular tissue, and sperm needs water for the *flagellated sperm* to swim to an egg.

Bryophyte is an embryophyte (i.e., land plant) non-vascular plant without vascular tissue (mosses, liverworts).

continued…

Bryophytes are distinct from other land plants because they do *not contain xylem*, the tissue used by vascular plants to transport water.

Instead, bryophytes receive *water and nutrients through their leaves.*

Mosses and liverworts (i.e., Bryophytes) are restricted to moist habitats; deserts have no mosses.

43. A is correct.

Seed formation is a reproductive process in seed plants (*spermatophytes*).

Seed is a structure that includes a plant embryo, endosperm (a food supply), and seed coat (protective covering).

Embryo is a young multicellular organism before it emerges from the seed.

Endosperm is the nutritive tissue of the seed, often a combination of starch, oil, and protein.

Seed coat consists of one or more protective layers encasing the seed.

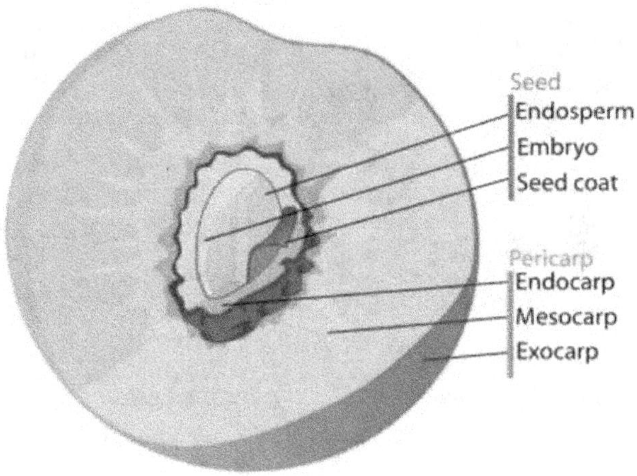

Flowering plant fruit anatomy with seed and pericarp components

B: *gametophyte* is the sexual phase with haploid (1N) male and female organs (or *gametangia*) developing and producing eggs and sperm gametes through mitosis as sexual reproduction.

C: *pollen grain* is a single cell containing two male gametes with small haploid (1N) cells encased in a protective coat to prevent desiccation (i.e., drying) and mechanical damage.

Pollen is an immature, endospore male gametophyte derived from male spores (or *microspores*) in seed plants.

Pollen grains are produced by seed plants (i.e., angiosperms and gymnosperms) and are male gametes by meiosis in the anther sac, producing four *microspores* (or pollen grains).

Microspores (or *pollen grains*) mature, the anther opens, and pollen is released.

D: *spore* is a haploid cell derived from sporangium via meiosis.

Spores are produced by ferns, lycopods, horsetails, mosses, and fungi.

44. C is correct.

When positioned horizontally, a plant's response to gravity is that auxins in the lower sides of stems cause cell elongation that bends the stem upright.

Tropism is a biological phenomenon indicating an organism's growth or turning movement (e.g., a plant) in response to an environmental stimulus.

Gravitropism is a movement or growth in response to gravity.

Gravitropism is a growth response directing shoots upward and roots downward, allowing each organ to reach good environments for performing its primary functions.

Auxins are hormones for cell elongation produced in rapidly growing regions near the tip of roots or stems and stimulate stem elongation by promoting cell growth and are a major growth factor in plants.

45. D is correct.

Ethylene is produced from almost all parts (e.g., leaves, stems, roots, flowers, fruits, tubers, seeds) of higher plants, and production is induced during certain stages of growth, such as germination, ripening of fruits, cutting off leaves, and senescence of flowers.

Developmental and environmental factors regulate ethylene production.

For example, ethylene can be induced by wounding, environmental stresses, and certain chemicals, including auxin and other regulators.

Fruit ripening uses *ethylene* for texture changes, softening, color, and other ripening processes.

For example, the ethylene level in under-ripe fruit is very low.

46. C is correct.

Stomata of leaves are usually open during the day if a plant has enough water.

Stomata are composed of a pair of specialized epidermal cells known as *guard cells.*

They regulate gas exchange (e.g., CO_2) and control water loss by changing the size of the stomatal pore.

Stomata are microscopic cell structures in the epidermis of tree leaves (and needles) involved in exchanging carbon dioxide and water between plants and the atmosphere, which is critical for photosynthesis.

47. B is correct.

Yellow, orange, and red colors of fall leaves result from *anthocyanin* and the *breakdown of chlorophyll.*

Chlorophyll breaks down, the green color disappears, and the yellow to orange colors become visible, giving leaves their fall splendor.

Other chemical changes may occur, forming additional colors by the development of red anthocyanin pigments.

Anthocyanins are colored water-soluble pigments responsible for the colors red, purple, and blue in fruits and vegetables. Berries, currants, grapes, and some tropical fruits contain high anthocyanins.

Carotene is a pigment that absorbs blue and indigo light and provides rich yellows and oranges.

Distinctive colors of mango, fall leaves, yams, yellow butter, and animal fats are due to carotene.

48. A is correct.

O_2 and CO_2 move in and out of a leaf through stomata.

Stomata (singular *stoma*) is a pore in the epidermis of leaves and stems used to control gas exchange.

Stoma pore is bordered by a pair of specialized *parenchyma cells* (i.e., *guard cells*) to regulate the size of the pore opening.

Stoma refers to a stomatal complex, the pore, and its accompanying *guard cells.*

49. D is correct.

Growth of plant seedlings is usually *phototropic* and *gravitropic.*

Tropism is a biological phenomenon indicating an organism's growth or turning movement (e.g., a plant) in response to an environmental stimulus.

Phototropism is the orientation of a plant or organism in response to light, either toward the source of light (*positive*) or away (*negative phototropism*).

Phototropism is observed in plants but can occur in organisms such as fungi.

Cells farthest from light contain auxin hormones that react during phototropism.

Gravitropism is a plant growth response directing shoots upward and roots downward, allowing each organ to reach environments adequate for performing its primary functions.

It is a collaborative process of differential growth by a plant or fungi in response to *artificial* or *natural gravity.*

It is a general feature of all higher and many lower plants.

> *negative gravitropism* is the upward growth of plant stems.
>
> *positive gravitropism* is the downward growth of plant roots.

Primary roots and portions of the root system show positive geotropism by growing towards gravity.

Thigmotropism is a movement or growth in response to touch or contact.

50. B is correct.

Vascular plants are larger than nonvascular plants because *vascular tissue carries water and nutrients* much more efficiently than diffusion. Nutrients and waste products are transported within larger plants over longer distances more efficiently.

The development of vascular tissue allowed plants to grow much taller than nonvascular plants, forming ancient swamp forests. Most of these large vascular seedless plants are extinct, but their smaller relatives remain.

Vascular plants evolved about 420 million years ago (probably) from moss-like bryophyte ancestors, but they had a life cycle dominated by the diploid sporophyte generation.

Early vascular plants evolved as more plant-like:

- Vascular plants evolved true *roots* made of vascular tissues.

  Compared with rhizoids, roots can absorb more water and minerals from the soil. They securely anchor plants in the ground to grow larger without toppling.

- Vascular plants evolved *stems* made of vascular tissues and *lignin.*

  Lignans stiffen the stems, so plants can grow high above the ground, where they can get more light and air. Because of vascular tissues, stems keep even tall plants supplied with water, so they do not dry out in the air.

- Vascular plants evolved *leaves* to collect sunlight.

  Initially, leaves were tiny and needle-like, which helped reduce water loss. Later, leaves were much larger and broader, so plants could collect more light and undergo photosynthesis to create food.

Early vascular plants had a selective advantage over nonvascular plants with vascular tissues and other adaptations to grow tall and outcompete others for sunlight.

Bryophytes were the photosynthetic pioneers onto land, but early vascular plants were the photosynthetic pioneers into the air.

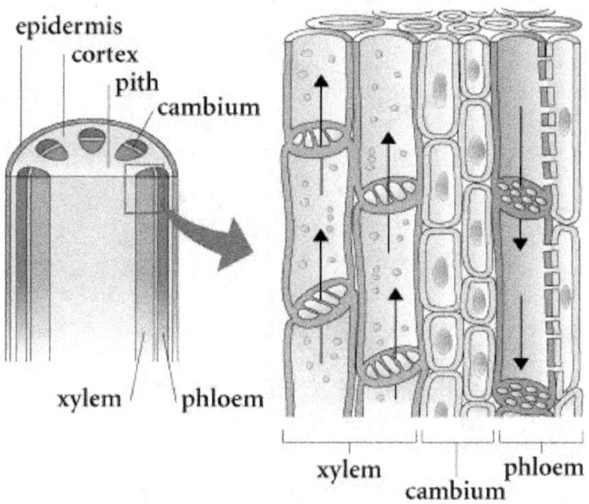

Vascular tissue: xylem (on left) functions for the transpiration of H_2O from roots with phloem (on right) for translocation of sugars from leaves during photosynthesis.

51. C is correct.

Photosynthesis in plants occurs mainly in *mesophyll tissue.*

Mesophyll tissue (i.e., chlorenchyma or parenchyma) makes up most of the leaf interior between the upper and lower layers of epidermis.

Assimilation (i.e., nutrient absorbing) tissue is the primary location of plant photosynthesis, with the products of photosynthesis, called *assimilates* (e.g., glucose, fructose).

Mesophyll tissue is internal ground tissue between two epidermal cell layers of leaves composed of:

> *palisade parenchyma* (upper layer of elongated chlorenchyma cells with many chloroplasts)
>
> *spongy parenchyma* (lower layer of spherical or ovoid cells with few chloroplasts and prominent intercellular air spaces).

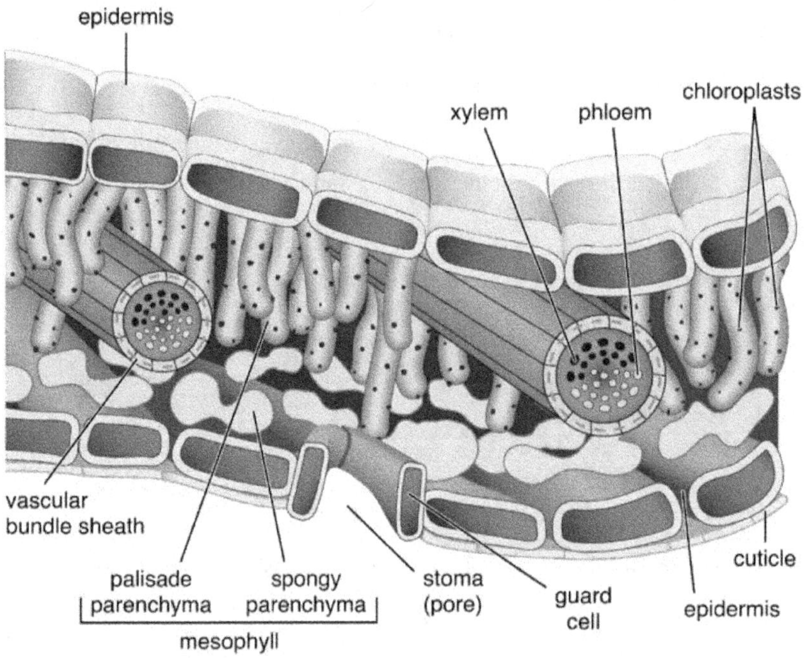

Leaf cross-section with organelles and cellular components

52. A is correct.

Stomata are only on the lower surface of leaves in many plants because water loss would be less on the shaded lower surface than in the direct sun.

Stomata (singular *stoma*) is a pore in the epidermis of leaves and stems for O_2 and CO_2 gas exchange.

The stomata pore is bordered by a pair of specialized *parenchyma cells* (i.e., guard cells) to regulate the size of the pore opening.

Stoma also refers to a *stomatal complex* (pore accompanying structures).

53. D is correct.

Growth of ivy tendrils wrapping around objects is *thigmotropism,* a growth response to touch or contact.

Tropism is a biological phenomenon of an organism's growth or turning movement (usually a plant) in response to an environmental stimulus.

There are many types of tropisms.

A: *photoperiodism* is not a type of tropism but the physiological reaction of organisms to the length of day or night. It occurs in plants and animals.

B: *chemotropism* is a movement or growth in response to chemicals.

C: *phototropism* is a movement or growth in response to lights or colors of light.

54. B is correct.

Seed-bearing plants differ from other plants because their gametes do *not* require water for fertilization.

Spermatophytes (i.e., seed plants) include *gymnosperms* (conifers) and *angiosperms* (flowering plants).

> *Seeds are produced by* flowering *plants* (angiosperms) and cone *plants (gymnosperms).*
>
> *S*pores are produced by ferns, mosses, and green algae and require water for fertilization.

Gymnosperms are a group of seed-producing plants (e.g., conifers, cycads) with seeds developing on the surface of scales or leaves, often modified to form cones or at the end of short stalks (e.g., Ginkgo).

Gymnosperm seeds are not enclosed within an ovary, unlike *angiosperms* (flowering plants).

Both male (*pollen*) gametes are involved in fertilization, forming a *zygote* (or fertilized egg) and an endosperm; double fertilization by two gametes is unique to flowering plants.

55. D is correct.

Xylem tissue in vascular plants is vital because it conducts water over long distances.

Xylem is in vascular bundles (non-woody plants and non-woody parts of woody plants), secondary xylem laid down by vascular cambium (meristem) in woody plants, and non-bundle arrangements (e.g., ferns).

Xylem transports water, soluble minerals, and nutrients from roots and replaces water lost during photosynthesis.

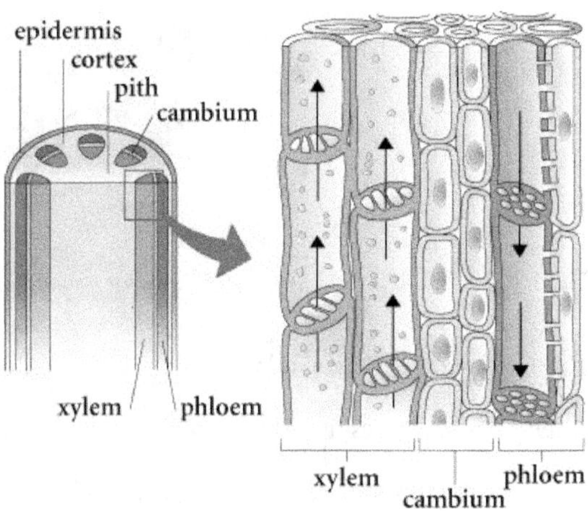

Xylem functions for transpiration of H_2O from roots

56. B is correct.

Corn, broccoli, and cabbage were developed by selective breeding.

Selective breeding is when humans use animal or plant breeding to selectively develop phenotypic traits by choosing representative male and female gametes to reproduce sexually and produce offspring.

Many fruits and vegetables have been bred through artificial selection.

For example, by selective breeding, broccoli, cauliflower, and cabbage were derived from wild mustard plants.

A: *auxins* are plant hormones for cell elongation produced in rapidly growing regions near the tip of roots or stems. They stimulate stem elongation by promoting cell growth and are a major growth factor in plants.

C: *plant propagation* is reproducing plants from a single parent plant.

Asexual propagation includes cuttings, layering, division, budding, and grafting.

> *cuttings* involve rooting a severed piece of the parent plant;
>
> *layering* involves rooting a part of the parent and then severing it; and
>
> *budding* and *grafting* join two plant parts from different varieties.

57. A is correct.

In the figure, the water pressure is high in *guard cells*, specialized cells in the lower leaf epidermis of plants.

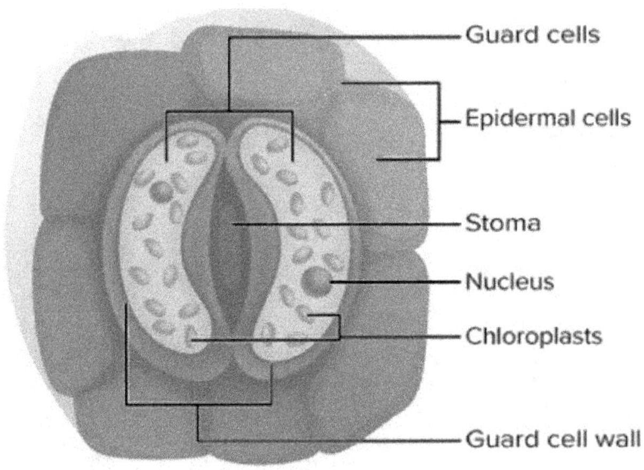

Leaf underside cross-section with an open stoma and support structures

Pairs of guard cells surround stomata pores and regulate CO_2 intake from the atmosphere for photosynthesis.

Guard cells also regulate how much water plants lose to the atmosphere.

Opening and closing the stomata pores uses changes in the turgor (i.e., osmotic) pressure of two guard cells.

When water flows into the guard cells, they swell, and the curved surface causes the stomata to open.

When the guard cells lose water, they shrink and become flaccid and straight, thus closing the stomata.

58. D is correct.

Xylem cells move water by *capillary pressure.*

Due to transpiration pull, when transpiration is high, xylem sap is under tension rather than pressure.

Capillary action is the movement of water within spaces of a porous material due to *adhesion*, *cohesion*, and *surface tension,* which is an elemental force for moving water and dissolved molecules.

> *Adhesion* is the attraction of molecules among different kinds of molecules and is strong for water, especially including positive or negative charges molecules.
>
> *Cohesion* (water sticking to each other) causes more water molecules to fill the gap in the xylem as the topmost water is pulled toward the stomata.
>
> *Surface tension* created by transpiration *pulls* water upward in the plant xylem.

A: *mesophyll cells* (i.e., chlorenchyma or parenchyma) are the internal ground tissue comprising most of the leaf interior between the upper and lower layers of the epidermis.

B: *albuminous cells* are specialized parenchyma cells and major component of ground tissue in vascular plants.

Albuminous cells and albumin protein are associated with phloem sieve tube cells in gymnosperm wood.

C: *phloem cells* are living tissue in vascular plants that transports (i.e., translocation) organic nutrients (e.g., sucrose) synthesized during photosynthesis to vascular plant parts.

59. C is correct.

Pollen grains are a haploid (1N) gymnosperm reproductive structure.

Spermatophytes (i.e., seed plants) include *gymnosperms* and *angiosperms* as two main groups.

Seeds are produced by flowering *plants* (*angiosperms*) and cone *plants (gymnosperms)*, while spores are produced by ferns, mosses, liverworts, and green algae and require water for fertilization.

Gymnosperms are seed-producing plants (e.g., conifers, cycads, Ginkgo) with the following characteristics:

> naked seeds,
>
> separate female and male gametes,
>
> pollination by wind, and
>
> tracheids to transport water and solutes in the vascular system.

Gymnosperm seeds develop on the surface of scales or leaves, often modified to form cones or at the end of short stalks, as in Ginkgo.

Gymnosperm seeds are not enclosed within an ovary, unlike flowering plants (angiosperms).

> *Gymnosperms* (or *naked seeds*) have seeds *exposed* at stalk ends (or *whorls*) or on *cone scales*.

continued…

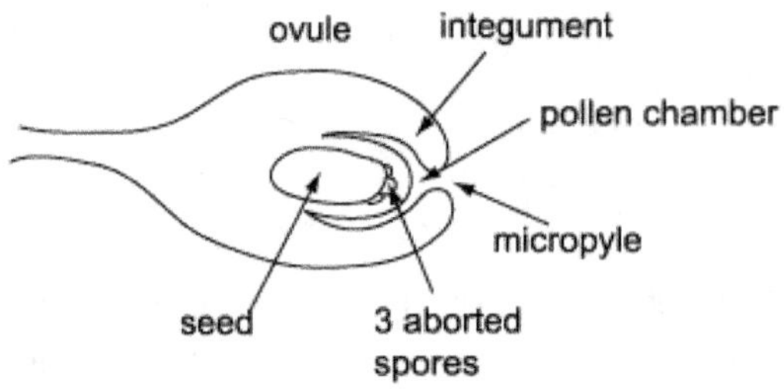

Gymnosperm with ovule, seed, pollen chamber, and aborted spores

Pollen grains mainly represent the (1N) male part of the reproductive process in plants and trees.

After forming four *haploid microspores*, they fuse to form *pollen grain,* and the pollen grain walls develop.

Therefore, a pollen grain is a haploid (1N) structure.

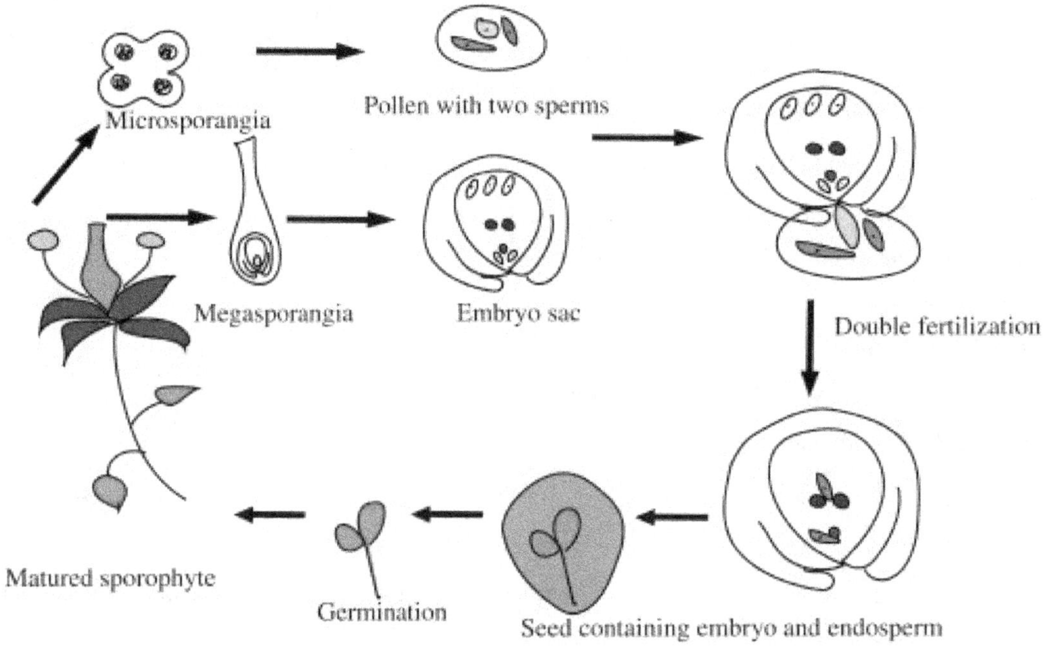

Sporophytes form megasporangia and four microsporangia. The embryo sac undergoes double fertilization to produce a seed containing embryo and endosperm

A: *seeds are produced by* flowering *plants* (*angiosperms*) and cone *plants (gymnosperms).*

Seed is the (2N) embryonic stage of the plant life cycle.

Most seeds consist of three parts:

embryo, *endosperm*, and *seed coat*

Endosperm is the nutritive tissue of the seed, often a combination of starch, oil, and protein.

B: *cones* are the reproductive structure in conifers with modified leaves densely arranged on a central stalk.

continued…

Female (or *seed cones*) bear ovules on the surface of their scales, while *male cones* produce pollen.

D: *seed cone* is the female fruit of conifers and terminal, cone-like, usually fleshy, and rarely dry.

Seed cones of conifers are compact branching systems composed of two separate structures: a bract (or modified leaf) and an ovuliferous scale (or modified seed-bearing shoot).

Seed-bearing cones are female, while the *pollen-filled cones are male.*

Both sexes of cones grow on the same tree, but the male cones typically grow on lower branches relying on the wind to blow pollen up to the female cones.

60. D is correct.

People use plants to provide food, raw materials, and medicines.

Medicinal plants (or *medicinal herbs*) have been used in traditional medicine practices since prehistoric times.

Medicinal plants and herbs (e.g., turmeric, ginger, basil leaves, mint, and cinnamon) are common in culinary dishes and offer health benefits (e.g., cold and flu remedies, relieve stress, improve digestion, and improve immune responses).

Plants synthesize chemical compounds for defense against insects, fungi, diseases, and herbivorous mammals.

61. A is correct.

Embryo develops in the ovule.

Carpel (or pistil) are female structures, while *stamen* are male structures.

Pistil (or *carpel*) is the female organ of a flower, with the stigma, style, and ovary.

Stamen is the male reproductive organ of flowering plants, consisting of the *anther* (site of pollen development) and often a stalk-like *filament* (transmits water and nutrients to the anther and positions it for pollen dispersal).

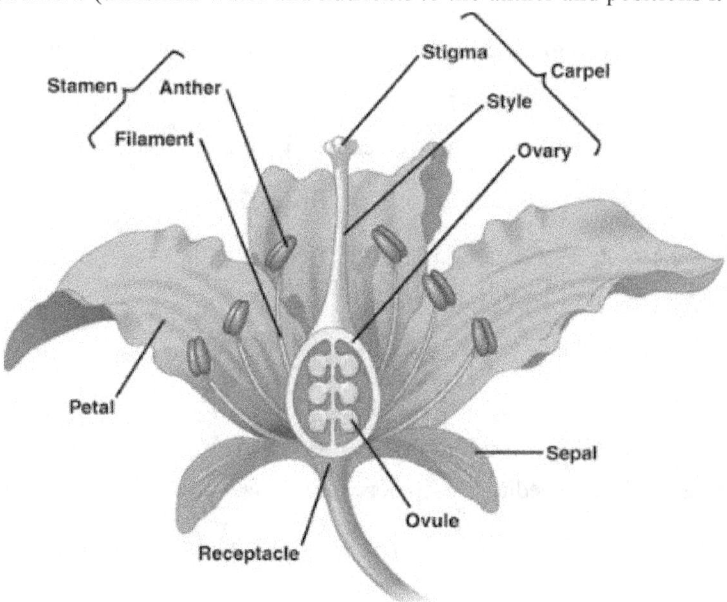

Carpel (or pistil) are female structures, while stamen are male structures

62. B is correct.

Pressure-flow hypothesis (or mass-flow hypothesis) is the best-supported theory to explain food movement through the phloem. German plant physiologist Ernst Munch proposed it in 1930.

Pressure-flow hypothesis proposes that water containing food molecules flows under pressure through the phloem. The pressure is created by the difference in water concentration of the solution in the phloem and the relatively pure water in the nearby xylem ducts.

A: *root pressure* is the transverse *osmotic pressure* within cells of a root system that causes sap to rise through a plant stem to the leaves.

Osmotic pressure is the minimum pressure for a solution to stop the flow of solvent molecules through a semipermeable membrane (osmosis). It is a colligative property and depends on the concentration of solute particles in the solution.

C: *transpiration pull* is a biological process in which the force of pulling is produced inside the xylem tissue. This force helps in the upward movement of water into the xylem vessels. In this process, water loss through leaves is observed as vapors.

D: *diffusion gradient* is a difference in the diffusion rates (net movement of a substance from higher to lower concentration) of molecules through a medium or substrate.

63. A is correct.

According to the pressure-flow hypothesis, the movement of H_2O into a nutrient-rich region of the phloem *increases* the pressure.

Pressure-flow hypothesis (or *mass-flow hypothesis*) explains the force for food movement through the phloem.

A high concentration of organic substances (e.g., sugars from photosynthesis) inside phloem cells at a source (e.g., a leaf) creates a diffusion gradient (osmotic gradient), drawing water into cells.

Sap originates in xylem cells of sapwood and is mostly water with minerals, sugar, nutrients, and hormones. *Sapwood* is the soft outer layer of recently formed wood between the heartwood and bark, with vascular tissue.

Phloem sap moves from sugar sources (leaves) to sugar sinks via turgor pressure (i.e., hydrostatic pressure).

A sugar source is any part of the plant producing or releasing sugar.

Tree sap can be in sapwood, which creates carbon dioxide. The sap is forced out in openings or wounds if the carbon dioxide creates pressure inside the tree.

During the plant's growth period (usually during the spring), storage organs (e.g., roots) are sugar sources, while many growing areas are sugar sinks.

Water moves sugar from roots to twigs when ambient temperatures rise, supplying the energy to grow new shoots and leaves. Pruning results in bleeding as sap oozes from the wound.

Movement in the phloem is bidirectional, while in xylem cells, it is unidirectional (upward).

64. C is correct.

Angiosperms (or *flowering plants*) enclose seeds in *ovaries* during development.

Spermatophytes (i.e., seed plants) include *gymnosperms* and *angiosperms* as two main groups.

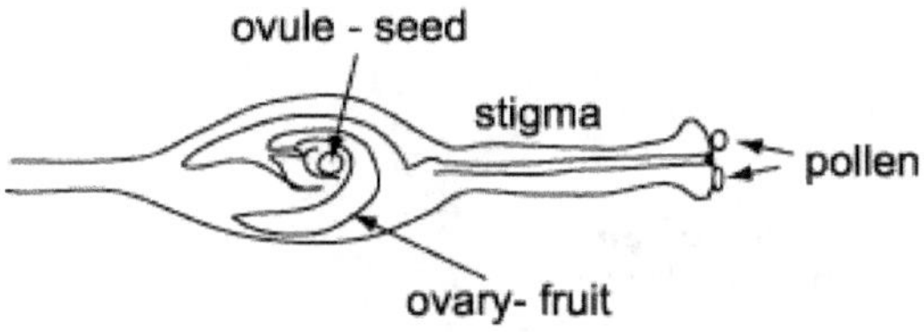

Angiosperm with ovule (seed), ovule (fruit), stigma, and pollen grains

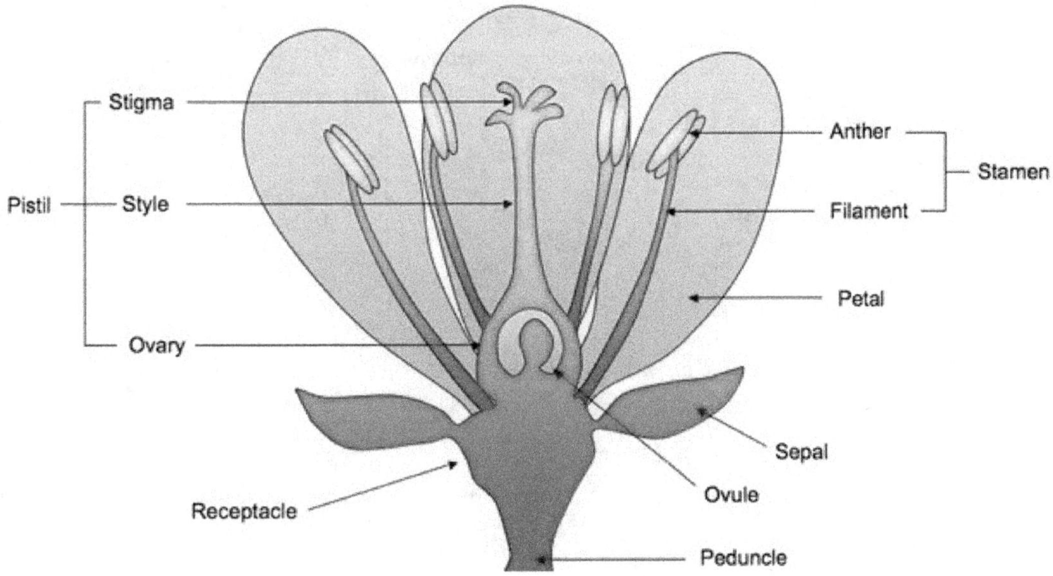

Carpel (or pistil) are female structures, while stamen are male structures

A: *petals* are sterile floral parts as visually conspicuous elements to attract specific pollinators to the flower.

B: *cones* are reproductive structures in conifers with modified leaves densely arranged on a central stalk.

Female (or *seed cones*) bear ovules on the surface of their scales, while male cones produce pollen.

Seed-bearing cones are female, while the pollen-filled cones are male.

Both sexes of cones grow on the same tree, but male cones typically grow on lower branches relying on wind to blow pollen up to the female cones.

D: *pollen grains* are produced by seed plants (i.e., angiosperms and gymnosperms) and are male gametes by meiosis in the anther sac, producing four microspores (or pollen grains).

Pollen grains carry male gametes with small haploid (1N) cells encased in a protective coat to prevent desiccation (i.e., drying) and mechanical damage.

continued…

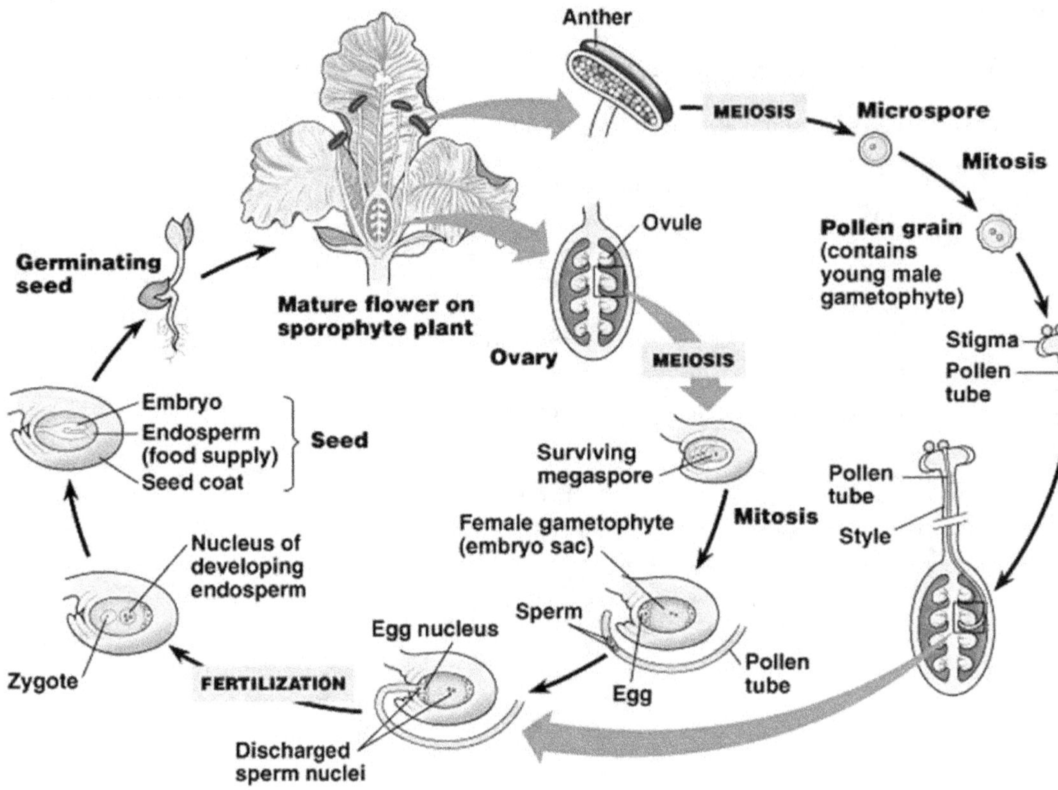

Angiosperm life cycle with microspore and megaspore undergoing fertilization

65. D is correct.

Ovaries develop into fruit in angiosperms. As the seed develops in an ovary, the walls of the ovary thicken and form the fruit.

A fertilized and fully grown, ripened ovary is a fruit.

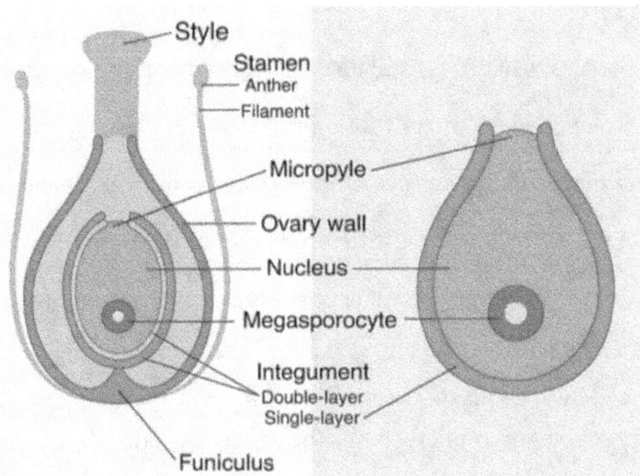

Angiosperm flowering plants (on left) and gymnosperm (on right) anatomy

66. D is correct.

Pollen grain landing on a stigma of a flower produces a *pollen tube*.

Pollination occurs when pollen lands on a compatible *pistil* or *female cone*. It germinates and produces a *pollen tube* that transfers the sperm to the *ovule* (contains female gametophyte).

Pollen grains are microscopic (require magnification to visualize) fine to coarse powder containing microgametophytes that produce male gametes (i.e., sperm cells).

Pollen grains have a hard coat that protects sperm cells moving from the:

> *stamens* (i.e., pollen-producing reproductive organ) to the *pistil* (i.e., a tube-like portion between stigma and ovary) of flowering plants, or
>
> *male cone* (i.e., organ with reproductive structure) to the *female cone* of coniferous plants.

Pollen is essential in forensics, archeology, paleoecology, and paleontology (i.e., the study of prehistoric life).

67. D is correct.

Annuals are flowering plants that grow, produce seeds, and die in one year.

A: *biennials* require two years to grow from seed to fruition and death.

B: *perennials* regrow every spring and generally have a shorter blooming period than annuals.

C: *dicots* (or *dicotyledons*) are one of two flowering plant (*angiosperms*) groups.

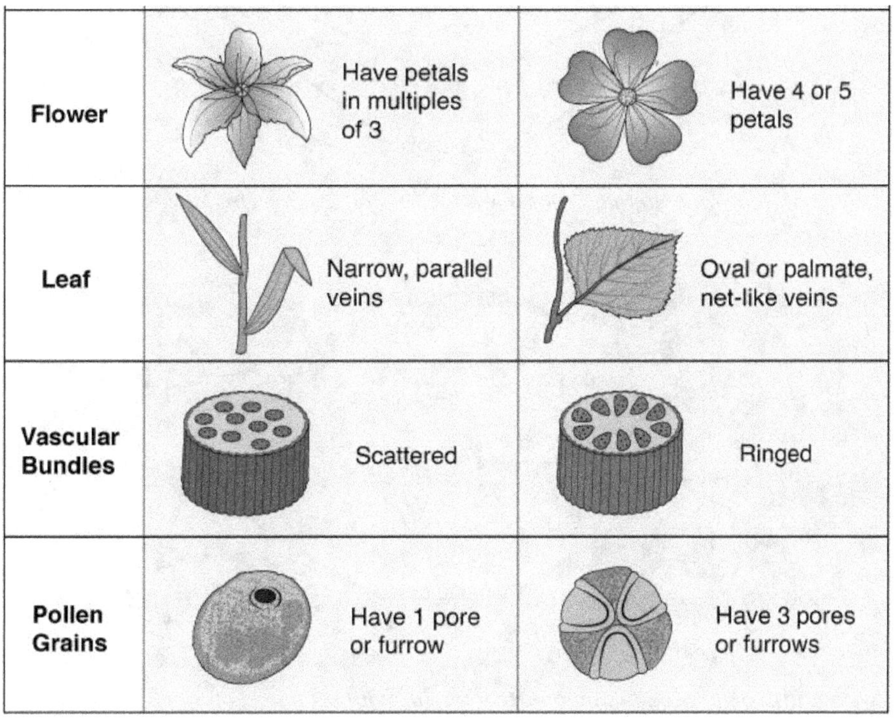

Monocots (on the left), compared to dicots (on the right), have distinctions in flowers, leaves, vascular bundles, pollen grains, cotyledons, roots, stems, and roots

continued...

Seed with two *embryonic leaves* (or *cotyledons*) is a characteristic of the dicots.

Cotyledon is part of the embryo within the seed and the embryonic leaf in seed-bearing plants, appearing first from a germinating seed.

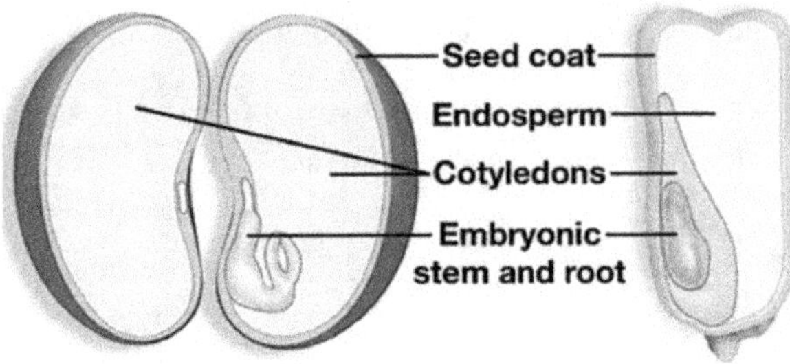

Bean seed (on left) and corn seed (on right) cotyledons and embryonic stem and root comparison

Cotyledons number is a characteristic used by botanists to classify flowering plants.

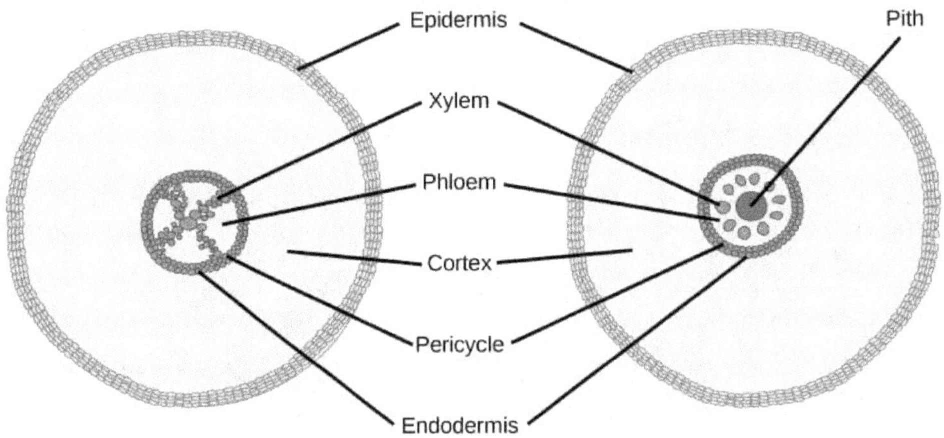

Dicot root (left) and monocot root (right) with scattered or ringed vascular bundles

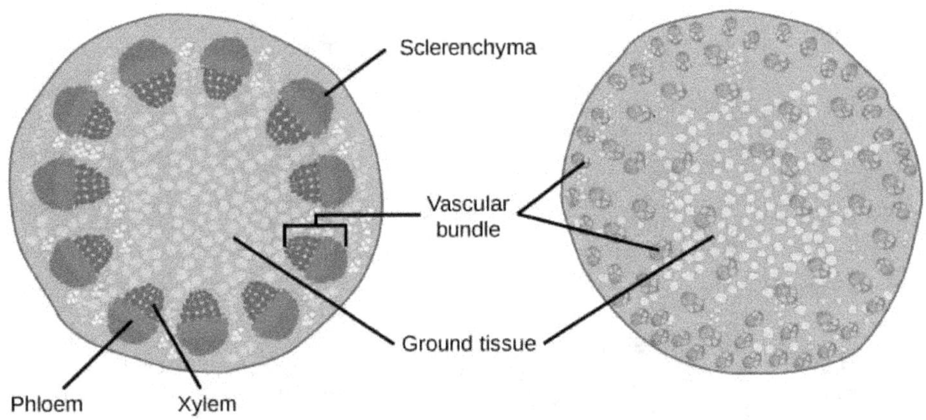

Dicot stem (left) compared to monocot stem (right) with scattered or ringed vascular bundles

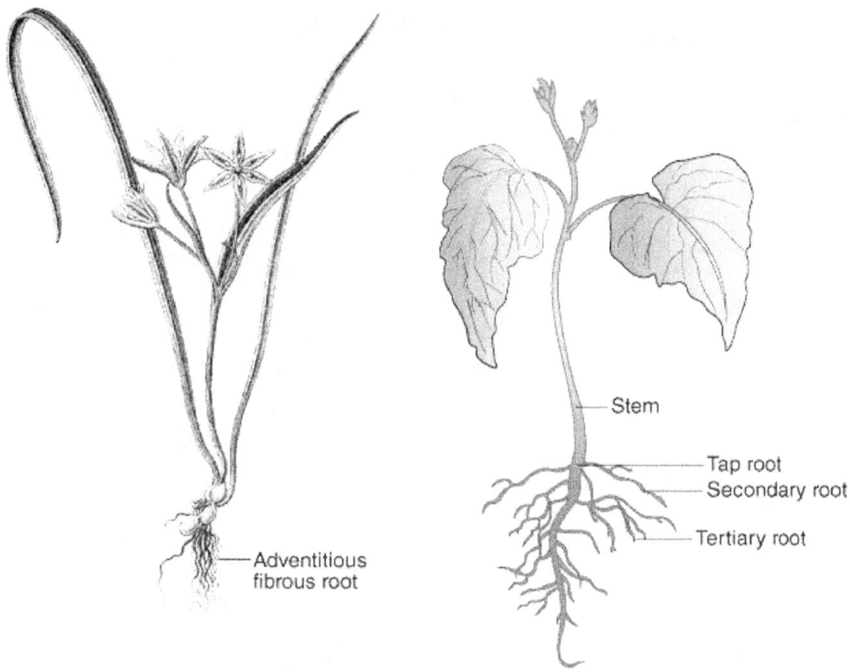

Monocot root (on the left) and dicot root (on right)

68. B is correct.

Ground tissue is in roots, stems, leaves, and all tissues that are neither dermal nor vascular.

Ground tissue by three types of cell walls.

> *Parenchyma cells* have *thin primary walls* and usually remain alive after they mature, which forms the *filler tissue* in the soft parts of plants.
>
> *Collenchyma cells* have *thin primary walls* with some areas of *secondary thickening* and provide extra *structural support* (i.e., regions of new growth).
>
> *Sclerenchyma cells* have *thick secondary walls* and often die when mature, providing the primary structural support to a plant.

69. D is correct.

Seed plants are held to the ground by *roots*.

A: *phloem* is living tissue in vascular plants that transports (i.e., translocation) organic nutrients (e.g., sucrose) synthesized during photosynthesis to vascular plant parts.

B: *epidermis* (or d*ermal tissue*) is the outer covering of a plant (roots, stems, leaves, flowers, fruits, and seeds); usually, one cell layer thick, and its cells lack chloroplasts.

C: *stems* are parts of plants that bear branches, leaves, and fruits.

Notes for active learning

Photosynthesis – Detailed Explanations

1. D is correct.

C_4 plants use C_4 carbon fixation.

There are three carbon fixation mechanisms: C_3, C_4 and CAM (*crassulacean acid metabolism*).

C_4 is named the 4-carbon molecule in carbon fixation compared to a 3-carbon molecule product in C_3 plants.

C_4 plants have a competitive advantage over plants that use the C_3 carbon fixation pathway under drought, high temperatures, and nitrogen or CO_2 limitation conditions.

About 8,100 plant species use C_4 carbon fixation, representing about 3% of terrestrial species of plants.

C_4 and CAM overcome the tendency of the RuBisCO to fix O_2 rather than CO_2 in photorespiration. They use a more efficient enzyme to fix CO_2 in mesophyll cells and shuttling this fixed carbon to bundle-sheath cells.

In bundle-sheath cells, RuBisCO is isolated from atmospheric oxygen and saturated with CO_2 released by decarboxylation of malate or oxaloacetate.

C_4 fixation is hypothesized to have evolved later.

Because these additional steps require more energy from ATP, C_4 plants can efficiently fix carbon in specific conditions, with the C_3 pathway being efficient in other conditions.

2. D is correct.

CAM (*crassulacean acid metabolism*) photosynthesis is a carbon fixation pathway that evolved in some plants to adapt to arid conditions.

In a plant using CAM, the stomata in the leaves remain shut during the day to reduce evapotranspiration but open at night to collect carbon dioxide (CO_2).

CO_2 is stored as the four-carbon acid malate and used for photosynthesis during the day.

The pre-collected CO_2 is concentrated around the RuBisCO enzyme, increasing photosynthetic efficiency.

3. A is correct.

Autotrophs (i.e., *producers*) synthesize complex organic compounds (i.e., carbohydrates, fats, proteins) from simple substances in their environment, generally using energy from light (i.e., photosynthesis) or inorganic chemical reactions (i.e., chemosynthesis).

Autotrophs are the producers in a food chain (plants on land and algae in water), in contrast to heterotrophs that consume autotrophs. They can make their food and do not need living energy or organic carbon source.

Autotrophs can reduce carbon dioxide to make organic compounds.

Most autotrophs use water as the reducing agent, but some can use other hydrogen compounds.

4. B is correct.

Adenosine triphosphate (ATP) is a *nucleotide* composed of adenosine (A), ribose, and three phosphate groups.

ATP has three phosphate groups and is produced by enzymes (e.g., ATP synthase) from adenosine diphosphate (ADP) or adenosine monophosphate (AMP) and phosphate groups.

Mitochondria are the site of most ATP molecules produced during aerobic respiration.

5. C is correct.

Photosynthesis reaction:

$$6\ CO_2 + 6\ H_2O \rightarrow C_6H_{12}O_6 + 6\ O_2$$

The reactants are six carbon dioxide (CO_2) and six water (H_2O) molecules, converted by light energy captured by chlorophyll pigments into one sugar ($C_6H_{12}O_6$) and six oxygen (O_2) molecules.

Chloroplasts are the site of photosynthesis and are only in algae and plant cells.

Chloroplasts contain DNA and ribosomes and may have similarly evolved via endosymbiosis.

6. D is correct.

ATP is a nucleoside triphosphate with a nitrogenous base (adenine), ribose sugar, and triphosphate.

The complete structure represents ATP.

Structure A represents *adenine* (nitrogenous base).

Structure B represents *ribose.*

Structure C represents *diphosphate.*

Structure D represents *phosphate*

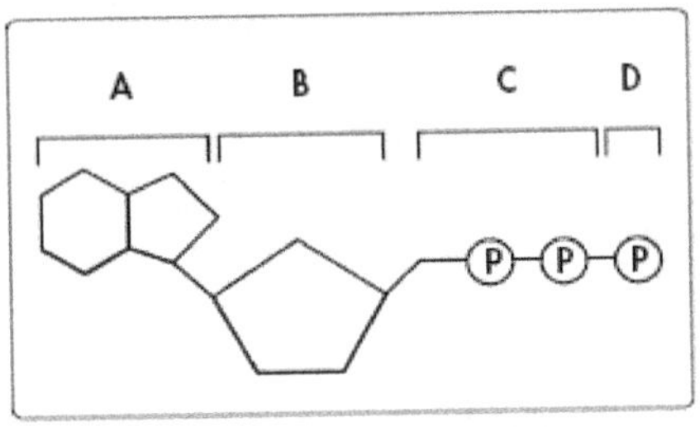

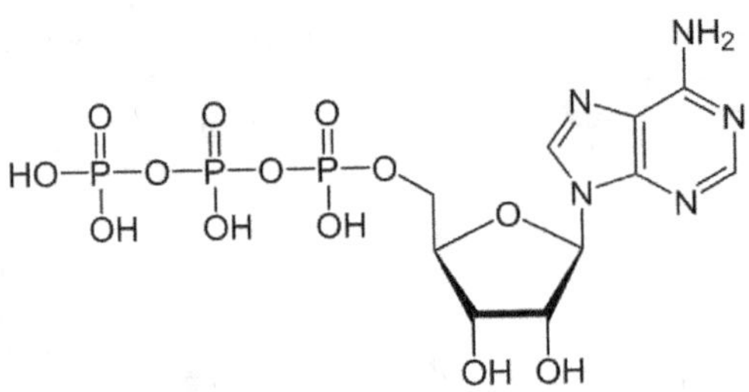

Adenosine triphosphate

7. A is correct.

Stroma is the colorless fluid surrounding the thylakoids within the chloroplast.

Within stroma are *grana* (i.e., stacks of thylakoids) – the sub-organelles, where photosynthesis begins before the chemical changes are completed in the stroma.

8. B is correct.

During photosynthesis, autotrophs produce carbohydrates (e.g., glucose).

Photosynthesis:

$$6\ CO_2 + 6\ H_2O \rightarrow C_6H_{12}O_6 + 6\ O_2$$

The reactants are six carbon dioxide (CO_2) and six water (H_2O) molecules, converted by light energy captured by chlorophyll pigments into one sugar ($C_6H_{12}O_6$) and six oxygen (O_2) molecules.

9. A is correct.

Stomata are composed of a pair of specialized epidermal cells known as *guard cells.*

Stomata regulate gas exchange (e.g., CO_2) and control water loss by changing the size of the stomatal pore.

10. D is correct.

Plants use photosynthesis to make food.

During photosynthesis, plants absorb photons (light energy) with their leaves.

Plants use energy from the sun to convert water (H_2O) and carbon dioxide (CO_2) into glucose ($C_6H_{12}O_6$), six-carbon chain sugar.

Plants use glucose ($C_6H_{12}O_6$) for cellular energy to make biomolecules (e.g., cellulose and starch in plants; glycogen in animals).

Cellulose builds cell walls, while starch is stored in seeds and plant parts as a food source.

Chlorophylls *a* and *b* are photosynthetic pigments produced in chloroplasts of photosynthetic tissues of leaves.

Chlorophyll molecules are water-repelling because of the long phytol tail in the molecule.

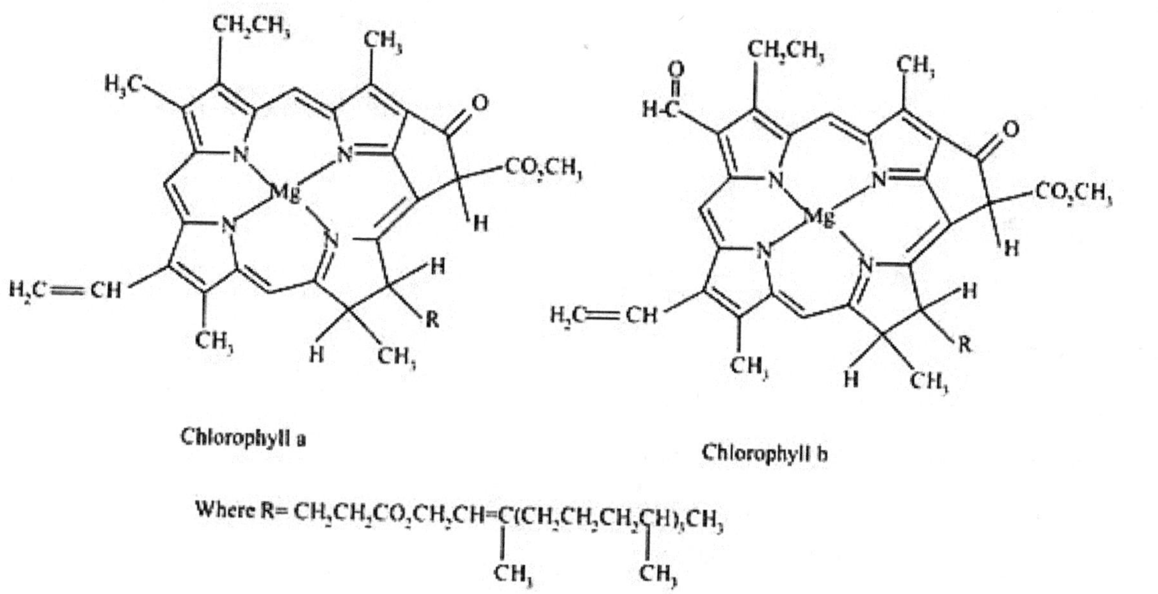

Photopigments chlorophyll a and chlorophyll b that absorb energy during photosynthesis

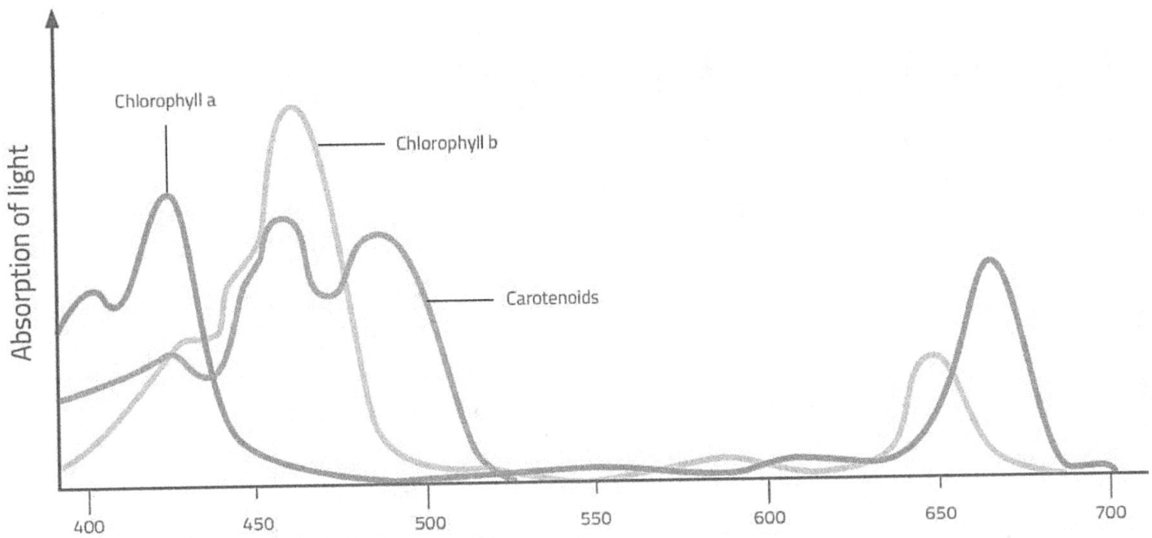

Photopigments absorb specific wavelengths of light with a visible light spectrum from 380 to 700 nm

11. C is correct.

Roots are used by most of their water plants to obtain.

12. C is correct.

Thylakoids are labeled in the figure.

13. B is correct.

Thylakoids are membrane-bound compartments inside chloroplasts and cyanobacteria. They are the site of photosynthesis.

Thylakoids consist of a thylakoid membrane surrounding a thylakoid lumen.

Thylakoids stack as disks (i.e., grana).

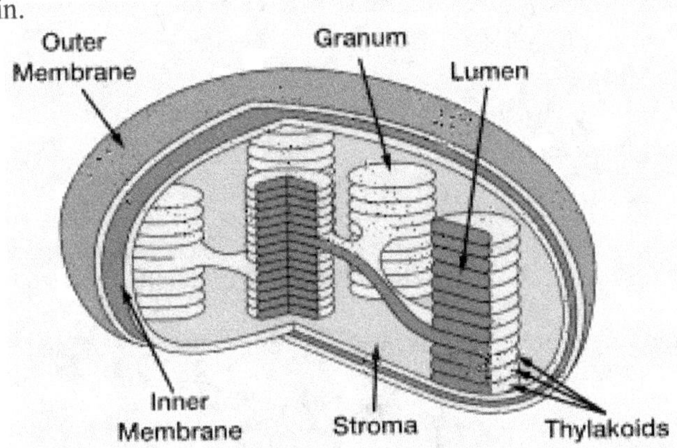

Chloroplast with internal structures

14. B is correct.

The chlorophyll is within the thylakoid membrane.

15. A is correct.

Granum (pl. grana) is a stack of thylakoid discs. Chloroplasts can have from 10 to 100 grana.

Grana are connected by stroma thylakoids (or lamellae).

Grana thylakoids and stroma thylakoids are distinguished by their protein composition.

16. A is correct.

Electron carrier molecule in the figure is $NADP^+$.

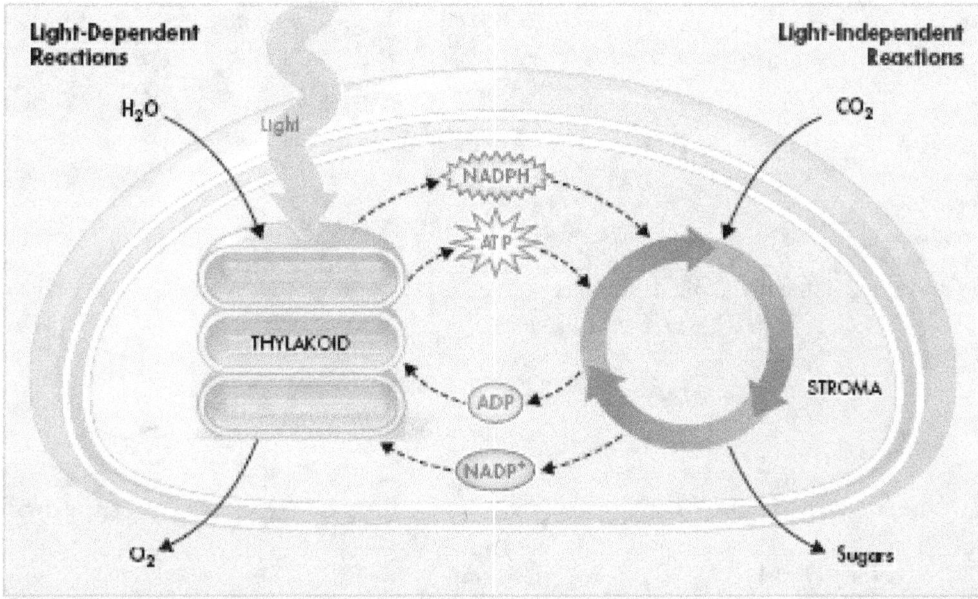

Chloroplast with light-dependent (ATP and NADPH) and light-independent reactions (sugars)

17. B is correct.

Low ATP production is likely if a shade-tolerant plant has minimal water intake.

18. C is correct.

High-energy electrons are highly reactive and are transported in the chloroplast by *electron carriers* ($NADP^+$).

19. D is correct.

Photosynthesis reaction:

$$6\ CO_2 + 6\ H_2O \rightarrow C_6H_{12}O_6 + 6\ O_2$$

The reactants are six carbon dioxide (CO_2) and six water (H_2O) molecules, converted by light energy captured by chlorophyll pigments into one sugar ($C_6H_{12}O_6$) and six oxygen (O_2) molecules.

20. A is correct.

The rate of photosynthesis *increases* and then levels off as the intensity of light increases.

21. A is correct.

In the redox reactions of photosynthesis, *electron transfer* is from $H_2O \rightarrow CO_2$

22. B is correct.

Steps in the light-dependent reaction include ATP synthase allowing H^+ ions to pass through the thylakoid membrane, high-energy electrons through the electron transport chain, and pigments in photosystem II absorb photons of light.

23. C is correct.

Oxygen is the product of the light reaction of photosynthesis.

24. C is correct.

Photosystem I and II are in the thylakoid membrane.

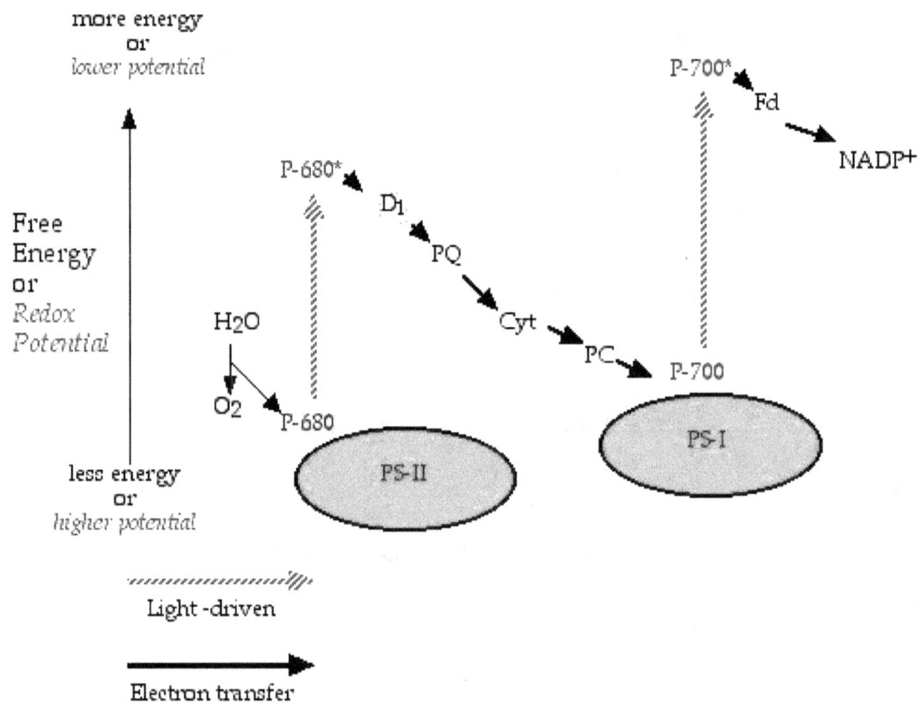

Photosystem II and I in the thylakoid membrane of the chloroplast. The light-driven reactions include electron promotions for P-680 and P-700. $NADP^+$ is the electron carrier.

25. B is correct.

Calvin cycle (i.e., C_3 cycle) is a series of biochemical redox reactions in the stroma of chloroplasts in photosynthetic organisms.

It is one of the light-independent reactions used for carbon fixation.

RuBisCO is the key enzyme of the cycle.

continued...

Photosynthesis occurs in two stages.

1) *Light-dependent reactions* capture the energy of light and use it to make energy storage and transport molecules ATP and NADPH.

2) *Light-independent reactions* (Calvin cycle) use energy from short-lived electronically excited electron carriers to convert carbon dioxide and water into organic compounds used by the organism (and animals that feed on it).

This set of reactions is *carbon fixation.*

Calvin cycle has four steps:

1) carbon fixation,
2) reduction phase,
3) carbohydrate formation, and
4) regeneration phase.

ATP and NADPH provide energy for the chemical reactions in this sugar-generating process.

Sunlight provides the energy for the reaction:

ATP + NADPH + carbon dioxide (CO_2) → sugar

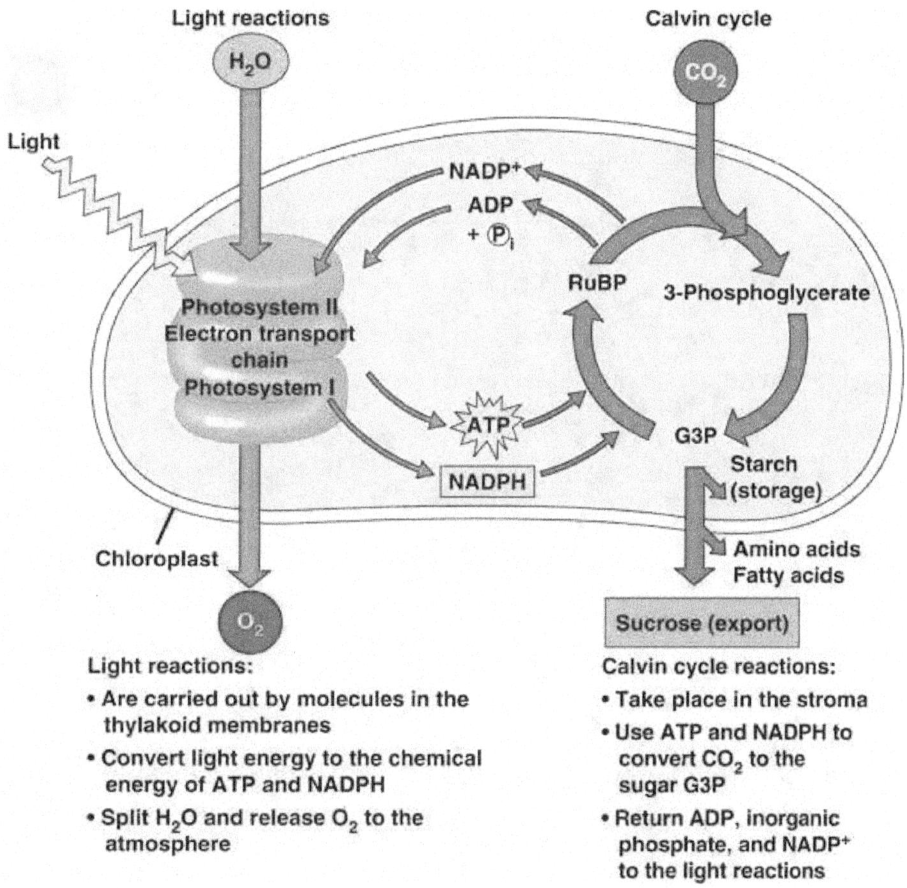

Products of light-dependent (ATP and NADPH) and light-independent reactions (sugars)

26. B is correct.

Electrons flow during photosynthesis is H_2O → $NADP^+$ → Calvin cycle.

27. D is correct.

Light-dependent reactions of photosynthesis produce NADPH and ATP.

28. D is correct.

Calvin cycle occurs in the stroma of the chloroplast.

29. C is correct.

Blue-colored plants reflect light of blue wavelength (380 – 500 nm); visible light spectrum (380 – 700 nm).

30. A is correct.

Three components of an ATP molecule are adenosine (*base*), ribose (*sugar*), and three *phosphate* groups.

31. C is correct.

The wavelength *reflected* determines the color of an object.

32. A is correct.

Energy is *released* in the hydrolysis of ATP (triphosphate) into ADP (diphosphate).

33. C is correct.

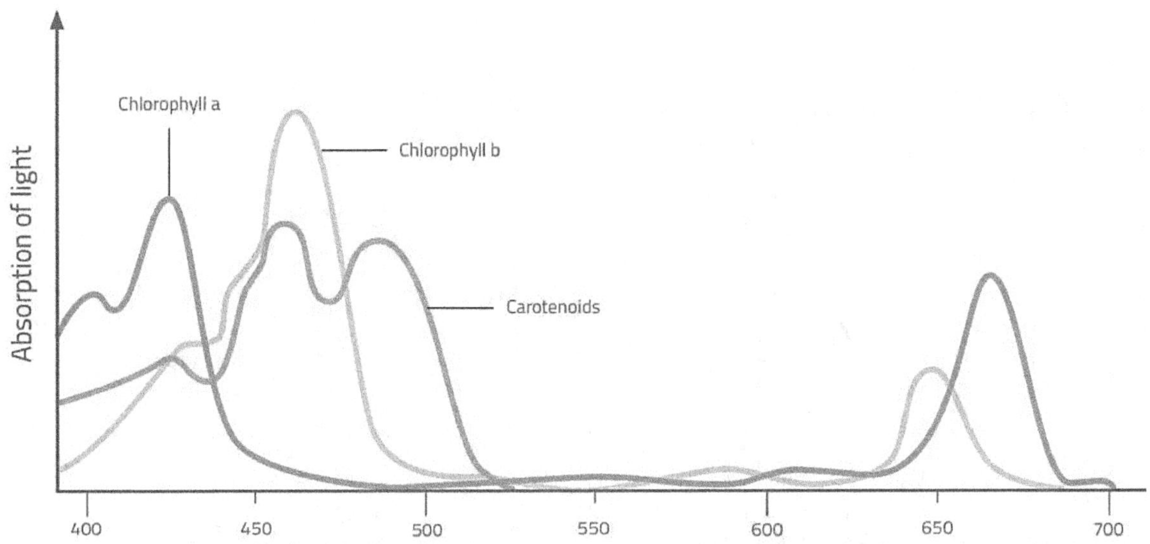

Photopigments absorb specific wavelengths of light with a visible light spectrum from 380 to 700 nm

34. C is correct.

Autotrophs (i.e., *producers*) synthesize complex organic compounds (i.e., carbohydrates, fats, proteins) from simple substances in their environment. They generally use energy from light (i.e., photosynthesis) or inorganic chemical reactions (i.e., chemosynthesis).

Autotrophs are the producers in a food chain (plants on land and algae in water), in contrast to heterotrophs that consume autotrophs. They can make their food and do not need living energy or organic carbon source.

Autotrophs can reduce carbon dioxide to make organic compounds. Most autotrophs use water to reduce (i.e., a gain of electrons), but some can use other hydrogen compounds.

Phototrophs (i.e., green plants and algae), a type of autotroph, convert physical energy from sunlight into chemical energy in the form of reduced carbon.

35. A is correct.

Carotenoids reflect light with yellow, red, and orange wavelengths.

36. D is correct.

Mushrooms are *heterotrophs* and derive nutrients from *complex organic molecules*.

37. C is correct.

Plants use chlorophyll *a*, chlorophyll *b*, and carotenoid accessory pigments to absorb light over a broad spectrum of wavelengths.

38. B is correct.

The *sun* is the *ultimate energy source* for living organisms.

39. A is correct.

Photons are discrete packets of light.

40. B is correct.

Most plants appear green because chlorophylls do not absorb wavelengths of green light (500 – 560 nm).

41. A is correct.

The electromagnetic spectrum (ROY G BIV) has higher energy light with shorter wavelengths.

42. B is correct.

Thylakoids are membrane-bound compartments inside chloroplasts and cyanobacteria and the site of photosynthesis.

43. B is correct.

The energy from photons is transferred to the electron and raises it to a higher energy (i.e., excited) level.

44. C is correct.

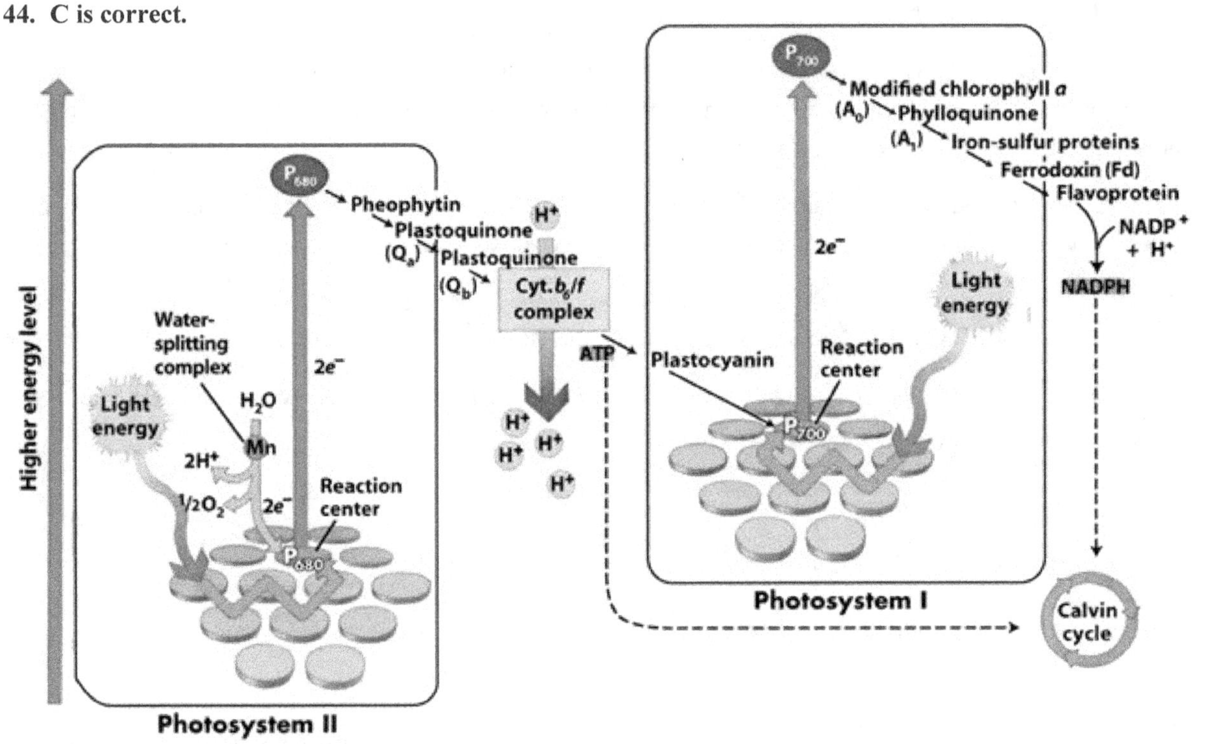

Photosystem II and I in the thylakoid membrane of the chloroplast. The light-driven reactions include electron promotions for P-680 and P-700 with $NADP^+$ as the electron carrier.

45. D is correct.

Molecules can *release energy* as fluorescence, heat, or light.

46. A is correct.

Photosynthesis reaction: $6\ CO_2 + 6\ H_2O \rightarrow C_6H_{12}O_6 + 6\ O_2$

The reactants are six carbon dioxide (CO_2) and six water (H_2O) molecules, converted by light energy captured by chlorophyll pigments into one sugar ($C_6H_{12}O_6$) and six oxygen (O_2) molecules.

47. D is correct.

Photosystems are in the *thylakoid* membrane of chloroplasts.

48. A is correct.

The light-dependent reaction reactions occur within the thylakoid membranes.

49. A is correct.

Chlorophyll *a* is a form of chlorophyll used in photosynthesis. It absorbs most energy from wavelengths of violet-blue and orange-red light. Chlorophyll *a* reflects green and yellow light and contributes to the green color of many plants. This photosynthetic pigment is essential for photosynthesis in eukaryotes, cyanobacteria, and prochlorophytes because it is a primary electron donor in the *electron transport chain*.

50. C is correct.

The products of the light-dependent reaction are ATP, NADPH, and O_2 gas.

51. A is correct.

Calvin cycle has four steps:

carbon fixation,

reduction phase,

carbohydrate formation, and

regeneration.

ATP and NADPH provide energy for the chemical reactions in this sugar-generating process.

Sunlight provides the energy for the reaction: ATP + NADPH + carbon dioxide (CO_2) → sugar

52. B is correct.

Calvin cycle (i.e., C_3 cycle) is a series of biochemical redox reactions in the stroma of chloroplasts in photosynthetic organisms. It is one of the light-independent reactions used for carbon fixation. The key enzyme of the cycle is RuBisCO.

Photosynthesis occurs in two stages in a cell.

Light-dependent reaction captures energy from light and uses it to make the energy storage and transport molecules ATP and NADPH.

Light-independent Calvin cycle uses energy from short-lived electronically excited electron carriers to convert carbon dioxide and water into organic compounds used by the organism (and animals that feed on it).

This set of reactions is *carbon fixation*.

53. D is correct.

Water is the source of *oxygen* and electrons released by a photosystem.

54. C is correct.

Light-independent reaction of photosynthesis used for carbon fixation is the *Calvin cycle*.

55. B is correct.

Water is the source of oxygen and *electrons* released by a photosystem.

56. A is correct.

Electron transport chain occurs in light reactions of photosynthesis (in chloroplasts) and *cellular respiration* (in mitochondria).

57. C is correct.

H^+ gradient is formed across the *thylakoid membranes* of chloroplasts during photosynthesis.

Notes for active learning

Macromolecules in a Biological System – Detailed Explanations

1. A is correct.

Primary structure of a protein is the *linear sequence* of amino acids.

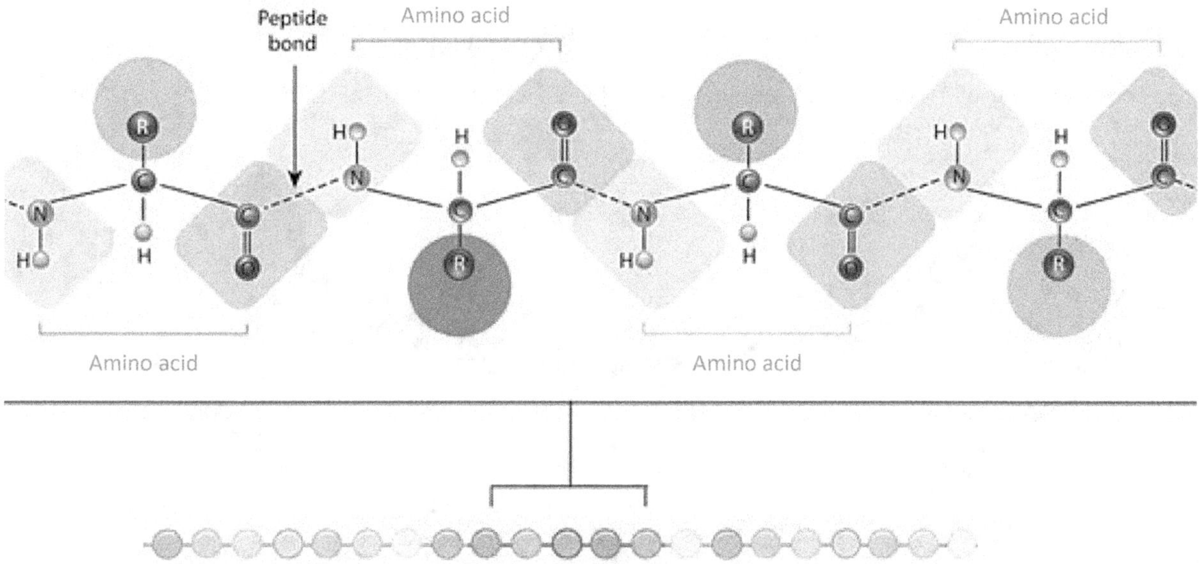

Peptide bonds link amino acids in the primary structure of proteins (circles) with amino acid side chains (R groups) facing opposite directions to reduce steric hindrance

2. B is correct.

Biomolecules are divided into macromolecules of *carbohydrates*, *lipids*, *nucleic acids*, and *proteins*.

3. A is correct.

Amino acids are linked as peptides by *amide bonds*.

The peptide (amide) bond is anti-periplanar and includes 4 atoms, as shown in the box.

The peptide forms in a condensation reaction via dehydration (loss of water) when the lone pair on the nitrogen of an amino group of one amino acid makes a nucleophilic attack on the carbonyl carbon of another.

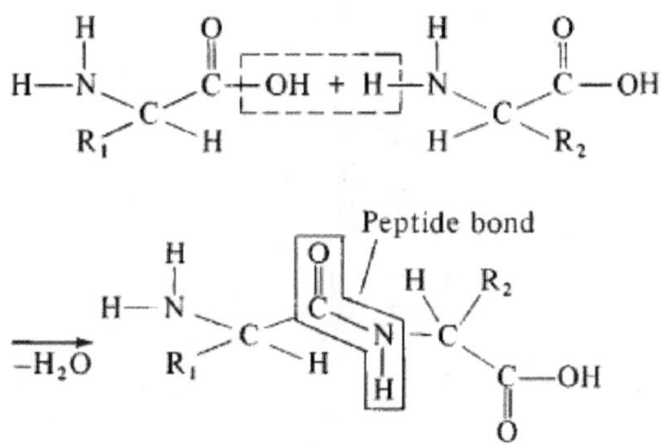

Peptide bonds form by a condensation (dehydration) reaction of 2 amino acids

4. A is correct.

Hormones are substances secreted by a gland and released into the blood to affect a target tissue/organ.

Insulin is a hormone composed of amino acids (i.e., peptide hormones) and is a protein.

Insulin has two peptide chains (A chain and B chain).

Two disulfide bonds link the chains, and an additional disulfide is formed within the A chain.

The A chain consists of 21 amino acids, and the B chain has 30 amino acids in most species.

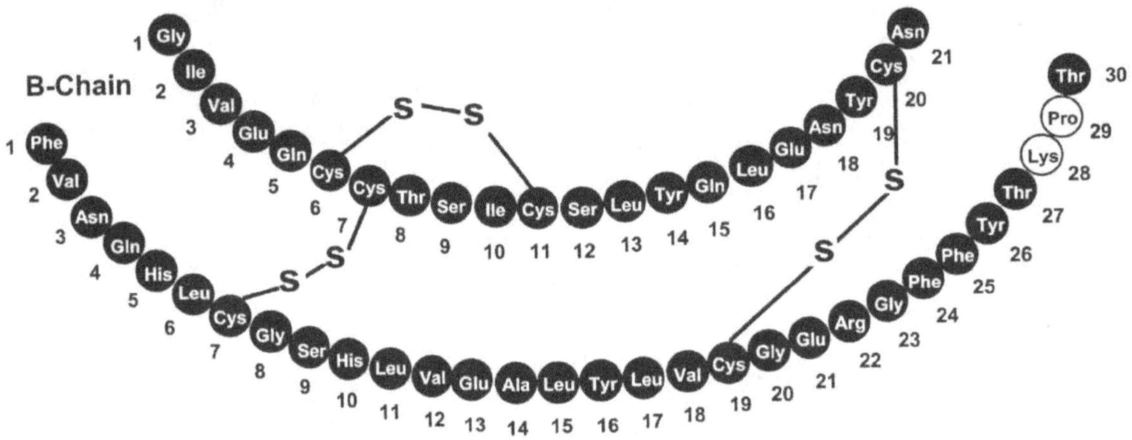

Insulin has an A chain and B chain linked by disulfide bonds

Hormones can be lipids, such as steroid derivatives (e.g., testosterone, progesterone, estrogen).

5. B is correct.

Alpha helix structure is one of two common secondary protein structures.

Hydrogen bonds hold the alpha-helix between every N–H (amino group) and the oxygen of C=O (carbonyl) in the next turn of the helix; four amino acids along the chain.

The typical alpha helix is about 11 amino acids long.

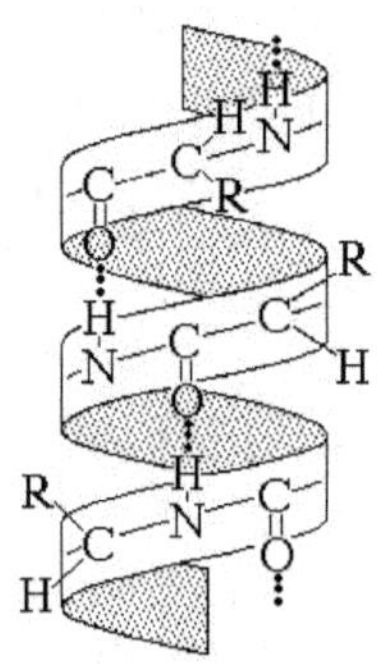

Hydrogen bonds hold the secondary structure of the alpha helix

Beta-pleated sheet is the other *secondary structure*.

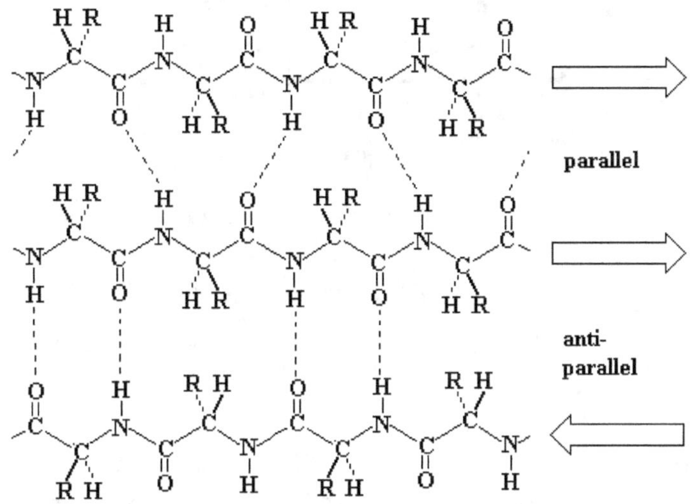

Beta pleated sheets are parallel or anti-parallel (a reference to the amino terminus)

6. C is correct.

Primary structure of a protein is the amino acid sequence, formed by covalent peptide linkages.

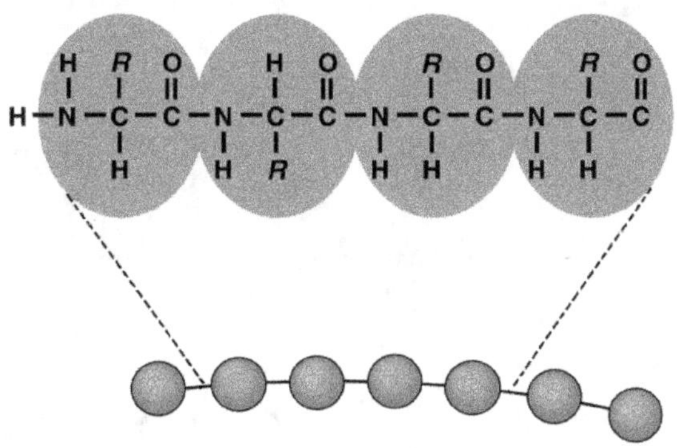

The amino acids (circles) are joined by covalent peptide bonds (lines). Peptide bonds link amino acids in the primary structure of proteins (circles) with amino acid side chains (R groups) facing opposite directions to reduce steric hindrance

A: only proteins containing more than one peptide subunit have a quaternary structure.

B: proteins are denatured by heating, and they lose their conformation above 35-40 °C.

D: many proteins contain more than one peptide chain (i.e., have a quaternary structure).

7. B is correct.

Amino acids are the building blocks of proteins.

Humans need 20 amino acids (see diagram), some are synthesized by the body (i.e., nonessential), and others must be obtained from the diet (i.e., essential).

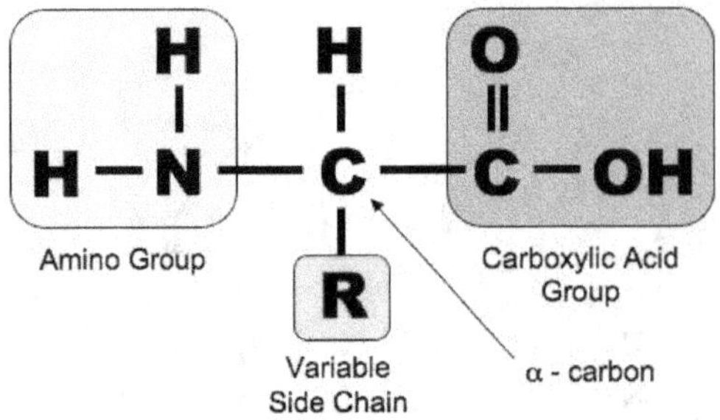

Amino acids contain an amine group, a carboxylic acid, an α-carbon, and an R group

Note: The 20 naturally occurring amino acids shown are not for memorization but for identifying characteristics (e.g., polar, nonpolar) for the side chains.

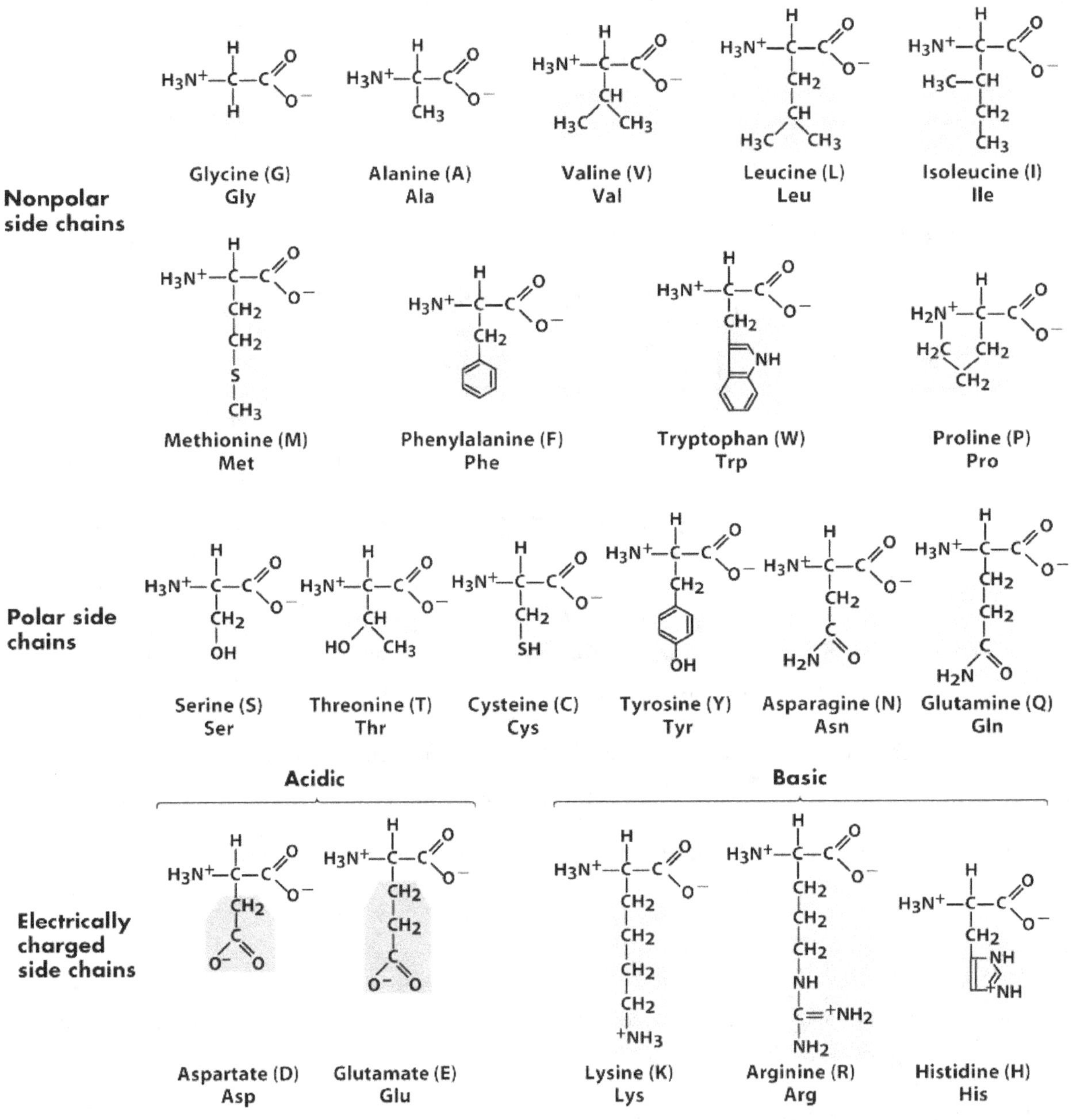

8. B is correct.

Peptide bonds link the primary sequence of amino acids in a protein between individual amino acids that form a peptide (*polypeptide*) chain.

Peptide bonds are not altered when a polypeptide bends or folds to form secondary structures (e.g., alpha helix).

Peptide bonds do *not* form when the polypeptide folds into a three-dimensional shape (tertiary structure).

A: interactions between charged groups (*electrostatic interactions*) can arise, especially in the tertiary structure.

C: *hydrogen bonds* are involved in the secondary and tertiary structures.

In the secondary structure, the polypeptide chain folds to allow the carbonyl oxygen and amine hydrogen to lie nearby. As a result, hydrogen bonding occurs to form sheets, helices, or turns.

Likewise, hydrogen bonding may stabilize the secondary or tertiary structure.

D: *hydrophobic interactions* also play an important role in the *tertiary structure.*

For example, in an aqueous environment, the hydrophobic side chains of the amino acids may interact to arrange themselves towards the inside of the protein.

9. C is correct.

Amino acids are the basic building blocks for proteins.

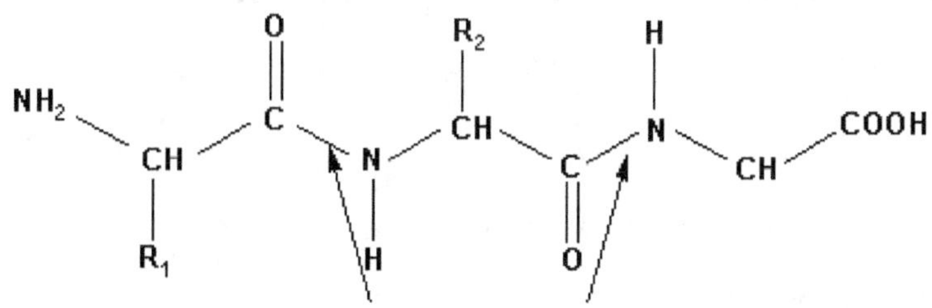

Two amino acids (dimer) with peptide bonds are indicated by arrows

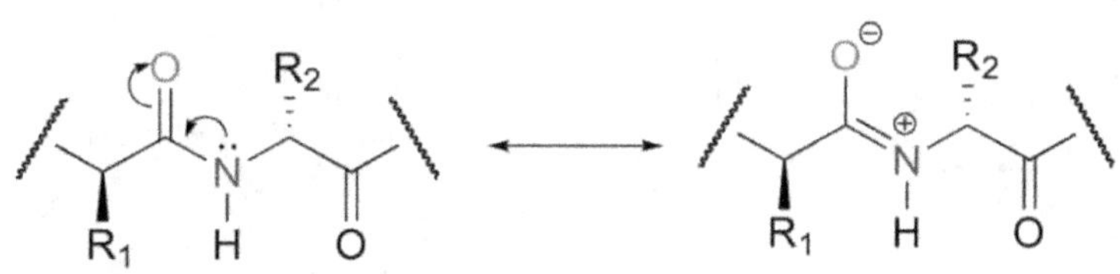

Peptide bonds are rigid with double bond character by lone pair nucleophilic attack on the adjacent carbonyl

The peptide bond is rigid due to the resonance hybrids involving the lone pair of electrons on nitrogen, forming a double bond to the carbonyl carbon (and oxygen develops a negative formal charge).

10. C is correct.

Protonation (or deprotonation) of an amino acid residue changes its *ionization state:* it may become positively charged, negatively charged, or neutral.

The process may lead to changes in the interactions among amino acid side chains, as some ionic bonds may be compromised from the lack of opposite charge pairing.

Specific hydrogen bonding interactions may be modulated if Lewis bases are protonated with Brønsted acids, impairing their ability to accept hydrogen bonds from nearby amino acid residues.

11. D is correct.

The *primary structure* of proteins refers to the *linear sequence* of amino acids.

Hydrogen bonding is essential for the secondary (alpha-helix and beta-pleated sheet) and tertiary (i.e., overall, 3-dimensional shape) structure of proteins.

Hydrophobic interactions involved in tertiary and quaternary (i.e., two or more polypeptide chains) structures arise from the hydrophobic side chains of the amino acid residues.

12. D is correct.

Collagen is a protein that supports hair, nails, and skin. It is composed of a triple helix, and the most abundant amino acids in collagen include glycine, proline, alanine, and glutamic acid.

Much excess protein consumed in an animal's diet is used to synthesize collagen.

13. B is correct.

Secondary structure for proteins involves localized bonding.

The critical intermolecular interaction for secondary structure is hydrogen bonding, which maintains the alpha helix and beta pleated (parallel and antiparallel) sheet structures.

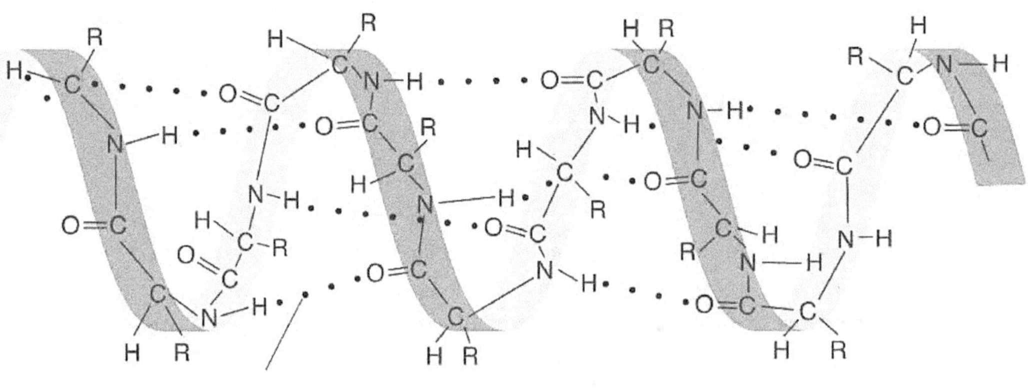

Alpha helix structure with hydrogen bonding shown as dotted lines

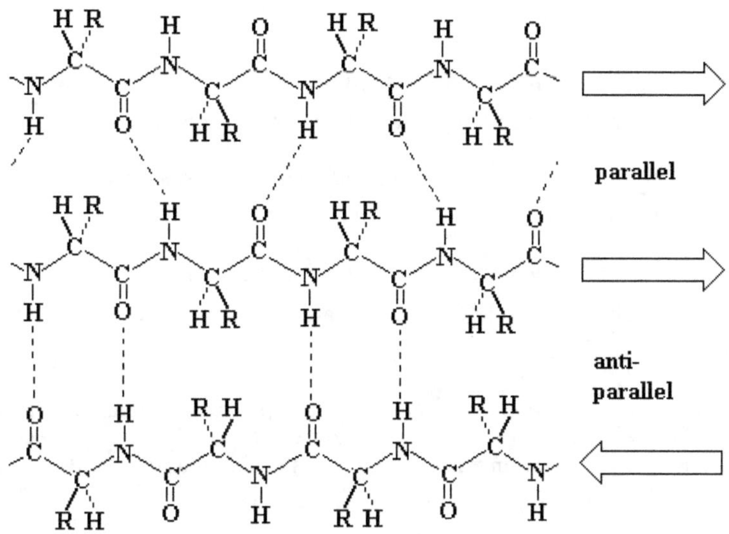

Beta pleated sheets (parallel and antiparallel) with hydrogen bonding shown

14. C is correct.

Standard conditions for breaking covalent bonds during peptide hydrolysis are concentrated HCl and several hours of reflux. The reaction time depends on partial or complete hydrolysis of the peptide.

15. B is correct.

Essential amino acids are those obtained from the diet.

Nonessential amino acids are synthesized by the body and do not need to be consumed.

Semi-essential (conditionally essential) amino acids are synthesized under special physiological conditions (e.g., in premature infants and under severe catabolic distress).

Nine essential amino acids for humans are histidine, isoleucine, leucine, lysine, methionine, phenylalanine, threonine, tryptophan, and valine.

Six conditionally essential amino acids are arginine, cysteine, glycine, glutamine, proline, and tyrosine.

Five nonessential amino acids are alanine, aspartic acid, asparagine, glutamic acid, and serine.

16. B is correct.

A polypeptide chain can undergo short-range bending and folding to form *β sheets* or *α helices*. These structures arise as the peptide bonds can assume a partial double-bond character and adopt different conformations.

The arrangement of groups around the relatively rigid amide bond can cause *R* groups to alternate from side to side and interact with one another.

The carbonyl oxygen in one region of the polypeptide chain could become hydrogen-bonded to the amide hydrogen in another region of the polypeptide chain. This interaction often results in forming a *beta-pleated sheet* or an *alpha helix*.

Localized bending and folding of a polypeptide do not constitute a protein's primary structure.

A: *primary structure* of a protein is the amino acid sequence; individual amino acids are linked through peptide (i.e., amide) linkages.

C: *tertiary structure* is the 3-D shape that arises by further folding the polypeptide chain. Usually, these nonrandom folds give the protein a particular conformation and associated function.

D: *quaternary structure* is the spatial arrangement between *two or more* polypeptide chains (often linked by *disulfide bridges* between *cysteine* residues).

17. B is correct.

Proteins are biological macromolecules of amino acids bonded with peptide (amide) bonds.

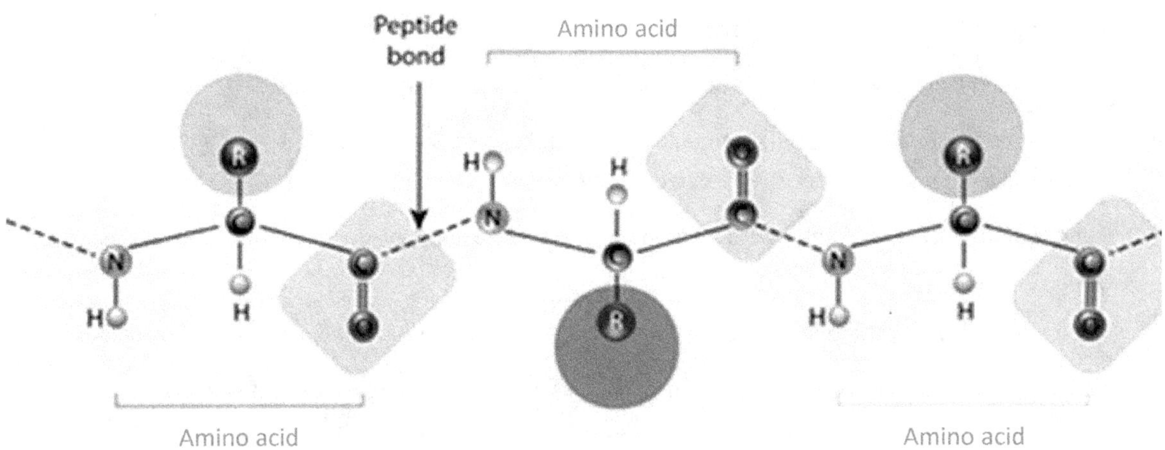

Three amino acid residues of a nascent (i.e., growing) polypeptide. Peptide bonds link amino acids in the primary structure of proteins (circles) with amino acid side chains (R groups) facing opposite directions to reduce steric hindrance.

18. D is correct.

The plasma membrane is made of lipids known as *phospholipids*. These molecules mainly possess nonpolar characteristics due to the long hydrocarbon chains, making the membrane permeable to nonpolar materials and semipermeable to polar or charged molecules.

19. D is correct.

Palmitic acid is a saturated acid, meaning that the molecule does not contain a double bond. Because this molecule lacks a double bond, the molecules stack better to form solids, and the fat has a higher melting point.

Unsaturated fats are frequently liquids at room temperature.

Linolenic acid (shown) is polyunsaturated with three *cis* double bonds.

Alkenes (i.e., unsaturation) introduce "kinks" in the chain that give the unsaturated fat an overall bent structure. This molecular geometry limits these fats from clustering closely to form solids.

Therefore, (relative to chain length), polyunsaturated molecules have the lowest melting point.

Position *omega* (ω) of the double bond(s) is the number of carbon atoms from the *terminal methyl group*.

Number of *carbon atoms* is noted from the carboxyl end

Palmitic Acid

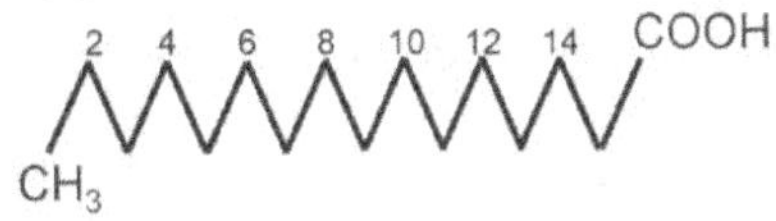

Oleic Acid

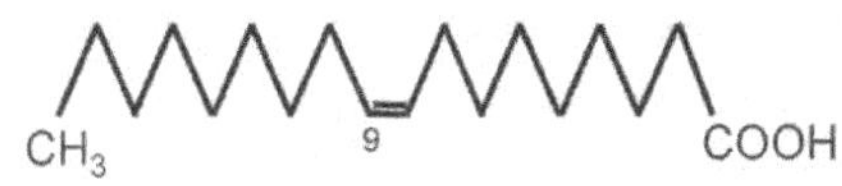

Saturated and unsaturated fatty acids – note the omega (ω) position 9 of the oleic acid double bond

Linoleic Acid

Arachidonic Acid

Omega-6 polyunsaturated fatty acids. Linoleic is ω-6,9, and arachidonic acid is ω-6,9,12,15 fatty acids.

α-Linolenic Acid

Eicosapentaenoic Acid

Docosahexaenoic Acid

Omega-6 polyunsaturated fatty acids. α-linoleic acid is ω-3,6,9, eicosapentaenoic acid is ω-3,6,9,12,15 and docosahexaenoic acid is ω-3,6,9,12,15,18.

20. D is correct.

Unsaturated fats become hydrogenated to form saturated fats, increasing the melting point.

An equivalent of hydrogen (H_2) is added across the double bonds of unsaturated fats during hydrogenation (i.e., increased hydrogen content), increasing the compound's molecular weight.

Saturation enhances the stacking ability of these compounds as solids.

The more saturated (*double bonds*) the compound is, the *higher* its melting point.

21. B is correct.

The ends of fatty acids are *hydrophobic*, nonpolar ends composed of hydrogen and carbon atoms.

The *hydrophilic*, polar end is composed of oxygen atoms and can hydrogen bond with water molecules.

22. B is correct.

Cholesterol plays a pivotal role in the synthesis of steroids and the integrity of cell membranes, but too much cholesterol in the blood results in plaque deposits in blood vessels.

23. D is correct.

Steroid molecules are one of two kinds of fat molecules with fused rings.

Triacylglycerides (or *triglycerides*) are long hydrocarbon chains with functionalized head groups.

24. C is correct.

Saponification is the base-promoted hydrolysis of esters.

This type of hydrolysis (i.e., saponification) is typically used to form soaps.

carboxylate ester *sodium hydroxide* *sodium carboxylate* *alcohol*

Base hydrolysis (saponification) of an ester to form a carboxylate salt and an alcohol

Soap is hard or soft depending on the counter-ion of the carboxylate salt.

Hard soap is produced when *sodium hydroxide* is used for the base hydrolysis reaction.

Soft soap is produced when *potassium hydroxide* is used for the base hydrolysis reaction.

Other bases can give rise to soaps.

25. C is correct.

Cholesterol is a lipid molecule known as a steroid compound.

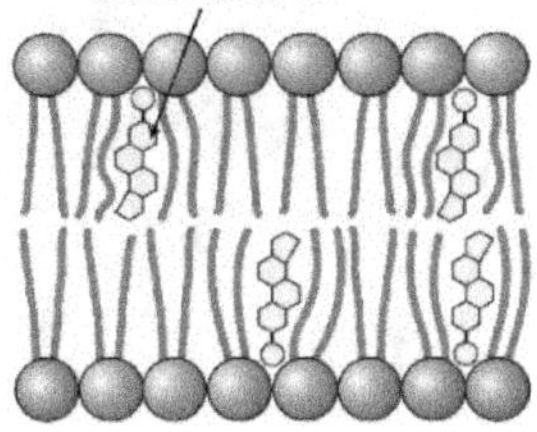

Cholesterol is embedded in the fluid mosaic phospholipid bilayer of membranes

Fused ring structure of steroid molecules makes them rigid with fewer degrees of motion due to the limited conformations available for cyclic molecules *vs.* acyclic molecules.

Molecules such as phospholipids lack fused-ring structures, and they exist as straight-chained molecules.

Cholesterol in the cell membrane acts as a *bidirectional regulator* of membrane fluidity: at high temperatures, it stabilizes the membrane and raises its melting point, whereas, at low temperatures, it intercalates between the phospholipids and prevents them from clustering and stiffening.

26. D is correct.

Saturated fats lack alkene double bonds.

Unsaturated fats convert to saturated through the process of hydrogenation. In this process, hydrogen gas is catalytically added to the alkene groups of the fatty acids to convert them to alkane groups.

27. D is correct.

The hydrogenation (i.e., H_2) of unsaturated fats adds hydrogen across the double bonds of the fat, causing the fats to be saturated and increasing their melting points.

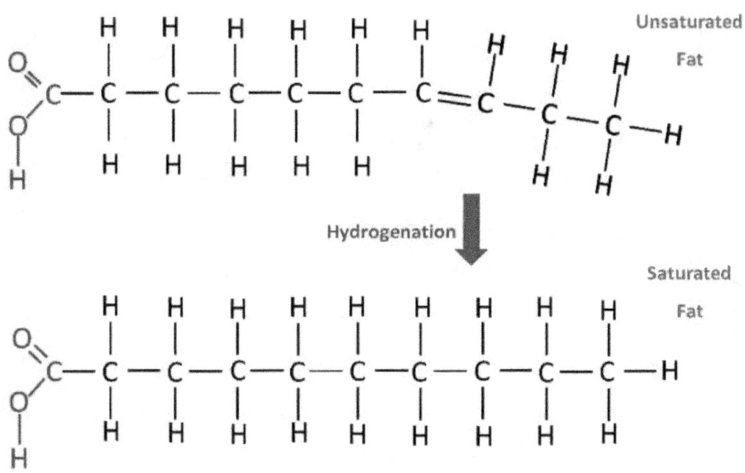

Hydrogenation adds H_2 across double bonds to form saturated fatty acid chains

Because they form solids easily, the consumption of hydrogenated fats should be limited for health concerns.

28. B is correct.

Triglycerides (or *triacylglycerides*) are used for storage and exist in the adipose tissue of animals.

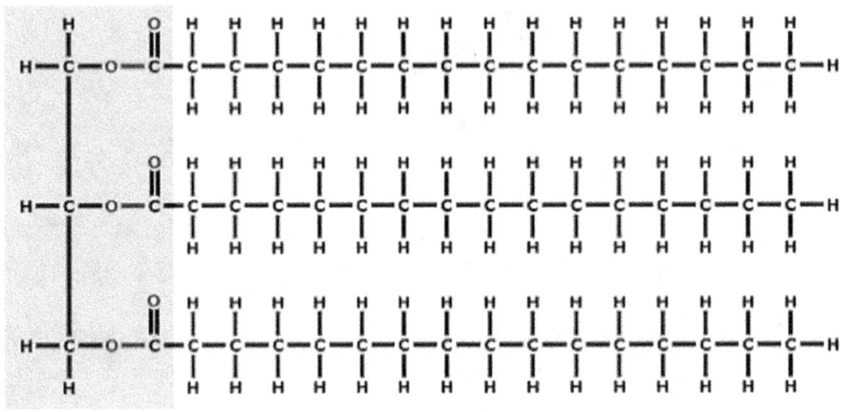

Triglyceride is glycerol backbone and three saturated fatty acid chains

Phospholipids are the largest component of semi-permeable cell membranes

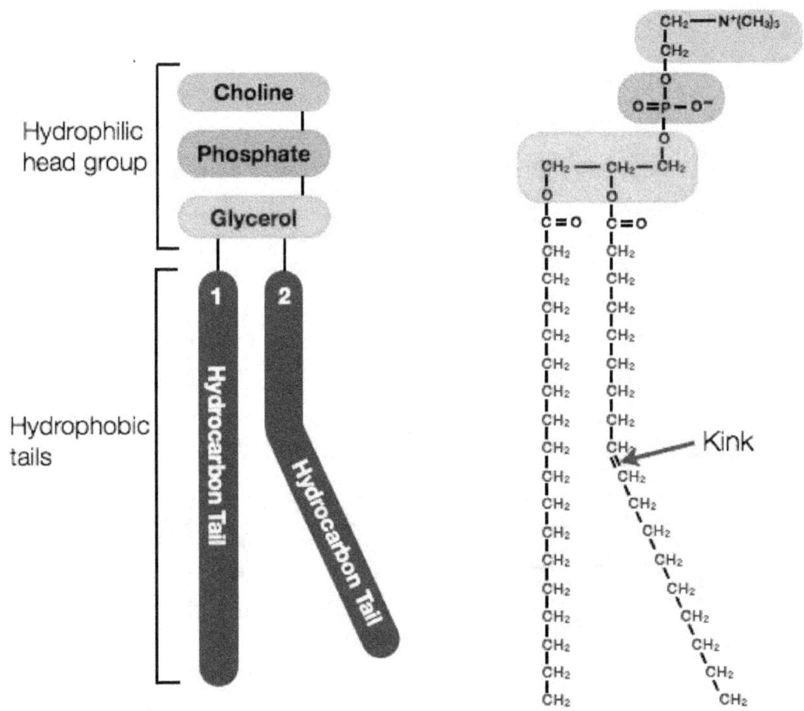

Phospholipids differ mainly in the composition of the polar head region. The hydrophobic region is the fatty acid chains (hydrocarbons) that are saturated (no double bonds) or unsaturated.

Steroids (see below) are lipids used for cell signaling.

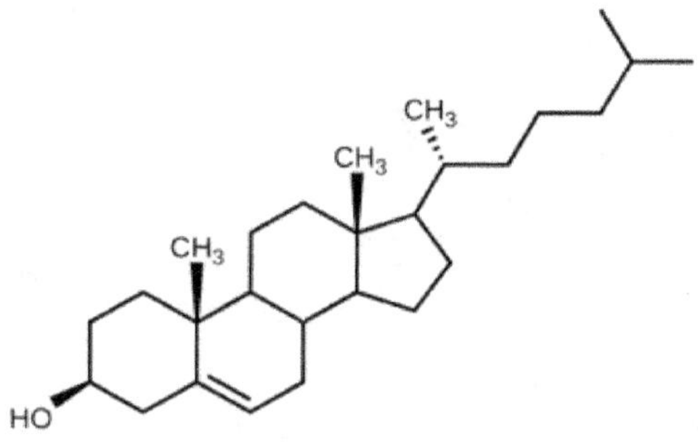

Cholesterol

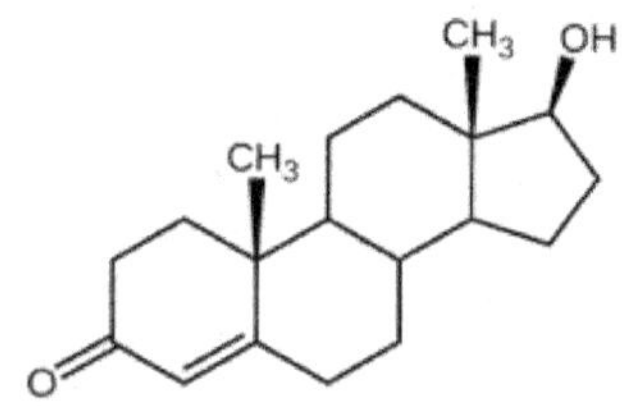

Testosterone

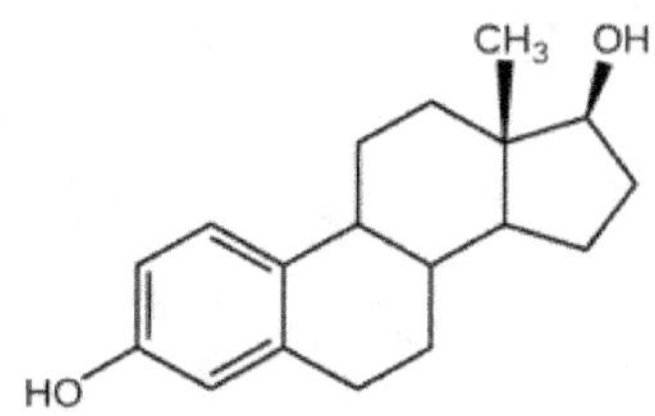

Estradiol

Cholesterol is the precursor molecule for several steroid hormones such as testosterone, estradiol, progesterone, and aldosterone

29. B is correct.

An omega-3 is when the alkene (E/Z) double bond is three carbon atoms from the methyl end.

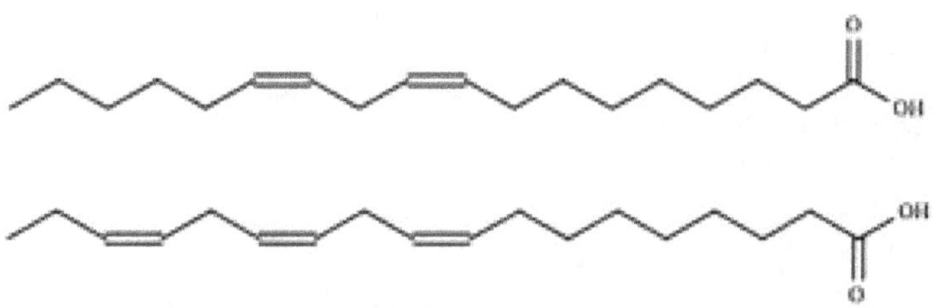

Linoleic acid (top)is ω-6,9 and linolenic acid (bottom) is ω-3,6,9. Note the positions of the double bonds and cis/trans (Z / E) relationships for double bonds

Linolenic acids are omega-3 fatty acids.

Linoleic acids are omega-6 fatty acids.

These molecules contain double bonds and are *unsaturated fats*.

30. D is correct.

Estradiol (shown) is a steroid hormone derived from cholesterol.

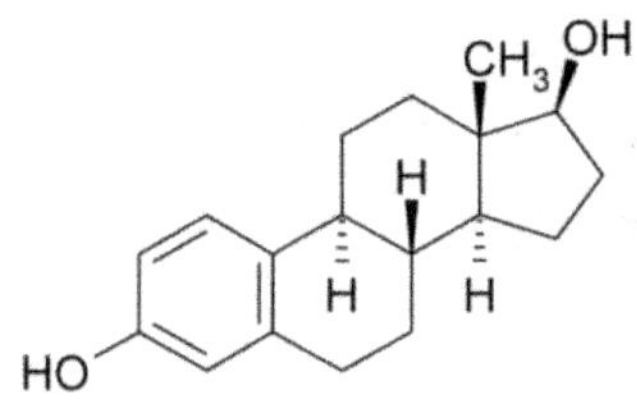

Estradiol is a derivative of cholesterol

Cholesterol is a lipid of four fused rings; three fused rings are six-membered, and the fourth is five-membered.

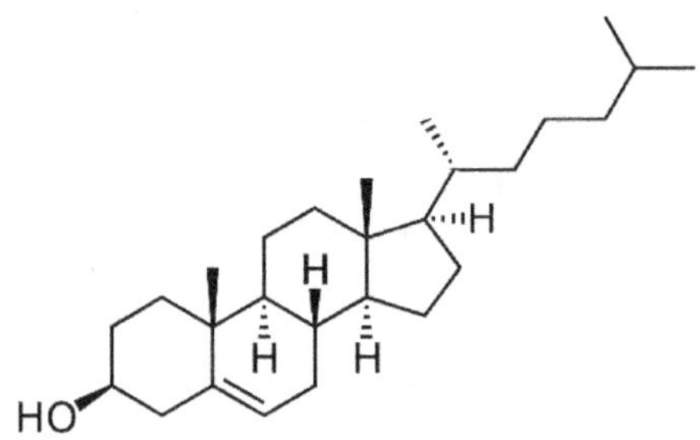

Cholesterol is a four-fused ring structure

Cholesterol is a steroid that makes up one of two types of lipid molecules.

Triglyceride (i.e., glycerol backbone with three fatty acid chains) is the other lipid (shown below).

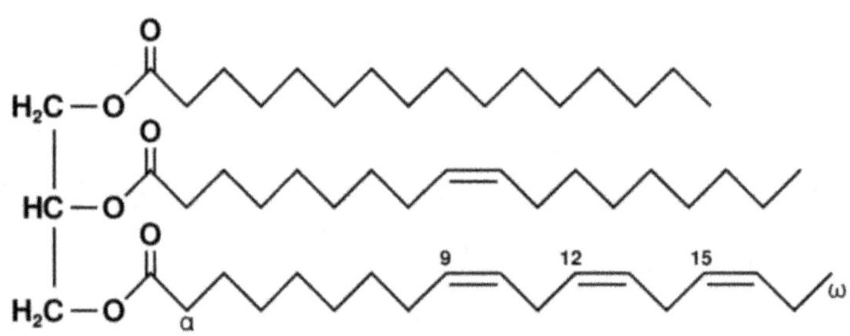

Triglyceride is glycerol backbone with three fatty acid chains

The carbon chain is numbered (example above) from the carboxyl end. Chemists number the double bonds as shown. Nutritionists specify the ω position from the terminal methyl group.

Position *omega* (ω) of the double bond(s) is the number of carbon atoms from the *terminal methyl group*.

31. C is correct.

When *micelles* form in aqueous environments, the hydrophobic tails cluster.

Polar head regions are exposed on the surface of the micelle and exposed to the aqueous environment.

32. B is correct.

Not all lipids are entirely hydrophobic.

The ionic and polar heads of soaps and phospholipids, respectively, enable the molecules to interact with aqueous or polar environments. These molecules are mainly hydrophobic because they primarily consist of hydrocarbon chains or rings.

33. A is correct.

The molecules are likely to exist as oils (i.e., liquids) for unsaturated fats at room temperature.

Saturated fats tend to be solid at room temperature because the reduced forms of these molecules have better stacking properties, forming solids.

34. B is correct.

The alkene molecules typically in fatty acids tend to be *Z* alkenes.

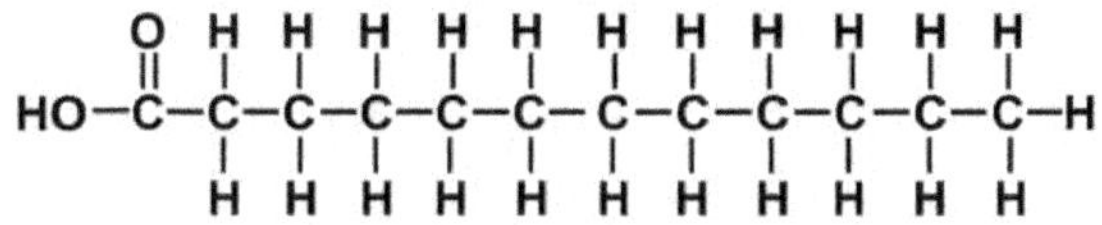

Saturated fatty acid lacks a double bond and is likely a sold at room temperature

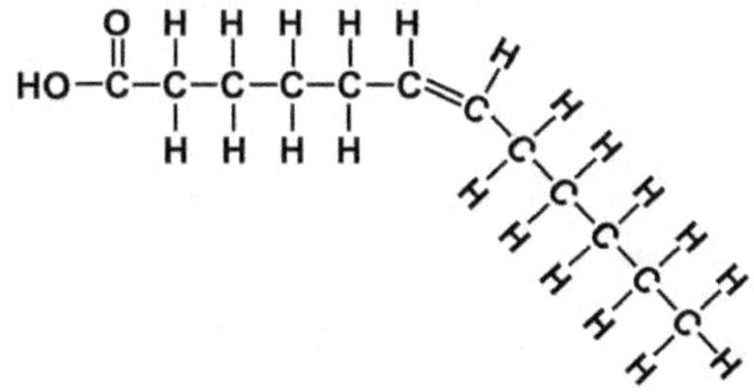

Unsaturated fatty acid has double bonds and is likely a liquid at room temperature

The alkene double bond (i.e., unsaturated) prevents the fatty acid molecules from stacking closely, lowering the melting point for the fat molecule.

The presence of double bonds may influence the state of matter of the oil, as unsaturated fats tend to be liquid at room temperature, and saturated fats tend to be solids.

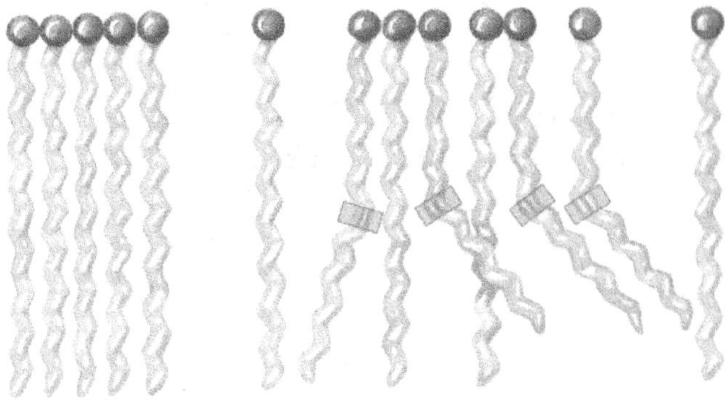

Saturated fatty acids — Mixture of saturated and unsaturated fatty acids

35. C is correct.

Triacylglycerols are composed of a *glycerol* substructure and *three fatty acids* condensed to form a triester.

36. A is correct.

Oils are isolated from plant sources and may consist of several different fatty acids.

These oils may contain saturated fat, but other fats in the mixture are unsaturated fat molecules.

Saturated fats include palmitic and stearic acid.

Unsaturated fats include oleic, palmitoleic, and linoleic acids.

37. B is correct.

There are two overall categories of lipids: long-chain lipids (e.g., triglycerides) and smaller, polycyclic lipids, such as steroids (e.g., cholesterol and its derivatives, such as estrogen and testosterone).

Lipid molecules are fat-soluble and are mainly soluble in organic/hydrophobic.

38. D is correct.

In hydrogenation reactions, *Z* and *E* alkenes are reduced (adding H_2 across double bonds) to alkanes, and this process is catalyzed by transition metals, such as nickel or palladium.

In this reaction, hydrogen (H_2) is added across the double bond of the alkene.

39. D is correct.

Saturated fats tend to be solid at room temperature because they lack alkene groups. The presence of alkene groups in fat molecules lowers the melting point for these compounds.

For example, butter is a dairy product made from the fat of cow's milk. It is solid at room temperature and is mainly composed of saturated fat molecules.

40. C is correct.

Although waxes are lipid molecules containing esters, waxes contain a single ester functional group.

Monoalcohols form waxes, whereas glycerol forms triglycerides and phospholipids.

41. D is correct.

Fatty acids make fat molecules known as *triglycerides*.

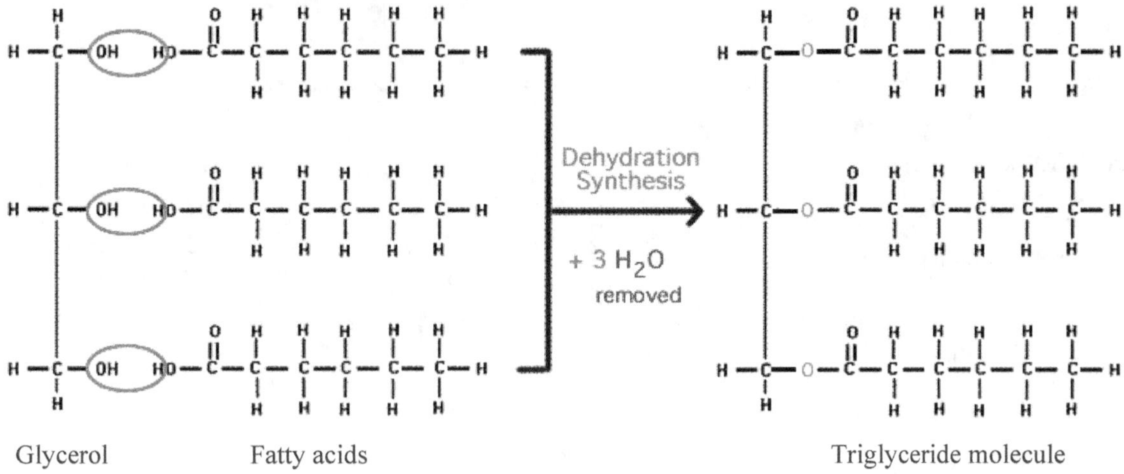

Dehydration (removal of H_2O) of glycerol and three fatty acids form triglycerides

Fatty acids are synthesized from one equivalent of a triol (glycerol and three equivalents of acid-containing groups) known as fatty acids.

42. A is correct.

Dietary triglycerides are composed of glycerol and three fatty acids.

The hydrolysis of triglycerides yields glycerol and three fatty acid chains.

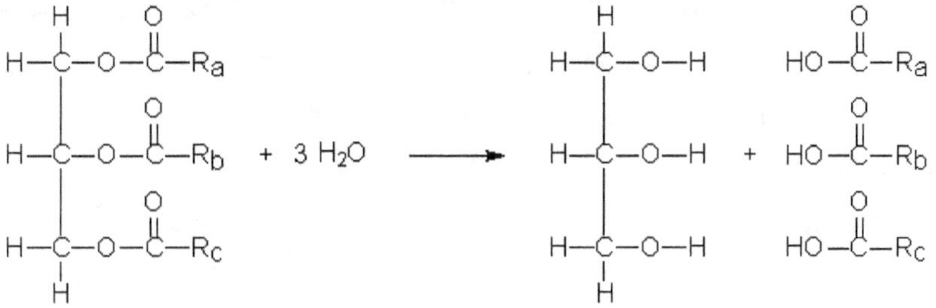

Hydrolysis of a triglyceride with three equivalents of water yields glycerol and three fatty acids

43. C is correct.

Phospholipids are essential lipids composing the bilayer structure of cell membranes, organelles, and other enclosed cellular structures.

Phospholipids are two (same or different) fatty acid molecules, a phosphate group and a 3-carbon glycerol backbone.

The phospholipid contains a hydrophobic (i.e., fatty acid tail) region and a hydrophilic (polar head) region.

Hydrophobic regions point toward each other in the membrane bilayer.

Hydrophilic polar heads point towards the inside (i.e., cytosolic) or outside (i.e., extracellular).

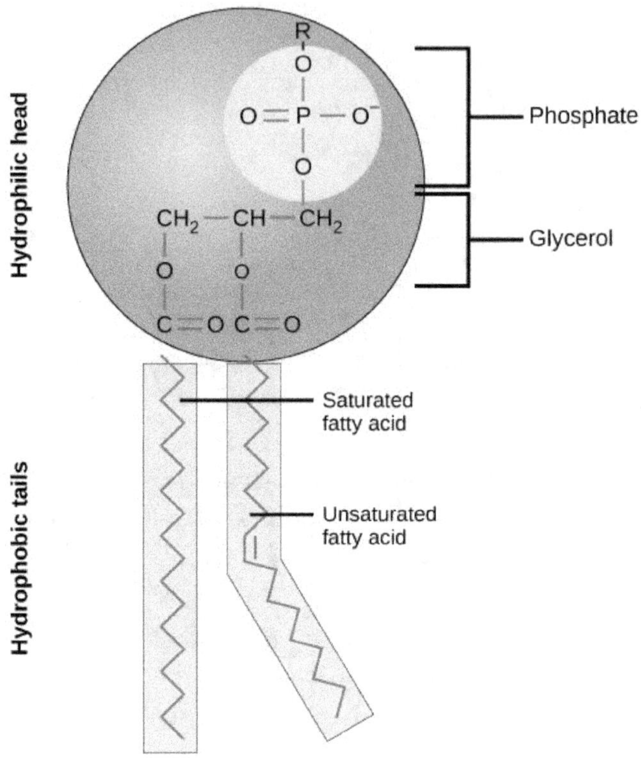

Phospholipids differ mainly in the composition of the polar head region. The hydrophobic region is the fatty acid chains (hydrocarbons) that are saturated (no double bonds) or unsaturated.

44. D is correct.

Fat can be synthesized from glucose, but glucose is *not* produced from animal fat.

Excess glucose consumption can lead to increased levels of fat in the body.

45. D is correct.

The two lipids categories are long-chain lipids (e.g., triglycerides) and smaller, polycyclic lipids, such as steroids (e.g., cholesterol and its derivatives such as estrogen and testosterone).

Terpenes are small alkene-containing hydrocarbon building blocks that can combine and cyclize to form steroids. Terpenes are simple lipids.

Examples of terpenes are shown:

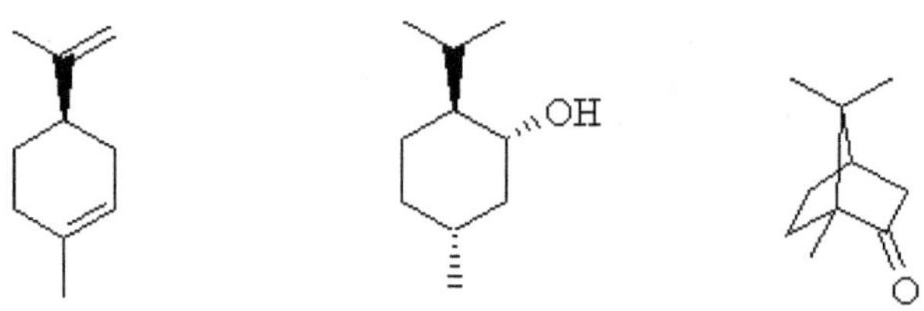

limonene menthol camphor

Limonene is in citrus fruit skins, menthol in peppermint, and camphor from camphor trees.

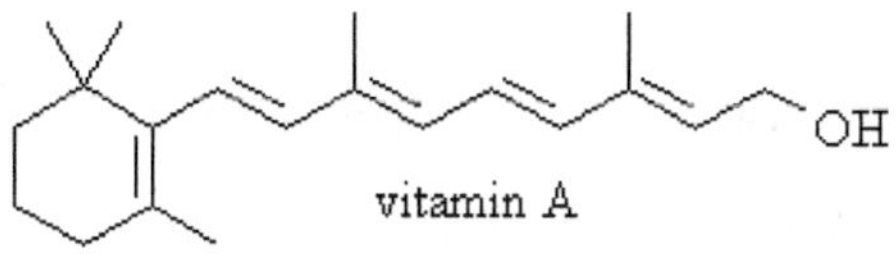

vitamin A

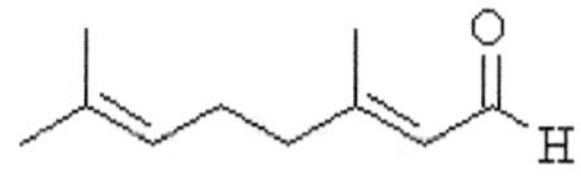

citral

(lemon grass)

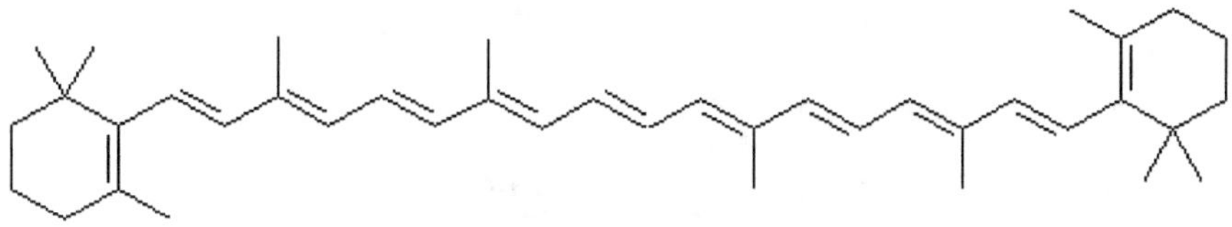

β carotene is a photopigment in carrots and vegetables converted to vitamin A

46. C is correct.

An exception to the tendency for sugars to have oxygen atoms linked to each carbon atom is deoxyribose (e.g., DNA). It is similar to a ribose (e.g., RNA) with one alcohol replaced with a carbon-hydrogen bond.

47. B is correct.

The ratio of atoms for sugars is typically 1:2:1 (carbon : hydrogen : oxygen).

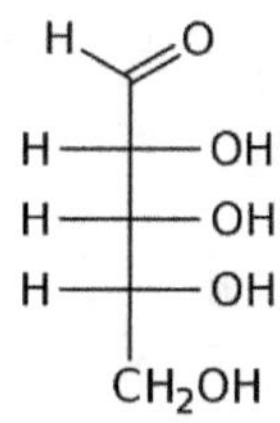

Fisher projections of an example D-ribose sugar with horizontal bonds projecting forward

48. A is correct.

Monosaccharides cannot be catabolized (i.e., broken down) into simpler sugar subunits.

Monosaccharides undergo oxidative degradation when treated with nitric acid to form carbon dioxide (CO_2) and carbon monoxide (CO).

49. A is correct.

Lactose is a disaccharide of glucose and galactose.

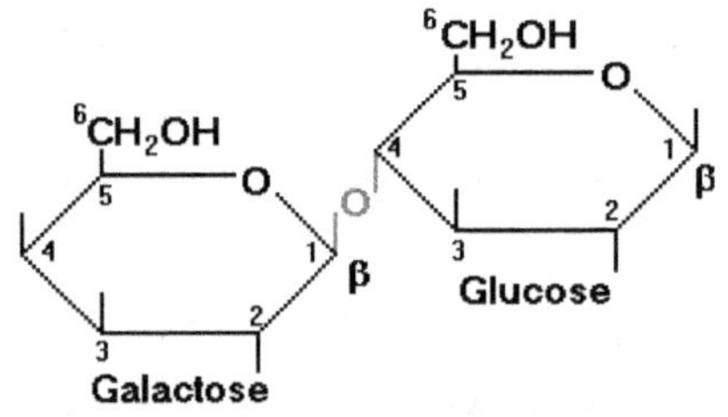

Lactose is a disaccharide of galactose and glucose

Lactose is a disaccharide formed by a β(1→4) linkage between galactose and glucose.

50. C is correct.

Glycogen is a polymer of glucose serving as an energy store of carbohydrates in animal cells (plants use starch), common in the liver, muscle, and red blood cells.

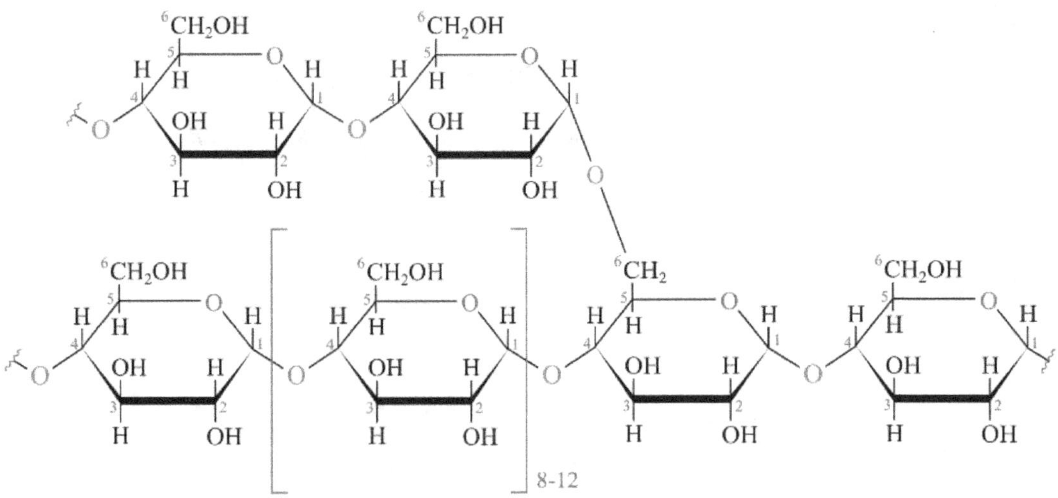

Glycogen is a large biomolecule consisting of repeating glucose subunits.
The anomeric position is carbon one and can be pointing upward (α) or down (β).

51. D is correct.

Disaccharides contain a glycosidic linkage that is an ether group.

The ether can be protonated with Brønsted acids and hydrolyzed in the presence of water.

Polysaccharides can be hydrolyzed to produce monosaccharides (i.e., individual monomers of the polymer).

52. D is correct.

Monosaccharides are the basic unit of carbohydrates.

Subjecting monosaccharides to acids or bases will not hydrolyze them; however, they undergo oxidative decomposition by treatment with periodic acid to form formaldehyde and formic acid.

53. D is correct.

Carbohydrates are organic compounds containing carbon, hydrogen, and oxygen.

The general molecular formula depends on carbohydrates, but many examples have the formula of $C_nH_{2n}O_n$.

54. C is correct.

The "di" prefix in the name indicates two smaller subunits.

Monosaccharides are linked by *glycosidic* (i.e., oxygen bonded to two ethers) functional groups.

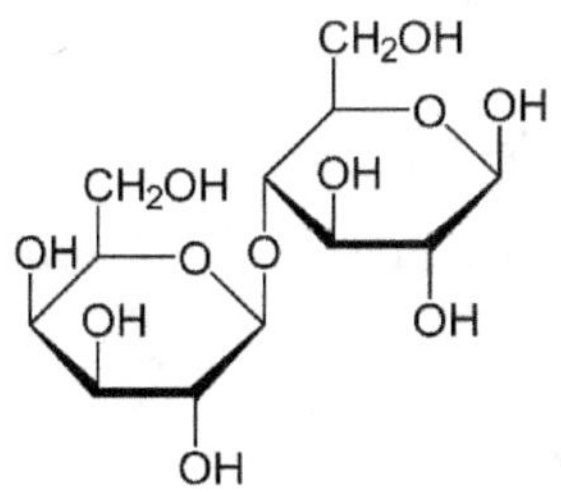

Lactose is a disaccharide of galactose and glucose

Lactose is a disaccharide formed by a β(1→4) linkage between galactose and glucose.

55. D is correct.

Proteins are denser than lipids, and lipoproteins with higher protein content are dense.

Lipoproteins with low-density transport mainly lipids and contain little protein (i.e., VLDL transport more lipids / less protein than LDL).

Chylomicrons are formed by the small intestine and have the lowest density and the highest lipid/lowest protein content of lipoproteins.

56. C is correct.

Lipids are the primary food storage molecule, and lipids release more energy per gram (9 kcal/gram) than carbohydrates (4 kcal/gram) or proteins (4 kcal/gram).

Lipids provide insulation and protection against injury as the major component of adipose (fat) tissue.

A: *proteins* are mainly composed of amino acids with C, H, O, and N but may contain S (i.e., cysteine).

B: α *helices* and β *pleated sheets* are secondary structures of proteins.

D: C:H:O ratio of carbohydrates is 1:2:1 ($C_nH_{2n}O_n$).

57. A is correct.

Carbohydrates (e.g., glucose, fructose, lactose, maltose) and proteins provide 4 calories per gram.

Fats (lipids) are energy-dense and provide 9 calories per gram.

58. B is correct.

The proper digestion of macromolecules is required to absorb nutrients from the small intestine.

Carbohydrates must be degraded into monosaccharides like glucose, fructose, and galactose.

Lactose is a disaccharide of glucose and galactose.

Sucrose is glucose and fructose.

Maltose is a disaccharide of two glucose units.

Disaccharide digestion into monomers occurs at the intestinal brush border of the small intestine via enzymes like lactase, sucrase, and maltase.

A: amino acids are the monomers of proteins.

Proteins must be hydrolyzed (catabolized) into monopeptides, dipeptides, or tripeptides for absorption in the duodenum of the small intestine.

D: lipids are degraded into free fatty acids and glycerol for absorption in the small intestine.

59. A is correct.

The addition of H_2O to break bonds (i.e., hydrolysis) is used during digestion when fats are catabolized (i.e., degraded) into fatty acids by detaching from glycerol or proteins that are digested (catabolized) into amino acids.

60. A is correct.

Dipole interactions joining complementary strands of DNA are *hydrogen bonds*.

These bonds form from the acid protons between the amides and imide functional groups and the carbonyl and amide Lewis basic sites of the matched nitrogen base pairs.

61. D is correct.

The DNA molecule (shown) has a deoxyribose sugar-phosphate backbone with bases (A, C, G, T) projecting into the center to join the antiparallel strand of DNA (i.e., double helix).

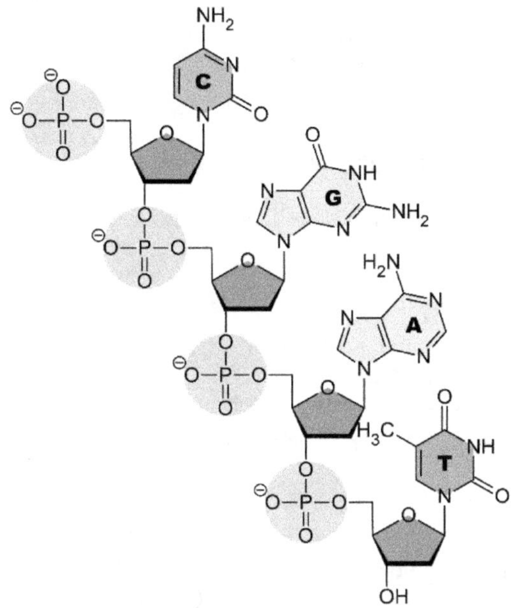

The deoxyribose sugar-phosphate backbone is negatively charged due to the formal charge of the oxygen attached to the phosphate group.

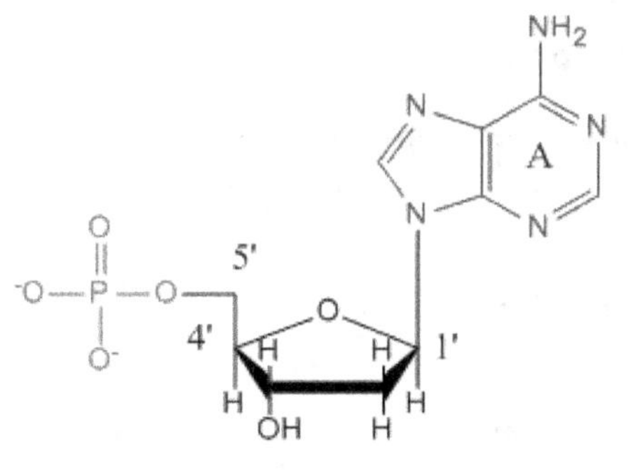

deoxyadenosine 5'-phosphate

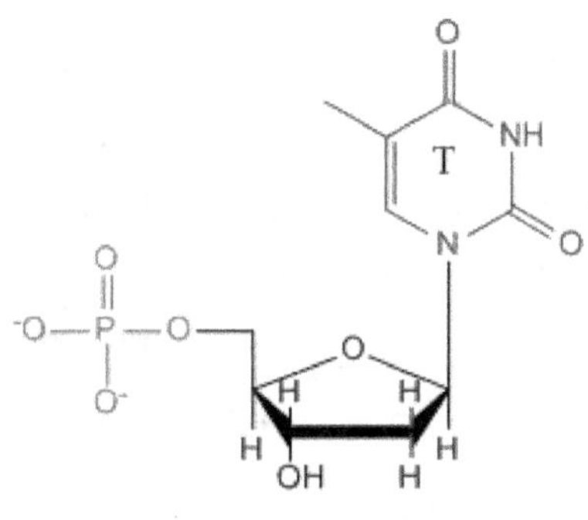

deoxythymidine 5'-phosphate

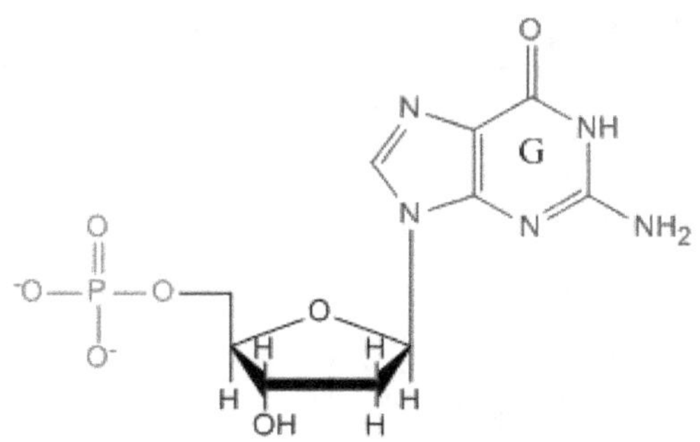

deoxyguanosine 5'-phosphate

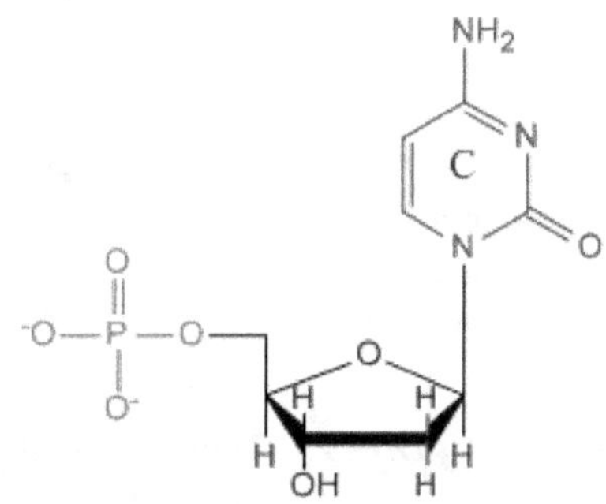

deoxycytosine 5'-phosphate

Nucleosides contain ribose sugar, phosphate, and base (A, C, G, T)

Purines (adenine and guanine) are double-ringed nitrogenous bases.

Pyrimidines (cytosine, thymine, and uracil-in RNA) are single-ringed.

*Nucleo**t**ides* have deoxyribose sugar, base, and phosphate.

*Nucleo**s**ides* have deoxyribose sugar and base without a phosphate group.

62. B is correct.

Transcription (DNA → RNA) is a biomolecular event in the nucleus.

During transcription, RNA molecules are synthesized using complementary base pairs of DNA single strands.

63. D is correct.

Amino acids are the monomers that comprise proteins.

Nucleotides have a nitrogenous base (i.e., adenosine, cytosine, guanine, thymine, or uracil), a phosphate group, and a five-carbon sugar (i.e., ribose for RNA or deoxyribose for DNA).

64. B is correct.

Four standard nucleotides are in DNA molecules: adenine (A), cytosine (C), guanine (G), and thymine (T)

Four standard nucleotides are in RNA molecules: adenine (A), cytosine (C), guanine (G), and uracil (U)

65. A is correct.

One strand is a template for a new strand synthesized at the replication fork during DNA replication.

The *continuously synthesized strand* is the *leading strand* and, when combined with one original (parental) strand of DNA, makes one new (daughter) DNA molecule. A new DNA molecule is created as the other parent strand is a template for its complementary strand.

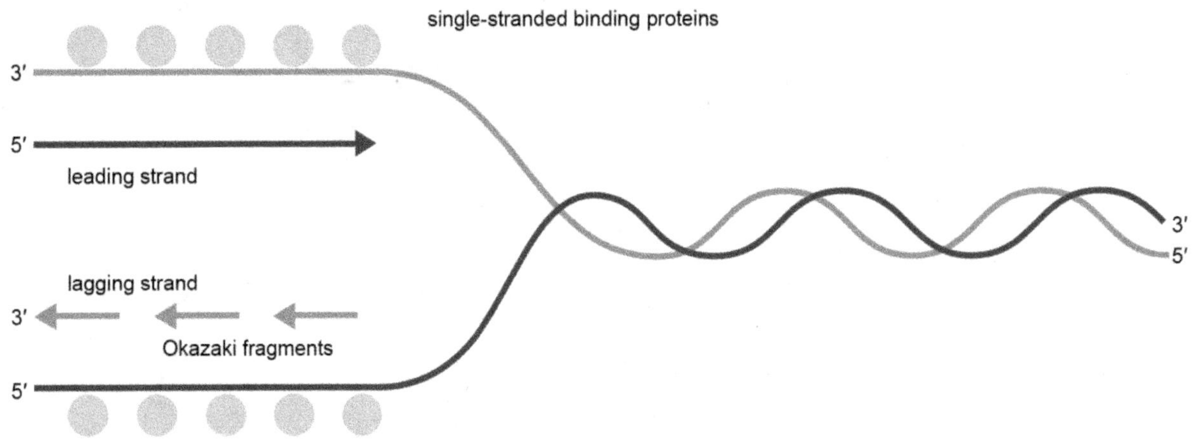

DNA replication with the leading and lagging strands indicated

Lagging strand is made of smaller segments called *Okazaki fragments* (about 150-200 nucleotides long).

Okazaki fragments are combined with DNA *ligase* (enzyme).

66. D is correct.

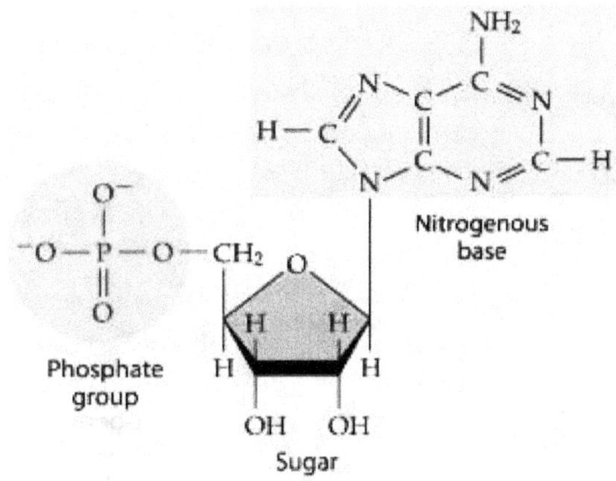

RNA molecules contain ribose as the carbohydrate component of the backbone

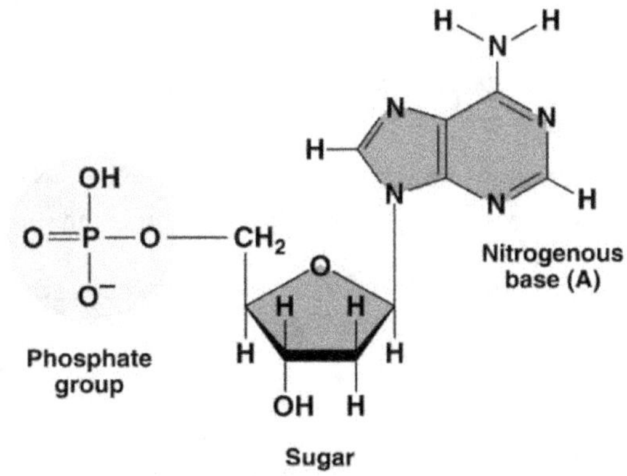

DNA contains deoxyribose (lacking a 2'-hydroxy) sugar as part of the backbone

67. D is correct.

Adenine forms two hydrogen bonds with thymine (A=T); cytosine forms three hydrogen bonds with guanine (C≡G).

A complementary nitrogen base is used by DNA polymerase (in the S phase of interphase) for strand synthesis.

68. C is correct.

DNA molecules hold genetic information.

RNA molecules are synthesized from the DNA strand to make proteins for the cell.

Genes are the sections of DNA responsible for synthesizing proteins in cells.

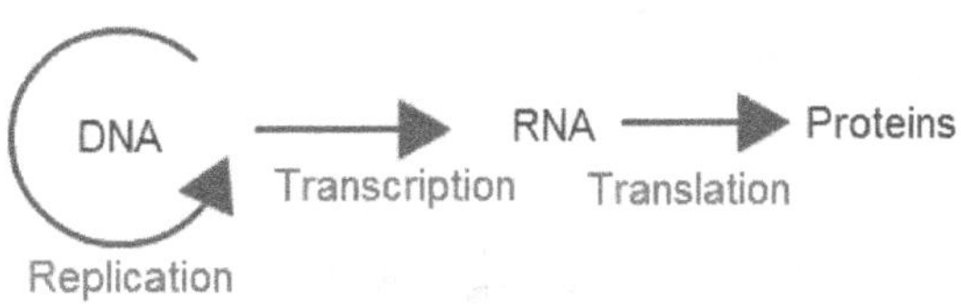

Central dogma of molecular biology designates information flow

69. D is correct.

There are five nucleotides, four of which appear in DNA.

These nucleotides are adenine (A), cytosine (C), guanine (G), and thymine (T).

In RNA molecules, thymine is replaced with the pyrimidine nucleotide of uracil.

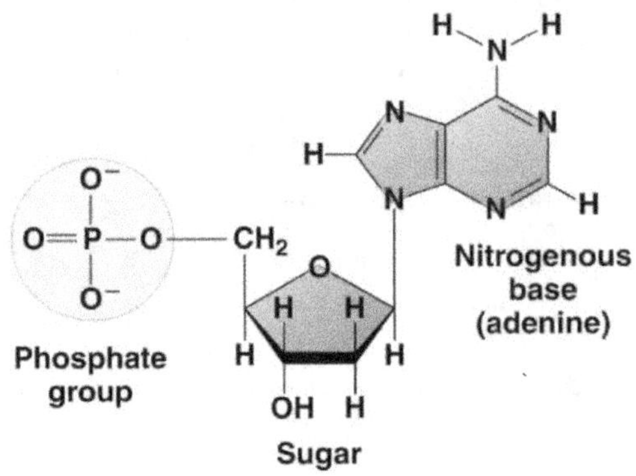

Nucleotides contain ribose sugar, phosphate, and base (adenine shown)

Nucleotides have a phosphate group (note the negative charge on oxygens), deoxyribose sugar (lack a 2'hydroxyl), and a nitrogenous base (adenine, cytosine, guanine, thymine, or uracil).

Nucleosides are nucleotides (sugar and base) with one or more phosphates added.

70. D is correct.

DNA	DNA	mRNA	tRNA
A	T	A	U
C	G	C	G
G	C	G	C
T	A	U	A

Complementary base pairing for nucleotides

DNA → DNA (replication); DNA → RNA (transcription); RNA → protein (translation)

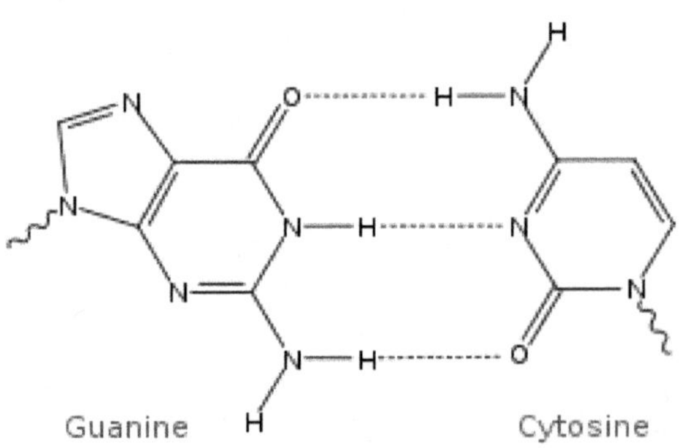

Guanine and cytosine with three hydrogen bonds

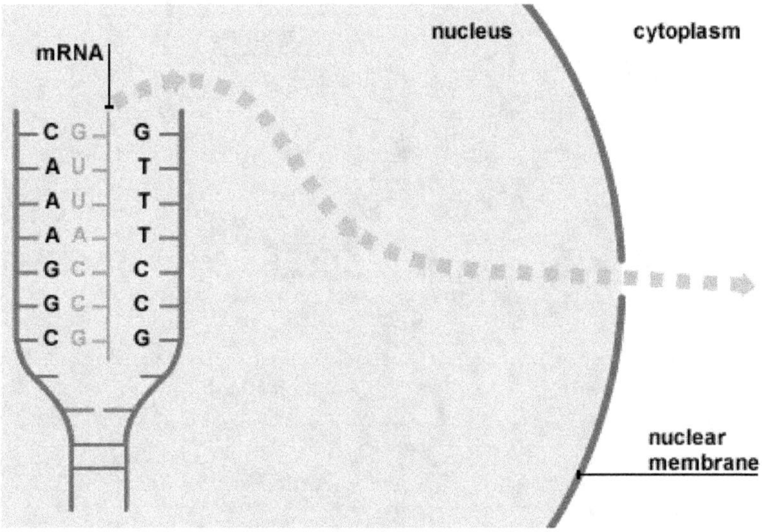

A single strand of DNA is the template for RNA synthesis during transcription

mRNA (after processing) is translocated to the cytoplasm for translation to proteins.

71. A is correct.

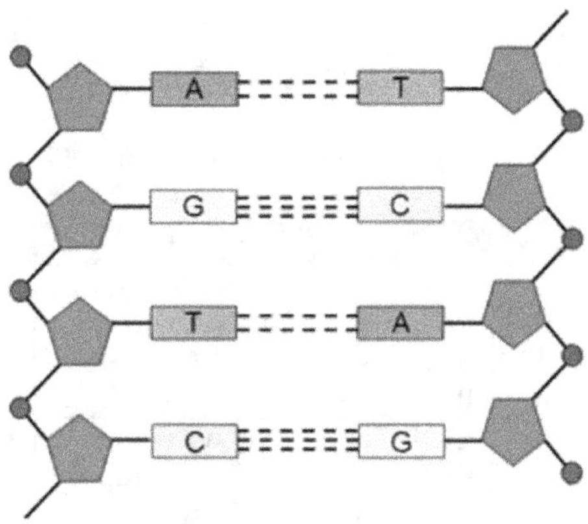

Double-stranded DNA with two hydrogen bonds between A=T and three between C≡G

DNA backbone is deoxyribose sugars (pentagons), and phosphates (circles) joined by covalent bonds.

72. D is correct.

Codons (schematically shown) are three-nucleotide sequences on the mRNA.

Anticodons are nucleotide sequences of three bases on the tRNA.

The codon-anticodon sequences hybridize by forming hydrogen bonds to the complementary base pair.

Relationship between codon (on mRNA), anticodon (on tRNA), and the resulting amino acid:

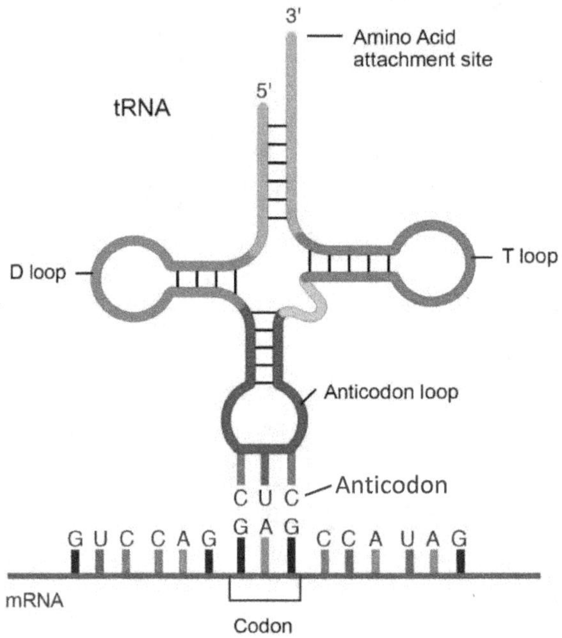

Three nucleotide codon/anticodon bonding with associated cloverleaf tRNA

Codon (mRNA): 5'–AUG–CAA–CCC–GAC–UCC–AGC–3'

Anticodon (tRNA): 3'–UAC–GUU–GGG–CUG–AGG–UAG–5'

Amino acids: Met–Gln–Pro–Asp–Phe–Ser

	U	C	A	G
U	UUU = phe	UCU = ser	UAU = tyr	UGU = cys
	UUC = phe	UCC = ser	UAC = tyr	UGC = cys
	UUA = leu	UCA = ser	UAA = stop	UGA = stop
	UUG = leu	UCG = ser	UAG = stop	UGG = trp
C	CUU = leu	CCU = pro	CAU = his	CGU = arg
	CUC = leu	CCC = pro	CAC = his	CGC = arg
	CUA = leu	CCA = pro	CAA = gln	CGA = arg
	CUG = leu	CCG = pro	CAG = gln	CGG = arg
A	AUU = ile	ACU = thr	AAU = asn	AGU = ser
	AUC = ile	ACC = thr	AAC = asn	AGC = ser
	AUA = ile	ACA = thr	AAA = lys	AGA = arg
	AUG = met	ACG = thr	AAG = lys	AGG = arg
G	GUU = val	GCU = ala	GAU = asp	GGU = gly
	GUC = val	GCC = ala	GAC = asp	GGC = gly
	GUA = val	GCA = ala	GAA = glu	GGA = gly
	GUG = val	GCG = ala	GAG = glu	GGG = gly

Genetic code of nucleotides and corresponding amino acids

Genetic code is the nucleotide sequence of the codon (on mRNA) that complementary base pairs with the nucleotide sequence of the anticodon (on tRNA).

Amino acid encoded by a gene (synthesized during translation into mRNA) derives from the *genetic code.*

73. A is correct.

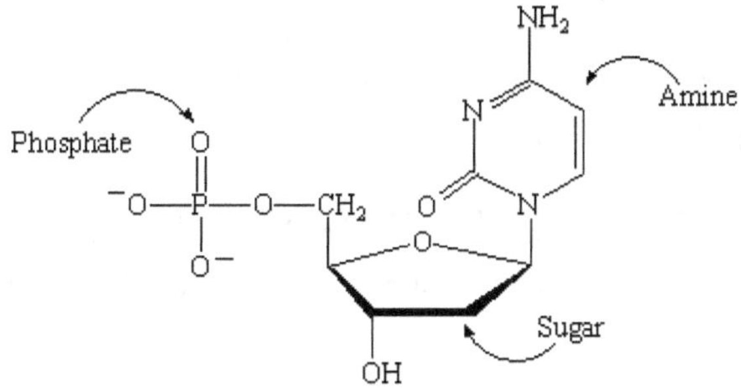

Nucleotide consisting of sugar, phosphate, and a nitrogenous base

The hexose sugar, nitrogen base, and phosphoric acid group comprise nucleotides, and these nucleotides are used to make larger molecules like *nucleic acids*.

74. D is correct.

Identical copies of DNA are necessary for the division of cells; these cells are *daughter cells*.

When the daughter strand is synthesized, a complementary nitrogenous base containing nucleotides (A ↔ T and C ↔ G) is incorporated into the growing strand.

Replication (i.e., synthesis of DNA) occurs during the S phase (i.e., within interphase) of the cell cycle.

75. B is correct.

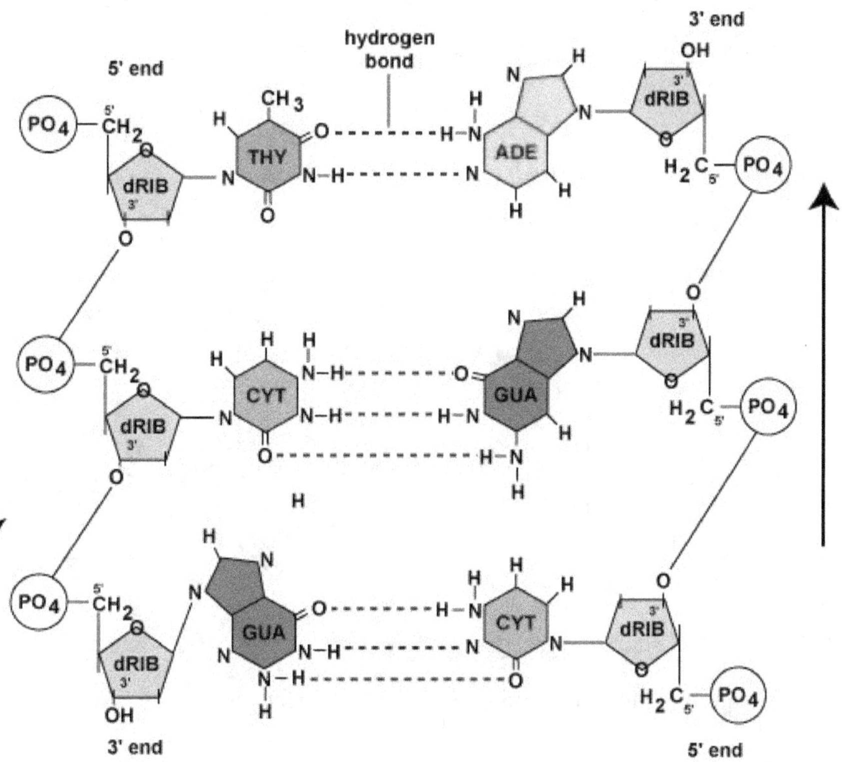

Hydrogen bonding between complementary strands of antiparallel DNA strands

Two antiparallel strands of DNA (schematically as vertical arrows) indicate the 3'-hydroxyl of the sugar (from the point of chain elongation).

The nucleotide is the building block of DNA and has sugar (i.e., deoxyribose), phosphate, and base (adenosine, cytosine, guanine, and thymine)

Two hydrogen bonds (A=T) hold the base pairs of adenine and thymine.

Cytosine and guanine nitrogen base pairs use three hydrogen bonds (C≡G) to support the nucleic acid structure.

76. B is correct.

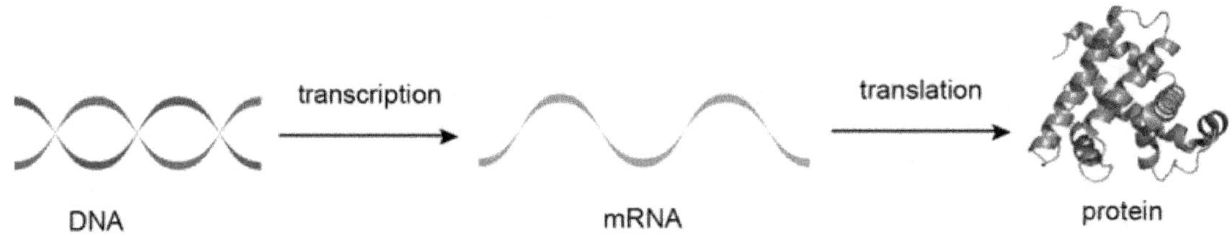

Central dogma of molecule biology: DNA → RNA → protein

tRNA and ribosomes are used to make peptide chains from mRNA.

Nucleotide triplets known as codons are combined with a complementary tRNA (i.e., containing the anticodon complementary in base pairing with the codon of the mRNA).

Each tRNA brings an appropriate (i.e., anticodon ↔ amino acid) to the growing polypeptide during translation.

Ribosomes combine *amino acid residue*s in the order of the codon sequence.

77. C is correct.

After transcription (DNA → mRNA), the synthesized mRNA is transported out of the nucleus and the ribosome to synthesize protein during translation (mRNA → protein).

RNA contains ribose, a sugar similar to deoxyribose of DNA molecules, except it has a hydroxyl group (~OH) at the 2' position on the sugar.

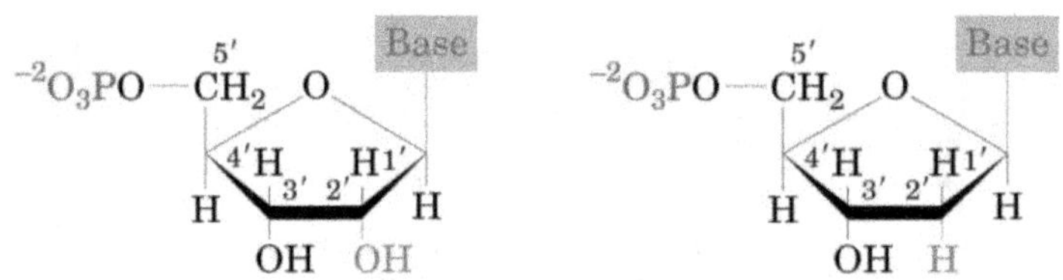

Ribonucleotide (left) and deoxyribonucleotide (right): note the 2' position of the sugars

Nucleic acids (DNA and RNA) bind via hydrogen bonds between bases (A=T/U and C≡G).

The sugar and phosphate groups form the backbone and do not affect (to any appreciable degree) the hydrogen bonding between bases.

continued…

A hydroxyl group (~OH) in RNA causes the backbone to experience steric and electrostatic repulsion.

Therefore, the grooves formed in helixes or hairpin loops between chains are larger.

Larger groove of DNA permits nucleases (i.e., enzymes that digest DNA or RNA) to efficiently bind to the RNA chain and digest the covalent bonds between the alternating sugar-phosphate monomers.

78. A is correct.

Nitrogen bases of nucleotides hydrogen bond one nucleic acid strand to the other.

The number of hydrogen bonds between the pairs varies between two (A=T) and three (C≡G) bonds.

Adenine and thymine nitrogenous base pairs bond with two hydrogen bonds (A=T).

Cytosine and guanine nitrogenous base pairs bond with three hydrogen bonds (C≡G).

79. A is correct.

Three nucleotides are combined to make a *codon* (on mRNA) or *anticodon* (on tRNA) required for translation.

A larger segment of DNA is a *gene*, and sections of different genes comprise nucleic acid strands.

Two single strands of DNA are used to make one DNA double helix.

80. B is correct.

Each thymine nitrogen base forms two hydrogen bonds with adenine in DNA molecules, and there is an equal number of nitrogen bases in the molecule.

81. D is correct.

Ribose differs from deoxyribose sugars because ribose lacks one alcohol (i.e., hydroxyl) group.

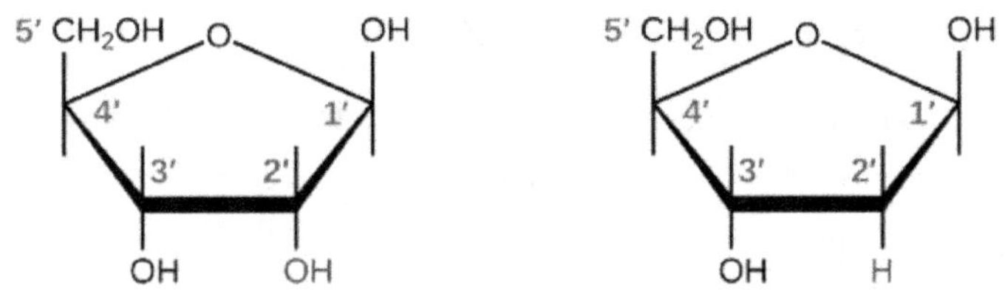

Ribose (left) and deoxyribose have a 3'-OH, but deoxyribose lacks a 2'-OH

Ribose sugar (RNA) and deoxyribose sugar (DNA) are used to synthesize nucleic acid polymers.

82. B is correct.

Replication is DNA → DNA during the S phase of the cell cycle.

A: *translation* is the process of synthesizing proteins from mRNA.

C: *transcription* is the process of synthesizing mRNA from DNA.

D: *complementation* is observed in genetics when two organisms with different homozygous recessive mutations that produce the same mutant phenotype (e.g., thorax differences in *Drosophila* flies), when mated or crossed, produce offspring with the wild-type phenotype.

Complementation only occurs if the mutations are in different genes.

Each organism's genome supplies the wild-type allele τo *complement* the mutated allele. Since the mutations are recessive, the offspring display the wild-type phenotype.

Complementation (i.e., *cis/trans*) test can test whether the mutations are in different genes.

83. D is correct.

Central dogma of molecular biology:

DNA → RNA → protein

DNA → RNA is transcription

RNA → protein is translation

DNA is the nucleic acid biomolecule that codes for nucleic acids and proteins.

DNA strands are synthesized from parental DNA strands during replication.

84. C is correct.

The peptide chain is assembled depending on the amino acid residue order, dictated by the mRNA sequence.

rRNA is the nucleic acid that comprises the ribosome used during translation (converting the codon into a corresponding amino acid in the growing polypeptide chains of the nascent protein).

Each *codon of RNA* has a corresponding *anticodon on tRNA*.

tRNA molecules have 3-nucleotide sequences of the anticodon and the appropriate amino acid corresponding to the anticodon at its 3' end.

The genetic code is the language for converting DNA (i.e., nucleotides) to proteins (i.e., amino acids).

DNA → mRNA → protein

DNA to RNA is transcription

mRNA to protein is translation

There are 20 naturally occurring amino acids with one start codon (i.e., methionine) and three stop codons (containing releasing factors that dissociate the ribosome).

85. D is correct.

RNA utilizes a uracil (U) nitrogenous base instead of thymine (T).

Therefore, thymine should not appear in the codon.

86. A is correct.

Thymine is a *pyrimidine nitrogenous base* that forms two hydrogen bonds with adenine (purine) in the base-paired structure of DNA.

In RNA molecules, the nitrogenous base thymine is replaced by uracil.

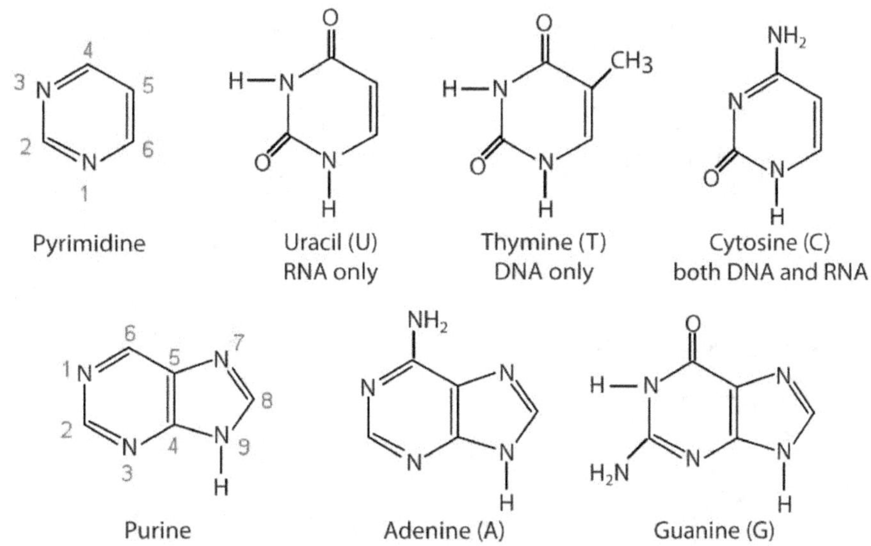

Purines (A, G) are single ring structures,
while pyrimidines (C, T, U) are double-ring structures

87. B is correct.

Three components of nucleotides are a phosphate group, cyclic five-carbon sugar, and a nitrogenous base.

Fat molecules are biomolecules in phospholipid membranes, storage fat molecules, and other lipid molecules.

88. A is correct.

There are three major nucleotide components in nucleic acids.

Nucleic acids have a nitrogen base used for hydrogen bonding, a hexose sugar (ribose or deoxyribose), and a phosphate group with a phosphate linkage with the sugar.

Ester linkages are in fats, *glycosidic linkages* in sugars, and *peptide linkages* in proteins.

89. B is correct.

The genetic code necessary for constructing peptide chains is the codon of three mRNA nucleotides complementary to the anticodon of tRNA molecules.

90. C is correct.

Because RNA contains uracil, these nitrogenous bases form hydrogen bonds with adenine.

Note that RNA molecules are single-stranded, and DNA molecules are double-stranded.

Adenosine (A) forms two hydrogen bonds with thymine (T) in DNA.

Cytosine (C) forms three hydrogen bonds with guanine (G) in DNA.

91. C is correct.

Thymine is a pyrimidine nucleotide base in DNA but not in RNA.

Instead, RNA has uracil.

92. C is correct.

In DNA, thymine hydrogen bonds with adenine.

In RNA, the thymine is exchanged for uracil.

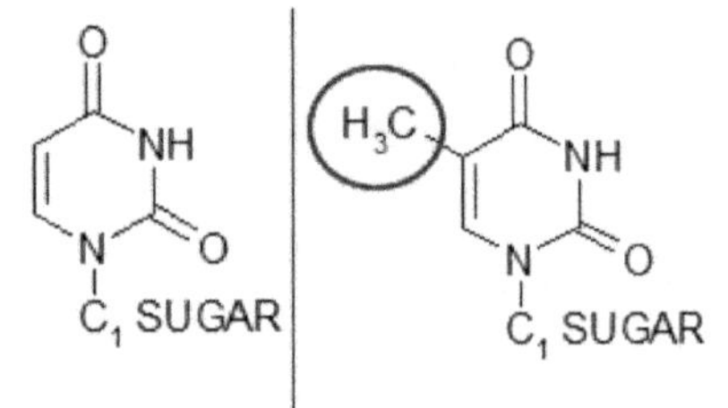

Uracil (RNA) Thymine (DNA)

93. C is correct.

Nucleic acids determine the sequences of amino acids because of groupings of nucleotides along a sequence corresponding to a particular amino acid.

The information of specific nucleic acids (i.e., mRNA) is translated on ribosomes by tRNA.

Prions are infectious, disease-causing agents of misfolded proteins.

94. C is correct.

DNA (deoxyribonucleic acid) is a long biological molecule composed of smaller nucleotides (i.e., sugar, phosphate, and base).

The sugar is deoxyribose (compared to ribose for RNA), and the bases are adenine, cytosine, guanine, and thymine (with thymine replaced by uracil in RNA).

The strands that these nucleotides are nucleic acids.

95. D is correct.

Ribose is the structural sugar of RNA, while deoxyribose is the sugar for DNA.

Uracil is a nucleotide (i.e., sugar, phosphate, and base) containing ribose sugar, similar to deoxyribose; however, one difference is that deoxyribose has one fewer alcohol group (2' position of the sugar) than ribose.

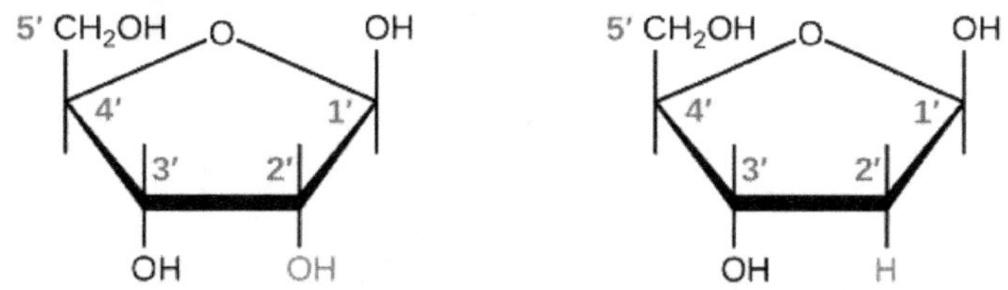

Ribose (left) and deoxyribose have a 3'-OH, but deoxyribose lacks a 2'-OH

Uracil is a *nucleotide base* of RNA and does not appear in DNA strands.

96. A is correct.

The two nucleic acid strands of DNA dissociate during transcription, and mRNA is synthesized (transcription) using the nitrogen bases as a template.

Therefore, the mRNA strand is a complementary strand to the DNA strand.

mRNA *exits the nucleus* to be used as a template to produce proteins (translation).

97. C is correct.

The bonds of every single strand of DNA are covalent.

The bonds linking the antiparallel strands of DNA are hydrogen bonds.

Two hydrogen bonds join adenine and thymine (A=T).

Three hydrogen bonds join cytosine and guanine (C≡G).

98. A is correct.

Protein function can be altered by *post-translational modifications* (often in Golgi) after synthesis.

99. D is correct.

Glucose and fructose have different chemical properties despite the same molecular formula

Notes for active learning

Chromosomes, Genes and DNA – Detailed Explanations

1. D is correct.

Histones are basic (i.e., positively charged) proteins associated with DNA within the nucleus to condense it into chromatin.

Nuclear DNA does not appear in free linear strands; instead, it is highly condensed and wrapped around histones to fit inside the nucleus and form chromosomes.

Three major types of RNA involved in gene expression:

1) messenger RNA (mRNA) molecules carry the coding sequences (i.e., "blueprints") for protein synthesis and are called transcripts;

2) ribosomal RNA (rRNA) forms the core of a cell's ribosomes (i.e., macromolecular cellular particles where protein synthesis takes place);

3) transfer RNA (tRNA) molecules transport amino acids (i.e., protein building blocks) to the ribosomes during protein synthesis.

2. C is correct.

A tumor suppressor gene is a gene that protects a cell from the path to cancer.

When this gene mutates to cause a loss (or reduction) in its function, the cell can progress to cancer, usually in combination with other genetic changes.

The loss of these tumor suppressor genes may be even more critical than proto-oncogene/activation to form many kinds of human cancer cells.

Apoptosis is the process of programmed cell death (PCD) that may occur in multicellular organisms.

Biochemical events lead to characteristic cell changes (morphology), including blebbing, cell shrinkage, nuclear fragmentation, chromatin condensation, chromosomal DNA fragmentation, and death.

In contrast to necrosis (i.e., traumatic cell death that results from acute cellular injury), apoptosis confers advantages during an organism's lifecycle.

For example, the separation of fingers and toes in a developing human embryo occurs because cells between the digits undergo apoptosis.

Unlike necrosis, apoptosis produces cell fragments called apoptotic bodies that phagocytic cells can engulf and quickly remove before the contents of the cell can spill out onto surrounding cells and cause damage.

A: telomerase is an enzyme that adds DNA sequence repeats (i.e., TTAGGG) to the 3' end of DNA strands in the telomere regions at the ends of eukaryotic chromosomes.

This region of repeated nucleotides called telomeres contains noncoding DNA and hinders the loss of essential DNA from chromosome ends.

When the chromosome is replicated during the S phase, 100–200 nucleotides are lost without damage to the coding region of the DNA.

Telomerase is a reverse transcriptase that carries its RNA molecule used as a template when it elongates telomeres shortened after each replication cycle.

Embryonic stem cells express telomerase, which allows them to divide repeatedly.

In adults, telomerase is highly expressed in cells that regularly divide (e.g., male germ cells, lymphocytes, and specific adult stem cells). In contrast, telomerase is not expressed in most adult somatic cells.

3. D is correct.

There are 4 different nucleotides (adenine, cytosine, guanine, and thymine/uracil).

Each codon is composed of 3 nucleotides, and therefore there must be 64 (4^3) possible variations of codons to encode for the 20 amino acids.

The genetic code is degenerate (or *redundant*) because several codons encode for the same amino acid.

Sixty-one codons encode for amino acids, and 3 stop codons terminate translation.

5. B is correct.

Chromosomes replicate during the synthesis (S) phase of interphase.

6. C is correct.

The replication fork opens by disrupting hydrogen bonds between complementary nucleotide base pairs (e.g., A bonded to T; C bonded to G).

Gyrase cuts one of the strands of the DNA backbone and relaxes the positive supercoil that accumulates as helicase separates the two strands of DNA.

Ligase seals the backbone of DNA (i.e., joins the Okazaki fragments) by forming phosphodiester bonds between deoxynucleotides in DNA.

7. B is correct.

The percent of adenine cannot be determined because RNA is a single-stranded molecule.

The base-pairing rules of DNA (i.e., Chargaff's rule) determining the possibility for double-stranded DNA do not apply to single-stranded RNA.

8. B is correct.

Adenosine (A) bonds with thymine (T) with two hydrogen bonds; guanine (G) bonds with cytosine (C) with three hydrogen bonds.

9. A is correct.

The addition of a 3' poly-A tail to mRNA (not proteins) is a *post-transcription* event.

The *post transcription* events include adding a 5' cap and splicing of exons (or removing introns) from the RNA molecule.

10. B is correct.

The *central dogma* of molecular biology refers to the direction of genetic information flow within a living system.

The *central dogma* states that DNA is transcribed into RNA, and then RNA is translated into protein:

DNA → RNA → protein

The central dogma of molecular biology was disrupted by the discovery of retroviruses that have RNA genomes using reverse transcriptase to make a copy of their RNA into DNA within the infected cell.

11. A is correct.

Heterochromatin is a tightly packed form of DNA.

Euchromatin refers to uncoiled regions of DNA actively engaged in gene expression because RNA polymerase binds to the relaxed region of DNA.

Heterochromatin mainly consists of genetically inactive satellite sequences (i.e., tandem repeats of noncoding DNA).

Centromeres and telomeres are heterochromatic (i.e., tightly coiled DNA), as is the Barr body of the second, inactivated X-chromosome in females.

12. D is correct.

The nucleolus is the organelle within the nucleus responsible for the synthesis of ribosomal RNA (rRNA).

A: the Golgi apparatus is the organelle responsible for processing, packaging, and distribution of proteins.

B: lysosomes are organelles with low pH that function to digest intracellular molecules.

C: mitochondrion is the organelle, bound by a double membrane, where the reactions of the Krebs (TCA) cycle, electron transport, and oxidative phosphorylation occur.

13. B is correct.

A polynucleotide (e.g., DNA) is the only macromolecule (i.e., nucleic acids, proteins, lipids, and carbohydrates) repaired rather than degraded. DNA repair is essential for maintaining cell function and is performed by biological repair systems (e.g., p53 tumor repressor protein).

Uncontrolled cell growth via disruption to the cell cycle regulation occurs if somatic cell mutations are not repaired.

14. B is correct.

A codon is a three-nucleotide segment of an mRNA molecule that hybridizes (via complementary base pairing) with the appropriate anticodon on the tRNA to encode for one amino acid in a polypeptide chain during protein synthesis.

The tRNA molecule interacts with the mRNA codon after the ribosomal complex binds to the mRNA.

A: tRNA molecule interacts with the mRNA codon after (not before) the mRNA is bound to the ribosomal complex.

C: translation involves the conversion of mRNA into protein.

D: operon regulates the transcription of genes into mRNA and is not involved in translating mRNA into proteins.

15. A is correct.

Ligase is an enzyme used by the cell during DNA replication (and other biochemical processes) that catalyzes the joining of two large molecules (e.g., DNA nucleotides) by forming a new chemical bond.

The newly formed bond is *via* a condensation reaction (joining) and usually involves dehydration with the loss of H_2O when the molecules (e.g., DNA, amino acids) are linked.

16. B is correct.

Gametes are formed via meiosis and are double-stranded haploids. A single chromosome consists of two hydrogen-bonded complementary DNA strands.

17. D is correct.

Genetic recombination is when two DNA strand molecules exchange genetic information (i.e., the base composition within the nucleotides), which results in new combinations of alleles (i.e., alternative forms of genes).

In eukaryotes, the natural process of genetic recombination during meiosis (i.e., the formation of gametes – eggs and sperm) results in genetic information passed to progeny.

Genetic recombination in eukaryotes involves pairing homologous chromosomes (i.e., a set of maternal and paternal chromosomes), which may involve nucleotide exchange between the chromosomes.

The information exchange may occur without physical exchange (a section of genetic material is copied – duplicated – without a change in the donating chromosome) or by the breaking and rejoining of the DNA strands (i.e., forming new molecules of DNA).

Mitosis may involve recombination, where two sister chromosomes form after DNA replication.

In general, new combinations of alleles are not produced because the sister chromosomes are usually identical.

For meiosis and mitosis, recombination occurs between similar DNA molecules (homologous chromosomes or sister chromatids, respectively).

In meiosis, non-sister (i.e., same parent) homologous chromosomes pair with each other, and recombination often occurs between non-sister homologs.

For both somatic cells (i.e., undergo mitosis) and gametes (i.e., undergo meiosis), recombination between homologous chromosomes or sister chromatids is a common DNA repair mechanism.

18. B is correct.

The pseudoautosomal regions get their name because genes within them are inherited like autosomal genes.

The function of these pseudoautosomal regions is that they allow the X and Y chromosomes to pair and properly segregate during meiosis in males.

Males have two copies of these genes: one in the pseudoautosomal region of their Y chromosome, the other in the corresponding portion of their X chromosome.

Typical females have two copies of pseudoautosomal genes, as each of their two X chromosomes contains a pseudoautosomal region.

Crossing over (during prophase I) between the X and Y chromosomes is usually restricted to the pseudoautosomal regions.

Thus, pseudoautosomal genes exhibit an autosomal, rather than sex-linked, pattern of inheritance.

Females can inherit an allele initially present on their father's Y chromosome, and males can inherit an allele initially present on the X chromosome of their father.

19. B is correct.

A primary spermatocyte has completed the synthesis (S) phase of interphase but has not completed the first meiotic division and is still diploid (2N) with 46 chromosomes (i.e., 23 pairs).

20. D is correct.

Klinefelter syndrome describes the set of symptoms resulting from additional X genetic material in males.

A: Turner syndrome describes the condition in females resulting from a single X (monosomy X) chromosome.

B: XYY syndrome is a genetic condition in which a human male has an extra male (Y) chromosome, giving him a total of 47 chromosomes instead of 46. 47, XYY is not inherited but usually occurs during the formation of sperm cells.

A nondisjunction error during anaphase II (of meiosis II) results in sperm cells with an extra copy of the Y chromosome.

If one of these atypical sperm cells contributes to the genetic makeup, the child will have an extra Y chromosome in each somatic cell.

C: Triple X syndrome is not inherited but usually occurs during the formation of gametes (e.g., ovum and sperm) because of nondisjunction in cell division that results in reproductive cells with additional chromosomes.

An error in cell division (nondisjunction) can result in gametes with additional chromosomes.

An egg or a sperm may gain an extra X chromosome as a result of nondisjunction, and if one of these gametes contributes to the genetic makeup of the zygote, the child will have an extra X chromosome in each cell.

21. D is correct.

DNA replication occurs during the synthesis (S) phase of interphase (i.e., G1, S, G2) to form sister chromatids joined by the centromere.

Replication occurs during the S phase of the cell cycle.

However, transcription (in the nucleus) and translation (in the cytoplasm) still occur during the S phase.

22. C is correct.

X-linked dominance is a mode of inheritance whereby a dominant gene is carried on the X chromosome.

X-linked dominant is less common than X-linked recessive.

For X-linked dominant inheritance, one copy of the allele is sufficient to cause the disorder inherited from a parent who has the disorder.

X-linked dominant traits do not necessarily affect males more than females (unlike X-linked recessive traits).

An affected father will give rise to all affected daughters, but no affected sons will be affected (unless the mother is also affected).

23. B is correct.

Meiosis is a particular process of cell division that occurs in sexually reproducing eukaryotes (animals, plants, and fungi), whereby the chromosome number is reduced by half, resulting in four genetically distinct haploid daughters.

In meiosis, DNA replication is followed by two rounds of cell division to produce four cells.

The two rounds of this meiotic division are Meiosis I and Meiosis II.

Meiosis I is known as *reductive division*, as the cells are reduced diploid to haploid.

Meiosis II (only G phase occurs) is known as *equational division*, as the cells begin and end as haploid cells.

24. D is correct.

Point mutations occur when a single nucleotide base (A, C, G, T) is substituted by another.

A silent mutation is a point mutation that either 1) occurs in a noncoding region or 2) does not change the amino acid sequence due to the degeneracy of the genetic code.

A frameshift mutation is either an insertion or deletion of some nucleotides.

These mutations have severe effects on the coded protein since nucleotides are series of triplets.

The addition or loss of nucleotides (except in multiples of three) changes the reading frame of the mRNA and often gives rise to premature polypeptide termination (i.e., nonsense mutation).

Even without a consequential frameshift mutation, insertions or deletions are often severe.

A missense mutation results from the insertion of a single nucleotide that changes the amino acid sequence of the specified polypeptide.

25. C is correct.

The DNA damage checkpoint is a signal transduction pathway that blocks cell cycle progression in G1, G2, and metaphase and slows down the rate of S phase progression when DNA is damaged.

It leads to a pause in the cell cycle, allowing the cell time to repair the damage before dividing.

Cell cycle progression:

G1 → S → G2 → prophase → metaphase → anaphase → telophase

Interphase (G1 → S → G2); Mitosis (PMAT) (prophase → metaphase → anaphase → telophase)

26. A is correct.

G-C base pairs are linked in the double helix by three hydrogen bonds.

Two hydrogen bonds join A-T base pairs.

Therefore, it takes more energy to separate G-C base pairs.

The less G-C rich a piece of double-stranded DNA is, the less energy is required to separate (i.e., denature) the two strands of the double helix.

Because of complementary base pairing, double-stranded DNA has equal quantities of G and C (and of A and T).

This is *Chargaff's rule*.

27. C is correct.

During meiosis I, homologous chromosomes separate. During meiosis II, sister chromatids (identical copies, except for recombination) separate.

In Klinefelter syndrome (XXY karyotype), sperm to cause the defect must contain an X and Y chromosome.

X and Y would be "homologous chromosomes" and would typically separate during meiosis I.

Failure to do so could create a sperm containing both an X and a Y, which causes Klinefelter syndrome.

Anaphase is when the centromere splits, and the homologous chromosomes / sister chromatids are drawn away (via spindle fibers) from each other toward opposite sides of the two cells.

The separation of homologous chromosomes occurs during anaphase I, while the separation of sister chromatids occurs during anaphase II.

Turner's syndrome is observed in females due to the single X karyotype (single X chromosome and no Y).

28. B is correct.

Only human gametes formed during meiosis are cells with a single copy (1N) of the genome.

Cells have a single unreplicated copy (i.e., devoid of a sister chromatid) after the second meiotic division.

29. A is correct.

During transcription, the two nucleic acid strands of DNA dissociate, and mRNA is synthesized (i.e., transcription) using the nitrogen bases of DNA as a template.

Therefore, the mRNA strand is a complementary strand to the DNA strand.

The mRNA then exits the nucleus to be used as a template for producing proteins (i.e., translation).

30. A is correct.

When DNA is replicated in the cell, one of the strands of DNA acts as a template for a new strand synthesized as the replication fork opens.

This continuously synthesized strand is known as the leading strand and, when combined with one of the old strands of DNA, makes up one new DNA molecule.

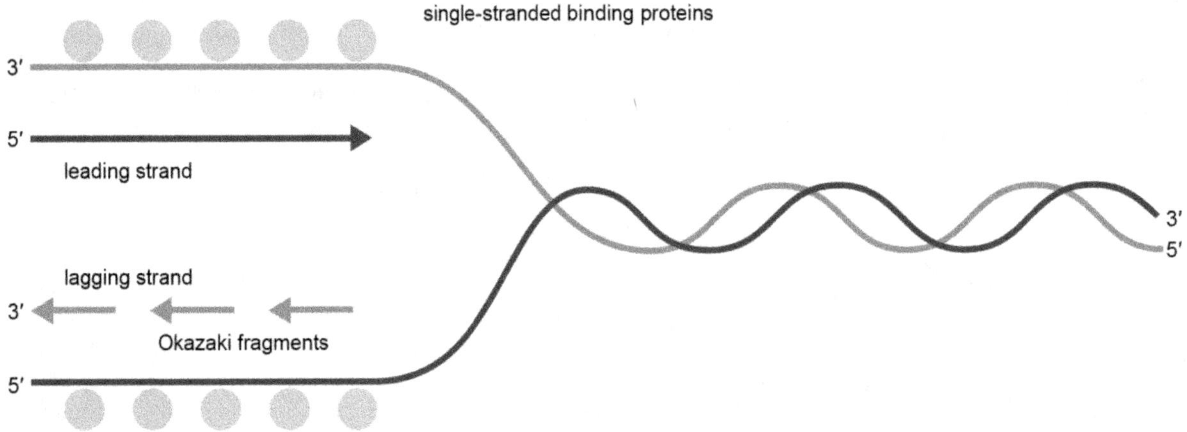

Furthermore, an additional DNA molecule is created in the process, as the other parent strand acts as a template for its complementary strand.

This lagging strand is made up of smaller segments called Okazaki fragments (about 150-200 nucleotides long).

Okazaki fragments are combined with the enzyme DNA ligase.

31. C is correct.

DNA molecules hold the genetic information of organisms.

RNA molecules are synthesized from the DNA strand to make proteins for the cell.

Genes are the sections of DNA responsible for the synthesis of proteins in cells.

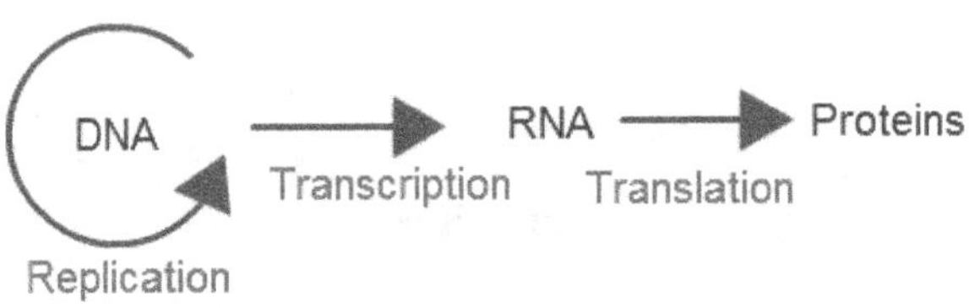

Central dogma of molecular biology (designates information flow)

32. D is correct.

There are five known nucleotides, four of which appear in DNA.

These nucleotides are cytosine, guanine, adenine, and thymine.

In RNA molecules, the thymine is replaced with another pyrimidine nucleotide known as uracil.

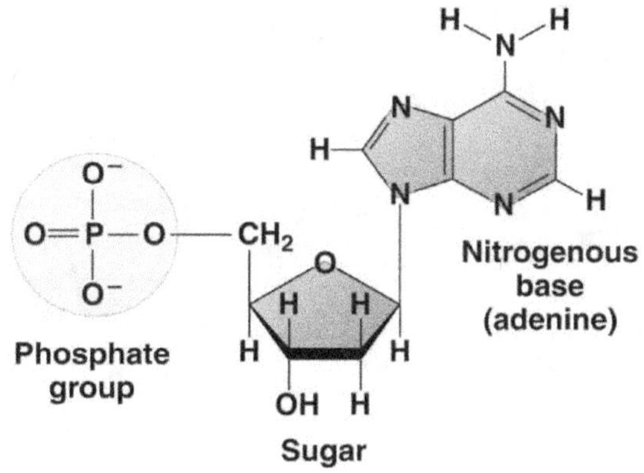

Nucleotides consisting of a phosphate group (note the negative charge on oxygens), deoxyribose sugar (lack a 2'hydroxyl) and a nitrogenous base (adenine, cytosine, guanine or thymine).

33. D is correct.

DNA	DNA	mRNA	tRNA
A	T	A	U
C	G	C	G
G	C	G	C
T	A	U	A

Complementary base pairing for nucleotides

DNA → DNA (replication); DNA → RNA (transcription); RNA → protein (translation)

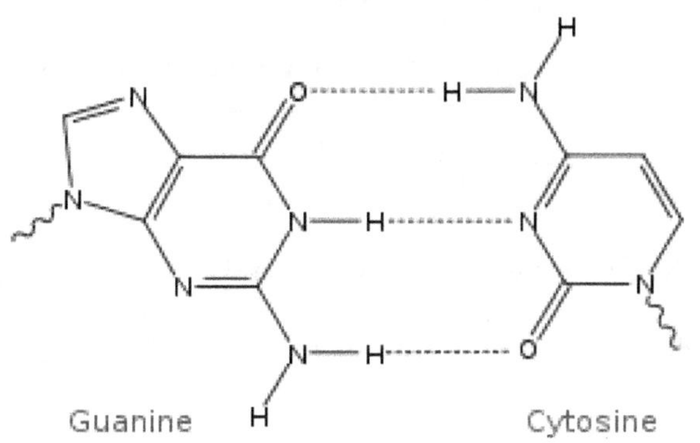

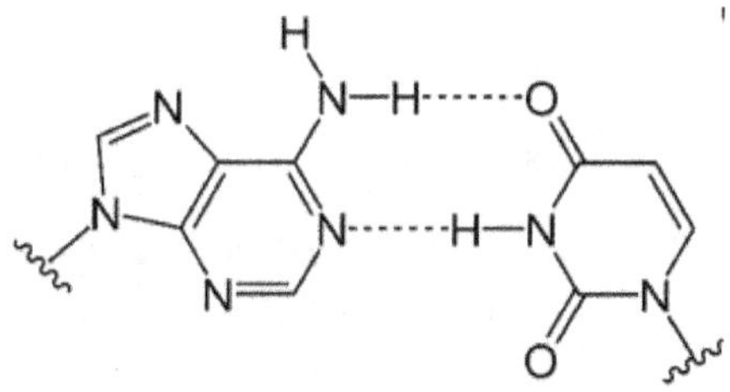

Adenosine (left) and uracil (replaces thymine in RNA)

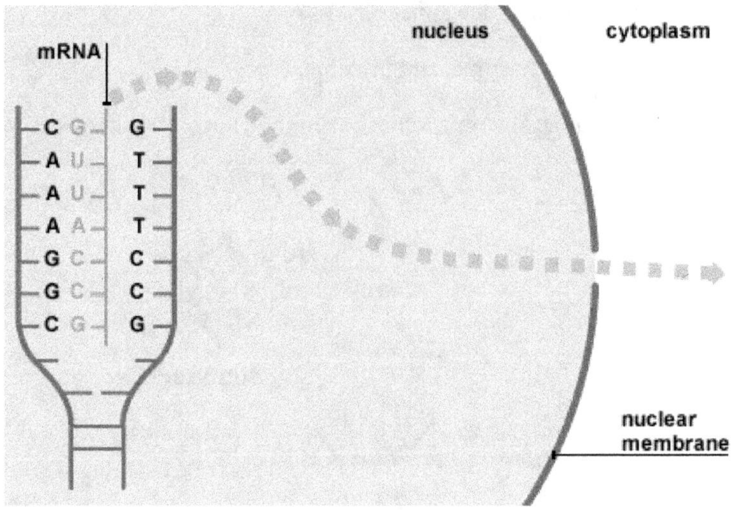

A single strand of DNA is the template for RNA synthesis during transcription

The mRNA (after processing) is translocated to the cytoplasm for translation to proteins.

34. A is correct.

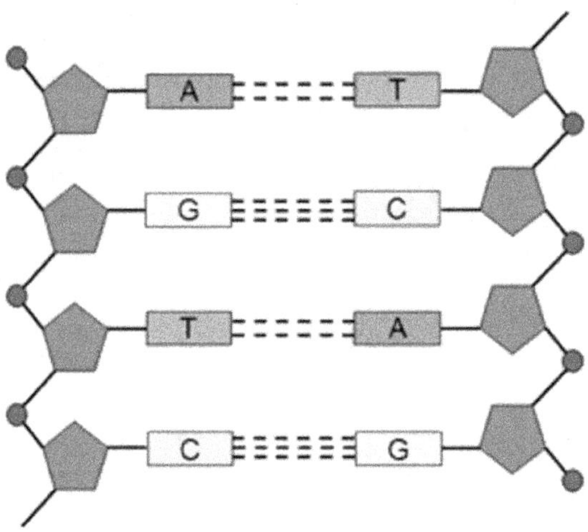

Double-stranded DNA with two hydrogen bonds between A=T and three hydrogen bonds between C≡G.

The backbone is comprised of deoxyribose sugar and phosphates (shown as circles between the pentose sugars).

35. D is correct.

Identical copies of DNA are necessary for cell division; these cells are known as daughter cells.

When the daughter strand is synthesized, a complementary nitrogenous base containing nucleotides (A↔T and C↔G) is incorporated into the growing strand.

Replication (i.e., synthesis of DNA) occurs during the S phase (i.e., within interphase) of the cell cycle.

36. B is correct.

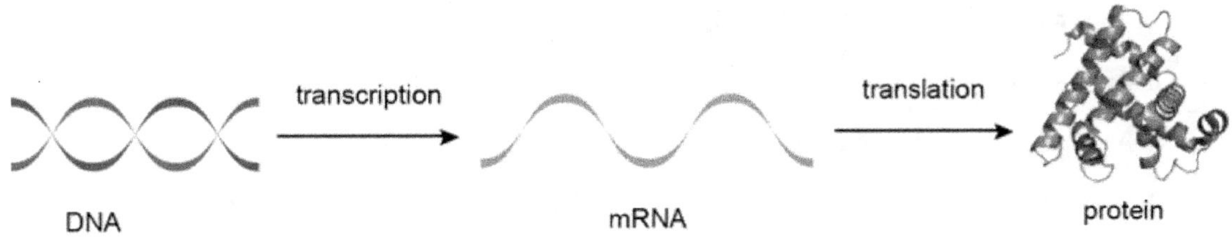

Central dogma of molecule biology: DNA → RNA → protein

tRNA and ribosomes are used to make peptide chains from mRNA.

Nucleotide triplets known as codons are combined with a complementary tRNA (i.e., containing the anticodon that is complementary in base pairing with the codon of the mRNA).

Each tRNA brings an appropriate (i.e., anticodon ↔ amino acid) to the growing polypeptide during translation.

The amino acid residues are combined in the order of the codon sequence.

37. A is correct.

Three nucleotides are combined to make a codon or anticodon required for translation.

A larger segment of DNA is called a gene, and sections of different genes make up nucleic acid strands.

Two single strands of DNA are used to make one DNA double helix.

38. B is correct.

Replication:

DNA → DNA during the S phase of the cell cycle.

A: translation is the process of synthesizing proteins from mRNA.

C: transcription is the process of synthesizing mRNA from DNA.

D: complementation is observed in genetics when two organisms with different homozygous recessive mutations produce the same mutant phenotype (e.g., thorax differences in *Drosophila* flies), when mated or crossed, produce offspring with the wild-type phenotype.

Complementation occurs if the mutations are in different genes.

Each organism's genome supplies the wild-type allele as the *complement* of the mutated allele of the other. Since the mutations are recessive, the offspring display the wild-type phenotype.

Complementation (i.e., *cis*/*trans*) test determines whether the mutations are in different (*trans*) genes.

39. D is correct.

Central dogma of molecular biology: DNA → RNA → protein

DNA → RNA is known as transcription.

RNA → protein is known as translation.

DNA is the nucleic acid biomolecule that can give rise to other nucleic acids and proteins.

DNA strands are synthesized from parental DNA strands during replication.

40. C is correct.

The peptide chain is assembled depending on the amino acid residue order, dictated by the mRNA sequence.

rRNA is the nucleic acid that comprises the ribosome used during translation (conversion of the codon into a corresponding amino acid in the growing polypeptide chains of the nascent protein).

Each codon of RNA has a corresponding anticodon located on the tRNA.

tRNA molecules have the 3-nucleotide sequence of the anticodon and the appropriate amino acid at its 3' end corresponding to the anticodon.

The genetic code is the language for converting DNA (i.e., nucleotides) to proteins (i.e., amino acids).

DNA → mRNA → protein

DNA to mRNA is transcription.

mRNA to protein is translation.

There are 20 naturally occurring amino acids.

There is one start codon (i.e., methionine) and three stop codons (containing releasing factors that dissociate the ribosome).

41. C is correct.

DNA (deoxyribonucleic acid) is a long biological molecule composed of smaller units called nucleotides (i.e., sugar, phosphate, and base).

The sugar is deoxyribose (compared to ribose for RNA), and the bases are either adenine, cytosine, guanine, and thymine (with uracil replacing thymine in RNA).

The strands that these nucleotides make up are called nucleic acids.

42. B is correct.

Western blotting separates proteins by electrophoresis and is commonly used to identify the presence of HIV antibodies (proteins).

All blotting techniques rely on gel electrophoresis to separate DNA, RNA, or proteins based on size.

After resolution by electrophoresis, the gel (containing the resolved products) is placed next to the blotting paper (e.g., nitrocellulose).

After the transfer of the macromolecules (i.e., DNA, RNA, or proteins) onto the gel by capillary action, the blotting paper is probed by specific markers that hybridize to regions with complementary sequences on the blotting paper.

A: Eastern blotting does not exist.

C: Northern blotting uses RNA in gel electrophoresis.

D: Southern blotting uses DNA in gel electrophoresis.

43. B is correct.

PCR requires knowledge of the sequence at the ends of the fragment being amplified.

From the ends, complementary base pairing is used to design the primers that anneal to the target fragment and permit amplification.

No knowledge of the region between the ends is required because the parent strands are the templates used by the DNA polymerase.

44. C is correct.

To show that DNA, rather than RNA, protein, or some other cell component, was responsible for transformation, Avery, McLeod, and McCarty used several biochemical tests.

Proteases and ribonucleases (i.e., degradation enzymes for protein and RNA) did not affect the transforming agent that caused disease.

However, a DNase treatment did degrade the DNA (genetic material) and destroyed the extract's ability to cause disease.

Streptococcus pneumoniae (i.e., pneumococcus) is a Gram-positive pathogenic bacterium.

S. pneumoniae was recognized as a significant cause of pneumonia in the late 19th-century had been the subject of humoral immunity (i.e., antibody-mediated) studies.

45. D is correct.

Hershey and Chase showed that when bacteriophages (i.e., viruses), composed of DNA and protein, infect bacteria, their DNA enters the host bacterial cell while their protein does not.

They grew separate populations of viruses and incorporated either radioactive sulfur (^{35}S to label protein) or phosphorus (^{32}P to label DNA) into the bacteriophages.

The viral ^{35}S protein coats remained outside the bacteria while the ^{32}P DNA entered the bacteria.

The experiment demonstrated that the DNA, not protein, was the *transforming* molecule that entered the bacteria from viral infection.

46. C is correct.

p53 is a *tumor suppressor protein* crucial for multicellular organisms, where it regulates the cell cycle and prevents cancer.

p53 is *the guardian of the genome* because of its role in conserving genetic stability by preventing genome mutations. The name p53 refers to its apparent molecular mass of 53 kDa.

Dalton is defined as 1/12 the mass of carbon and is a unit of a convention for expressing (i.e., often in kilodaltons or kDa) the molecular mass of proteins.

Notes for active learning

Notes for active learning

Heredity – Detailed Explanations

1. B is correct.

Only human gametes formed during meiosis have a single copy (1N) of the genome.

Cells have a single unreplicated copy (i.e., devoid of a sister chromatid) after the second meiotic division.

2. A is correct.

Centrioles are the organizational sites for microtubules (i.e., spindle fibers) that assemble during cell division (e.g., mitosis and meiosis).

The four phases of mitosis are prophase, metaphase, anaphase, and telophase, followed by cytokinesis.

Cytokinesis physically divides the cell into two identical daughter cells.

The condensed (heterochromatin) chromosomes are aligned along the equatorial plane in mitotic metaphase before the centromere (i.e., heterochromatin region on the DNA) splits.

The two sister chromosomes migrate to the respective poles of the cell.

3. B is correct.

For a recessive trait to appear in a phenotype of an offspring (e.g., long hair), offspring inherit a recessive allele (Mendel called it *traits*) from each parent (i.e., two copies of the recessive gene).

The short-haired parent carries one copy (i.e., heterozygous) of the recessive long-haired allele (or *gene*) passed to some offspring.

Combined with the second copy from the other long-haired parent, it produced the long-haired offspring.

4. C is correct.

Eye color is sex-linked in *Drosophila.*

Determine the phenotype of the parents.

A red-eyed fly with red-eyed and sepia-eyed parents must be heterozygous because a sepia-eyed parent only contributes to the recessive sepia allele.

When the heterozygous (Rr) red-eyed fly is crossed with a homozygous recessive (rr) sepia-eyed fly, ½ the offspring are red-eyed (Rr) because of the dominant (red) allele from the heterozygous fly.

The Punnett square is:

Red eyed parent

Sepia eyed Parent	R	r
r	Rr (red)	rr (sepia)
r	Rr (red)	rr (sepia)

Since the question does not assign the gender to the sepia and red-eyed parents, the Punnett squares for two possible combinations for sex-linked traits are:

		Red-eyed female (♀) R	Red-eyed female (♀) r
Sepia eyed male (♂)	r	Rr (red)	rr (sepia)
	y	Ry (red)	ry (sepia)

		Red-eyed male (♂) R	Red-eyed male (♂) y
Sepia eyed female (♀)	r	Rr (red)	ry (sepia)
	r	Rr (red)	ry (sepia)

5. B is correct.

Hardy-Weinberg law states that the gene ratios (P + Q =1) and the allelic frequencies ($p^2 + 2pq + q^2 = 1$) remain constant between generations in a population that is not evolving.

The Hardy-Weinberg law requires five criteria: 1) a large population, 2) random mating, 3) no gene flow, 4) no mutations, and 5) no selection.

These conditions are required for the gene frequency to remain constant.

If all of the criteria are met, the gene frequencies remain constant.

If one or more of these criteria are not met, the gene frequencies (and allele frequencies) change and evolution occurs.

6. C is correct.

The cross of spherical-seeded and wrinkled-seeded pea plants in Mendel's experiment inherited alleles (or *gene variants*) from both parents.

However, only spherical-seeded plants resulted from the cross.

The wrinkled-seed gene is a recessive allele compared to the spherical-seed (i.e., dominant gene).

A dominant gene is a gene variant, which for many reasons, is expressed more strongly than other variants (or *alleles*) of the gene (i.e., recessive) present.

7. B is correct.

Notation 2N indicates that a given cell line is diploid, two homologous versions of each chromosome.

Human somatic cells are diploid and have 23 different chromosome pairs (N = 23) for 46 chromosomes (2N = 46).

Gamete cells are haploid (i.e., 1N).

Mitosis is the mode of cell division used by somatic cells, which results in two diploid daughter cells genetically identical to the diploid parent cell.

8. D is correct.

The degree of genetic linkage measures the physical distance of two genes that are on the same chromosome.

The probability of a crossover and corresponding exchange between gene loci (location on the chromosome) is generally directly proportional to the distance between the loci.

Pairs of genes that are far apart from each other on a chromosome have a higher probability of being separated during crossover than genes physically close.

Thus, the frequency of genetic recombination between two genes is related to the distance between them.

Recombination frequencies can be used to construct a genetic map.

One map unit (Morgan units) is defined as a 1 percent recombinant frequency.

Recombination frequencies are roughly additive but merely an approximation for small percentages.

The percentage of recombinants cannot exceed 50%, which results when the two genes are at the opposite ends of the same chromosome.

In this situation, a crossover event results in an exchange of genes.

However, only an odd number of crossover events (a 50/50 chance between an even and odd number of crossover events) results in a recombinant product.

9. D is correct.

From Mendel's law of independent assortment, the probability of a cross resulting in a particular genotype is equal to the *product* of the individual probabilities.

This cross involves two heterozygous individuals for the three genes (A, B, and C) to produce an offspring that is homozygous dominant for each trait.

Both parents are heterozygous for A, genotype = Aa.

The ratio of their offspring = 1/4 AA, 1/2 Aa, and 1/4 aa; typical 1:2:1 ratio for heterozygous crosses.

Both parents are heterozygous for genes B and C; the probability of their offspring being BB is 1/4, and CC is 1/4.

Therefore, the probability that a particular offspring will have the genotype AABBCC.

AABBCC = the product of the individual probabilities: $1/4 \times 1/4 \times 1/4 = 1/64$.

10. B is correct.

The basic laws of inheritance explain the patterns of disease transmission.

The inheritance patterns of single-gene diseases are referred to as "Mendelian" after Gregor Mendel, who first observed the different patterns of gene segregation for specific traits of garden peas.

Mendel calculated probabilities of trait recurrence in the next generations.

Most genes have one or more versions (alleles) because of mutations or polymorphisms.

Individuals can carry a normal, mutant or rare allele, depending on the impact of the mutation/polymorphism and the frequency of the allele within a population.

Single-gene diseases are usually inherited in one of several patterns depending on the location of the gene and whether one or two normal copies of the gene are needed for the disease to manifest (i.e., affected individual).

The expression of the mutated allele is characterized as dominant, co-dominant, or recessive.

The five basic patterns of inheritance for single-gene diseases:

Autosomal dominant:

every affected person has an affected parent

manifests in every generation

Autosomal recessive:

both parents of an affected person are carriers (i.e., unaffected)

typically, NOT seen in every generation

X-linked dominant:

females affected more frequently

can affect males and females in the same generation

X-linked recessive:

males affected more frequently

often affects males in each generation

Mitochondrial:

both males and females can be affected but passed by females

can appear in every generation

When a family is affected by a disease, accurate family history is essential to determine a pattern of inheritance.

11. D is correct.

Let T = tall and t = short; B = brown eyes and b = blue eyes.

The father is homozygous tall and homozygous blue-eyed with a genotype of TTbb.

The mother is heterozygous tall and heterozygous brown-eyed with a genotype of TtBb.

Determine the probability that these parents could produce a tall child with blue eyes (T_bb).

The genes for height and eye color are unlinked.

The father (TTbb) contributes T and b alleles, so his gametes have both T and b alleles.

The mother (TtBb) contributes T or t and B or b, so her gametes are (in equal amounts): TB, tB, Tb, or tb.

Possible genotypes of the offspring: TTBb, TTbb, TtBb, Ttbb.

Half the offspring are tall and brown-eyed (T_B_), and half are tall and blue-eyed (T_bb).

Therefore, the probability of a tall child with blue eyes is ½.

A faster method is calculating phenotype ratios for height and eye-color separately and then combining them.

The mating of TT × Tt = 100% tall.

The mating of Bb × Bb = ½ blue and ½ brown.

Multiplying 1 tall × ½ blue = ½ tall blue.

12. A is correct.

Recombination is the exchange of genetic information between homologous chromosomes during prophase I.

In meiosis, crossing over occurs between non-sister homologs and results in a new combination of alleles (e.g., AB / ab can yield Ab / aB).

Recombination has been reported in eukaryotes during mitosis, but between sister chromatids, which are copies (replicated during the S phase), do not lead to novel genotypes.

Recombination is a DNA repair mechanism between homologous chromosomes.

B: recombination frequency would be the same for *cis*- and *trans*-heterozygotes.

The distance would be the same between the genes regardless of whether they are on the same chromosome (i.e., *cis*) or homologous chromosomes (*trans*).

Recombination frequency is not a completely random event.

Specific regions within the chromosome have differences in the propensity to undergo recombination; the presence of hotspots and architectural features increase or decrease recombination frequencies.

C: recombination frequency increases (not decreases) with distance.

D: different genes have different distances from each other.

Therefore, with different distances, the recombination frequency changes.

13. C is correct.

There are two possible alleles for each of the three genes.

If the genes assort independently (not linked), then there are $2^3 = 8$ possible combinations.

14. A is correct.

In the *autosomal dominant* inheritance pattern, every affected person has an affected parent.

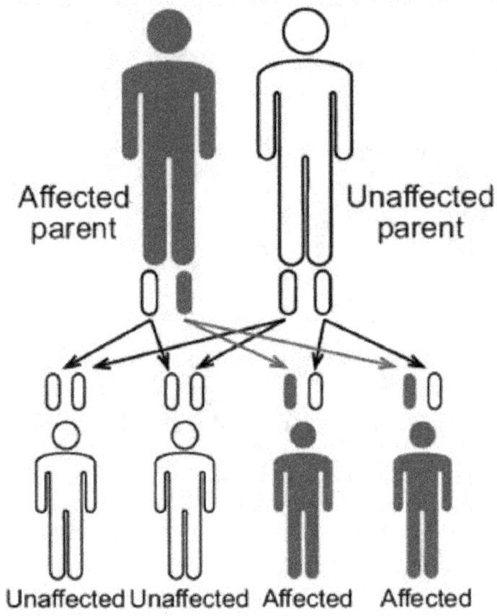

15. C is correct.

Parents are Aa × Aa (carriers but not afflicted with the disease).

The children (i.e., progeny) could be AA, Aa, Aa, or aa.

From the Punnett square, eliminate aa because this is the disease state.

The question asks what the chance is she is heterozygous (Aa) but not homozygous (AA) = 2/3

16. D is correct.

Afflicted children are either AA, Aa or Aa (but not aa), representing 3 of 4 possibilities.

Probability = ¾ or 75%.

	A	a
A	AA	Aa
a	Aa	aa

17. D is correct.

Autosomal dominant is the most common type of inheritance.

For autosomal dominant traits, a person needs a single copy of the mutant gene to exhibit the disease.

Usually, equal numbers of males and females are affected, traits do not skip generations, and father-to-son transmission is observed.

18. C is correct.

In males, spermatogonia are (2N) cells that undergo mitosis to produce (2N) cells called 1° spermatocytes, which undergo the first round of meiosis to yield 2° spermatocytes (1N).

The 2° spermatocytes undergo another round of meiosis to produce four spermatids (1N) that mature into sperm (i.e., male gamete).

In females, the 1° oocyte is (2N), whereby the 2° oocyte undergoes the second meiotic division to produce two (1N) cells – a mature oocyte (i.e., ovum – female gamete) and another polar body.

A: primary oocyte is a diploid cell. During fertilization, a (1N) ovum and a (1N) sperm fuse to produce a (2N) zygote.

B: spermatogonium is a diploid (2N) cell.

D: zygote is a fertilized egg.

19. D is correct.

The probability that a child will be male (or female) is ½.

Each event is independent; the first (or second or third) child is a particular gender that does not affect future events.

Therefore, the probability for 4 children is calculated as $\frac{1}{2} \times \frac{1}{2} \times \frac{1}{2} \times \frac{1}{2} = 1/16$.

The same would be the probability that the first and third child is female, with the second and fourth as male (or other combinations).

20. C is correct.

Let C designate the wild type and c designate the colorblind allele.

Mother is Cc, and father is CY (normal allele with a single copy of X gene).

From mating, the mother's gamete (as a carrier due to her dad) can be either C or c with a 50% probability of the gamete inheriting either the C or c allele.

Assuming the child is a boy (i.e., father transmits Y and not X), so the father's allele of Y (boy) is 100%, and the probability of the mother passing a c (colorblind) is 50%.

The probability of the son being colorblind is 50% or ½.

If the question had asked, "what is the probability that they will have a colorblind child?" the analysis changes to determine the probability among all children (not just among boys in the original question).

The same gametes are produced by the mother: C and c with 50% probability each.

The affected child is a boy.

What is the probability that the father passes the X or Y gene to the offspring; the probability is 50%.

The individual probabilities are multiplied to determine the overall probability: $\frac{1}{2} \times \frac{1}{2} = \frac{1}{4}$

21. D is correct.

Color blindness is a sex-linked trait because the gene is on the X chromosome.

The mother is a carrier (not afflicted with the condition) and is heterozygous for the recessive allele (color blindness).

The father has the allele on his X chromosome (the Y chromosome lacks the gene).

The genotype and phenotype of a son depend entirely on the mother (afflicted vs. carrier) since the afflicted father transmits the gene on his X.

Since the mother is heterozygous, a son would have a 50% chance of receiving the color blindness allele from her.

22. D is correct.

Loss of heterozygosity is a chromosomal event that results in loss of the entire gene and the surrounding chromosomal region.

Most diploid cells (e.g., human somatic cells) contain two copies of the genome, one from each parent.

Each copy contains approximately 3 billion bases, and for most positions in the genome, the base present is consistent between individuals.

However, a small percentage may contain different bases.

These positions are called single nucleotide polymorphisms (or SNP).

When the genomic copies from each parent have different bases for these regions, the region is heterozygous.

Most of the chromosomes within somatic cells of individuals are paired, allowing SNP locations to be potentially heterozygous.

However, one parental copy of a region can sometimes be lost, which results in the region having one copy.

If the copy lost contained the dominant allele, the remaining recessive allele appears as the phenotype.

23. A is correct.

A *gene* is a fundamental physical and functional unit of heredity transferred from a parent to offspring and determines some characteristics of the offspring.

Genes are DNA sequences encoding for proteins.

Alleles are forms of the same gene with slight differences in their sequence of DNA bases.

A *genome* is a complete set of genetic information of an organism.

A genome contains the genetic information needed for an organism and allows it to develop, grow and reproduce.

24. A is correct.

The desired phenotype is green smooth peas, and both green and smooth are dominant phenotypes.

Therefore, the genotypes selected for the cross must avoid the two recessive alleles (g and s).

For GgSs × GGSS, one parent (GGSS) is a double dominant, and therefore all offspring have the dominant phenotype (G and S) regardless of the genotype of the other parent.

B: Gg × gg yields 1/2 yellow (g) phenotype offspring.

C: ss × Ss yields 1/2 wrinkled (s) phenotype offspring.

D: Gg × Gg yields 1/4 yellow (g) phenotype offspring.

25. D is correct.

True breeding means that the organism is homozygous (e.g., AA or aa) for the trait.

All progeny are heterozygous Aa (shown below) and exhibit the dominant phenotype.

	A	A
a	Aa	Aa
a	Aa	Aa

26. C is correct.

Microscopy is the laboratory technique of magnifying objects that cannot be seen with the unaided eye.

Gregor Mendel performed crossbreeding experiments with pea plants to study inheritance patterns and introduced the terms *dominant* and *recessive* for phenotypic traits (or *alleles*).

27. A is correct.

Autosomal recessive inheritance is the product of mating two carriers (i.e., heterozygous parents).

In the mating of two heterozygotes for an autosomal recessive gene, there is a:

1) 25% (1/4) probability of a homozygous unaffected child

2) 25% (1/4) probability of a homozygous affected child

3) 50% (1/2) probability of a heterozygous (carrier) child

Overall, 75% of children are phenotypically normal (25% AA and 50% Aa).

Out of all children, 50% are phenotypically normal but carry the mutant gene (Aa).

28. A is correct.

Maternal inheritance involves all the progeny exhibiting the phenotype of the female parent.

B: not maternal inheritance because the progeny exhibit the phenotype of only the male parent.

C and D: normal Mendelian 1:1 segregation and not maternal inheritance.

Maternal inheritance is a type of uniparental inheritance by which all progeny have the genotype and phenotype of the female parent.

29. B is correct.

A recessive trait is expressed only when present in both copies or is the only copy of the gene present.

A recessive X-linked allele would only be expressed in unaffected women with two (homozygous) copies of the allele. If one X is present, then all recessive (single copy) alleles on the X are expressed (e.g., hemophilia).

The presence of a Y chromosome (in humans) confers maleness on the organism.

30. A is correct.

A model organism for genetic studies (e.g., pea plants, Drosophila, zebrafish) has common features. Genetic studies rely upon statistics that favor large sample numbers. Therefore, a short generation time (i.e., the span between birth and fecundity) is preferred. The organism must be bred in large numbers, and ease of cultivation favors viable offspring that transmit genetic information between generations.

There should be discreet phenotypic differences among alleles (alternative forms of the gene). For example, among the seven traits that Mendel observed, he inventoried tall *vs.* short plants, round *vs.* smooth seeds, and green *vs.* yellow seeds. Mendel was able to control the crosses by manually transferring pollen from the anther of a mature pea plant of one variety to the stigma of a separate mature pea plant of the second variety.

The organism should have a genome well characterized (gene identity and function have been studied). Increasing the number of chromosomes increases the genes that can influence the observable phenotypic outcomes. An organism with fewer chromosomes makes statistical analysis direct and supports causation when the genome is manipulated.

Notes for active learning

Notes for active learning

Cardiovascular System – Detailed Explanations

1. B is correct.

The capillaries are the most extensive capillary network of blood vessels, having the greatest cross-sectional area and blood flow resistance.

Blood pressure is highest in the aorta and drops until the blood returns to the heart via the inferior and superior venae cavae.

2. A is correct.

Ventricles are the heart's pumping chambers, while the atria are collecting chambers.

Oxygenated blood returns from the lungs (i.e., pulmonary circulation) and feeds into the heart's left atrium.

From the left atrium, the oxygenated blood flows into the left ventricle that pumps it through the systemic (i.e., body) circulation.

The right atrium receives deoxygenated blood via the inferior and superior vena cava.

The blood is pumped from the right ventricle to the lungs.

3. C is correct.

At the arteriole end of the capillary bed, the hydrostatic pressure is approximately 30-36 mmHg, while the opposing osmotic pressure is approximately 25-27 mmHg.

The higher hydrostatic pressure forces fluid out of the capillaries and into the interstitial space.

At the venule end of the capillary bed, the osmotic pressure is higher than the hydrostatic pressure, which has dropped to 12-15 mmHg.

The higher osmotic pressure draws fluid into the capillaries from the interstitial space and permits the liquid to return to the heart.

Most of the fluid (blood) is forced from the capillaries to the interstitial space at the arteriole end (via hydrostatic pressure). The capillaries reabsorb it at the venule end (via osmotic pressure).

4. A is correct.

Veins are thin-walled inelastic vessels that usually transport deoxygenated blood (pulmonary vein is the exception) *toward* the heart.

Veins do not have a steady pulse because of the fluid exchange that occurs in the capillaries. Because of their low pressure, blood flow along veins depends on compression by skeletal muscles rather than on the elastic, smooth muscles that line the arteries.

B: arteries transport blood *away* from the heart and are thick-walled, muscular elastic circulatory vessels.

Blood in the arteries is usually oxygenated. The pulmonary arteries are the exception whereby deoxygenated blood from the heart, returning from the body, is pumped to the lungs.

C: mammals have a four-chambered heart.

D: lymphatic system is an open circulatory system that transports excess interstitial fluid (lymph) to the cardiovascular system and maintains constant fluid levels.

Lymph enters the bloodstream at the thoracic duct that connects to the superior vena cava.

5. D is correct.

Blood enters the heart through the inferior and superior vena cava.

The right atrium receives the oxygen-poor blood and contracts, bringing the blood to the right ventricle.

When the ventricle contracts, blood enters the pulmonary artery and lungs, where it is oxygenated.

The pulmonary vein empties the blood into the left atrium, after which it flows into the left ventricle.

Finally, the ventricle contracts and the blood leaves through the aortic valve into the aorta.

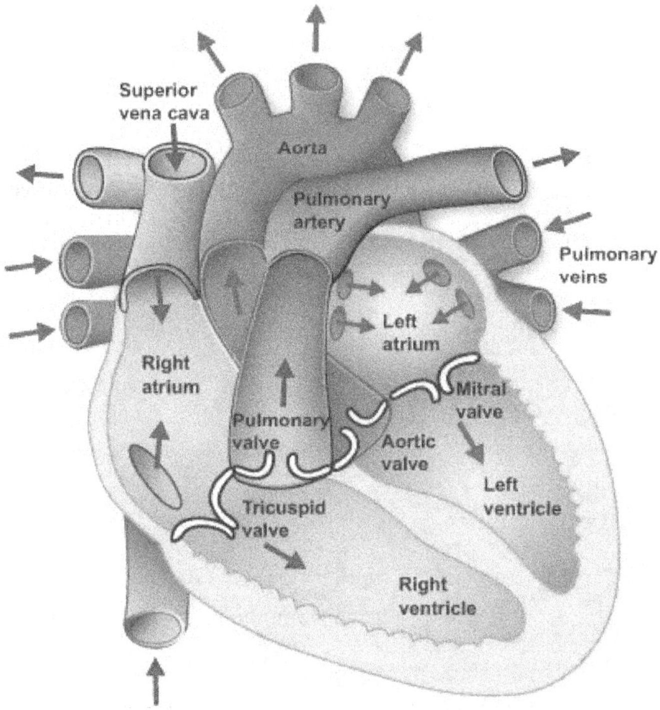

6. C is correct.

Blood has three leading roles in the body: transport, protection, and regulation.

In its transport role, blood transports the following substances:

- Gases (oxygen and carbon dioxide) between the lungs and tissues
- Nutrients from the digestive tract and storage sites to the cells
- Waste products to be removed by the liver and kidneys
- Hormones from where they are secreted to their target cells

Blood's protective role includes:

- Leukocytes (i.e., white blood cells) destroy invading microorganisms and cancer cells
- Antibodies and other proteins destroy pathogenic substances
- Platelet factors initiate blood clotting and minimize blood loss

Blood helps regulate:

- pH by interacting with acids and bases
- Water balance by transferring water to and from tissues

7. C is correct.

Platelets are cell fragments that lack nuclei and are involved in clot formation.

Platelets (i.e., thrombocytes) are small, disk-shaped clear cell fragments (i.e., do not have a nucleus), 2–3 μm in diameter, derived from the fragmentation of precursor megakaryocytes.

The lifespan of a typical platelet is 5 to 9 days.

Platelets are a natural source of growth factors circulating in the blood and are involved in hemostasis (formation of blood clots to stop bleeding).

A low platelet count may result in excessive bleeding, while too high of a number may result in the formation of blood clots (thrombosis), which may obstruct blood vessels and result in stroke, myocardial infarction, pulmonary embolism or the blockage of blood vessels, for example, in the extremities of the arms or legs.

A: erythrocytes (i.e., red blood cells) are the blood's oxygen-carrying components and contain hemoglobin that binds up to four molecules of oxygen.

B: macrophages perform phagocytosis of foreign particles and bacteria by engulfing them, digesting the material, and then presenting the fragments on their cell surface.

D: T cells lyse virally infected cells or secrete proteins that stimulate B cells' development or other T cell types.

8. D is correct.

Capillaries have higher hydrostatic pressure at the arteriole end and lower pressure at the venule end.

As blood flows from arterioles to capillaries, blood pressure gradually drops due to friction between the blood and the walls of the vessels and the increase in cross-sectional areas provided by numerous capillary beds.

Blood plasma in the capillaries has a higher osmotic pressure than interstitial fluid pressure. This results from the higher number of dissolved solutes in the capillaries' blood plasma.

Hydrostatic pressure is defined as the force per area that blood exerts on the blood vessels' walls.

The pumping force of the heart through the blood vessels creates hydrostatic pressure.

9. C is correct.

In an adult, hematopoiesis (i.e., the formation of blood cellular components) occurs in the bone marrow.

In the fetus, hematopoiesis occurs in the fetal liver.

The spleen acts as a reservoir for red blood cells (red blood cells) and filters the blood.

Erythrocytes are produced in the bone marrow to replace erythrocytes during their 120-day life span.

A: erythrocytes (mature red blood cells) are not nucleated to create more space for hemoglobin.

B: blood platelets are crucial for the clotting of blood.

D: leukocytes (white blood cells) such as macrophages and neutrophils engulf foreign matter.

Neutrophils are the most abundant white blood cells and account for approximately 50-70% of white blood cells.

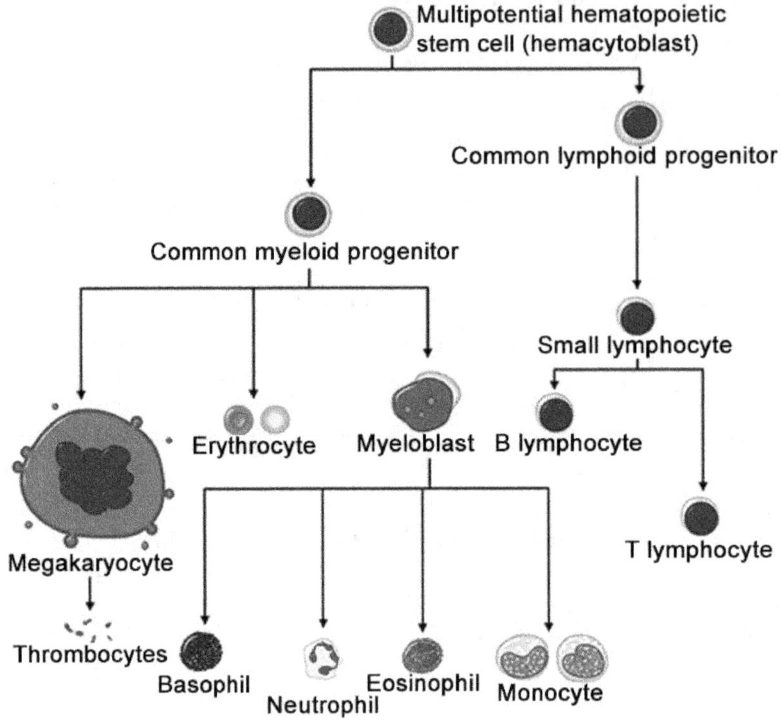

10. D is correct.

Blood traveling from the left ventricle enters the aorta and then flows to the body areas (i.e., systemic circulation) except the lungs (i.e., pulmonary circulation).

Examples of systemic circulation include blood in the brachiocephalic artery (travels to the head and the shoulders) and blood in the renal artery (travels to the kidney to be filtered).

Superior/inferior vena cava → right atrium → right ventricle → pulmonary artery (*pulmonary circulation*) → lungs → pulmonary vein → left atrium → left ventricle → aorta (*systemic circulation*) → superior/inferior vena cava

For systemic circulation, blood moves from arteries into arterioles, then capillaries, where nutrients, waste, and energy are exchanged.

Blood then enters venules, collects in veins, is transported to the superior and inferior venae cavae, and then enters the right atrium and flows into the right ventricle.

For pulmonary circulation, blood is transported to the lungs via the pulmonary artery.

Capillary beds around the alveoli exchange gas (O_2 into the blood and CO_2 from the blood) in the lungs.

The pulmonary veins bring blood back to the left atrium to start the systemic circulation.

11. C is correct.

The right atrium is the upper chamber of the heart's right side. The blood returned to the right atrium by the superior, and inferior vena cava is deoxygenated (low in O_2) and passes into the right ventricle to be pumped through the lungs' pulmonary artery for O_2 and removal of CO_2.

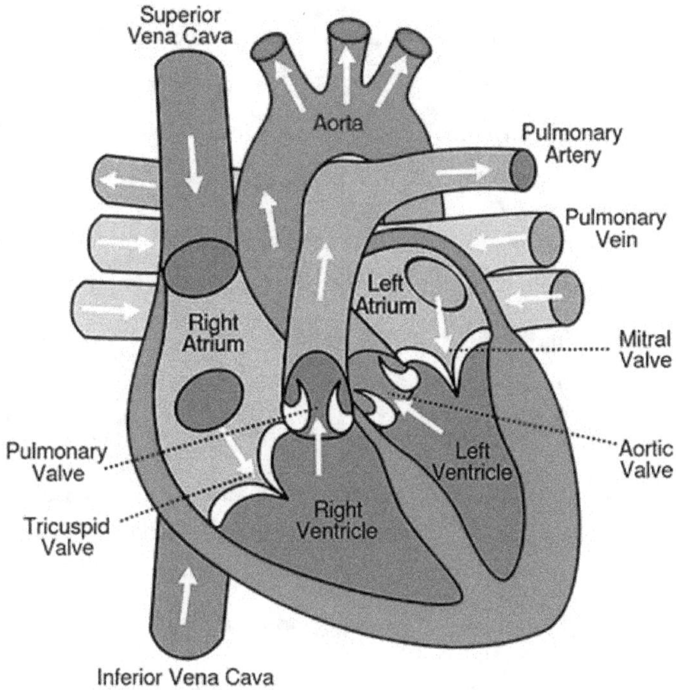

The left atrium receives newly oxygenated blood from the lungs by the pulmonary vein. The blood is passed into the muscular left ventricle to be pumped through the aorta to the body's organs.

In a healthy heart, the right ventricle pumps blood through the pulmonary arteries to the lungs during pulmonary circulation. The left ventricle pumps blood through the aorta to the body by systemic circulation.

If blood mixes between the ventricles, some deoxygenated blood in the right ventricle mixes with the left ventricle's oxygenated blood for systemic circulation. This mixing lowers the O_2 supplied to the tissues.

12. A is correct.

Blood plasma is the liquid part of blood, consisting primarily of water (up to 95%), suspending blood cells and acting as the extracellular matrix.

13. A is correct.

The aorta carries oxygenated blood from the heart through the systemic (i.e., body) circulation and (like the pulmonary vein) has the highest partial pressure of O_2.

B: coronary veins carry deoxygenated blood to the heart's right side.

C: the superior vena cava carries deoxygenated blood from the body's upper regions to the heart. The inferior vena cava carries deoxygenated blood from the lower half of the body into the right atrium.

D: pulmonary arteries carry deoxygenated blood *away* from the heart towards the lungs. The pulmonary artery is the only artery in the adult carrying deoxygenated blood.

14. C is correct.

Deoxygenated blood returns from the systemic circulation and drains into the right atrium from the inferior and superior vena cava.

From the right atrium, blood is pumped through the tricuspid valve into the right ventricle, pumps blood to the lungs via the pulmonary arteries.

CO_2 is exchanged for O_2 in the alveoli of the lungs.

Oxygenated blood is returned to the left atrium via the pulmonary veins.

The left atrium pumps the blood through the mitral valve into the left ventricle, which ejects it into the aorta for systemic circulation.

If a tracer substance were injected into the superior vena cava, it would take the longest time to reach the left ventricle.

A: the tricuspid valve separates the right atrium from the right ventricle. It prevents blood's backflow into the right atrium when the right ventricle contracts, forcing blood into the pulmonary artery.

B: blood with the tracer would travel from the right ventricle to the lungs via the pulmonary arteries and return through the pulmonary veins to the left atrium.

D: from the systemic circulation, the tracer first enters the right atrium and then passes through the tricuspid valve as it is pumped into the right ventricle.

15. D is correct.

Semi-lunar valves prevent blood backflow as it leaves the heart via the aorta (systemic circulation to the body) and pulmonary arteries (pulmonary circulation to the lungs).

The ventricles have negative pressure once they start to relax before receiving blood from the atria.

16. B is correct.

The atrioventricular (AV) node is located at the junction between the atria and the ventricles and functions to delay the ventricles' conduction impulse for a fraction of a second.

A wave of excitation spreads out from the sinoatrial (SA) node through the atria along specialized conduction channels.

This activates the AV node, which delays conduction impulses by approximately 0.12s.

This delay in the cardiac pulse ensures that the atria have ejected their blood into the ventricles before the ventricles contract.

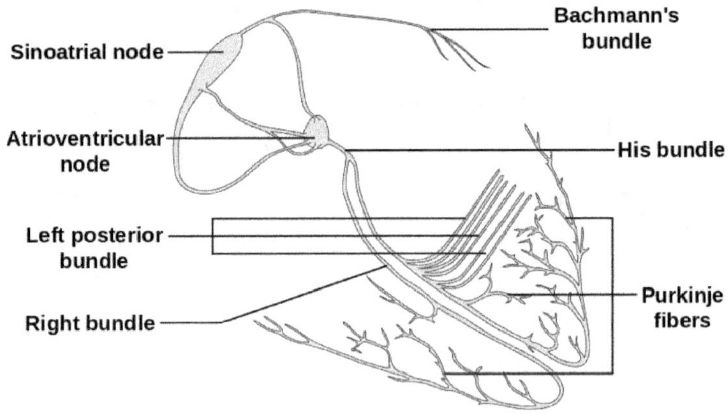

17. D is correct.

The mitral valve is between the left atrium and the left ventricle and prevents the backflow of blood into the left atrium during the contraction of the left ventricle.

A patient suffering from stenosis of the mitral valve has impeded blood flow from the left atrium into the left ventricle.

Therefore, the effect is an increase in blood volume in the left atrium and a reduced net movement of blood from the left atrium into the left ventricle, which results in increased left atrial pressure and decreased left ventricular pressure.

A: the right atrium receives deoxygenated blood by venous circulation and is too removed for the circulatory process to be affected by mitral valve stenosis.

B: the left ventricle ejects blood into the aorta as it begins systemic circulation.

Since volume/pressure in the left ventricle is reduced due to impeded flow from the left atrium, the patient experiences decreased pressure in the aorta.

18. C is correct.

A hemorrhage is a loss of blood from the circulatory system.

Hemorrhage can be external (i.e., when blood is lost to the outside of the body through a wound or a natural opening) or internal (i.e., when blood escapes from the blood vessels into other structures within the body).

Moderate blood loss results in decreased cardiac output (volume of blood pumped by the heart each minute).

19. C is correct.

The cardiac conduction pathway begins with the SA node in the right atrium.

The conduction signal then travels to the AV node located between the atria and the ventricles before moving down the bundle of His, which then splits into the right/left Purkinje fibers.

The Purkinje fibers spread the conduction signal to the two ventricles, which contract simultaneously to eject blood.

The blood from the right ventricle enters the pulmonary artery. Blood from the left ventricle enters the aorta.

SA node → AV node → bundle of His → Purkinje fibers

20. B is correct.

Blood in the left ventricle returned via the pulmonary vein from the lungs, where it was oxygenated.

The pulmonary vein is the only vein in the adult that carries oxygenated blood.

The superior vena cava returns deoxygenated blood to the heart from the body's upper regions.

A: the right atrium receives blood via the inferior and superior venae cavae returning from systemic circulation.

Blood returning from the body has low O_2 content as it is transported to the lungs.

C: pulmonary artery blood is pumped to the lungs to be oxygenated and is the least oxygenated blood in the circulatory system.

D: lymph vessels, such as the thoracic duct, return lymphatic fluid to the venous circulation.

The lymph fluid has low O_2_partial pressure.

21. C is correct.

The pulmonary artery's blood is pumped to the lungs to be oxygenated and is the least oxygenated blood in the circulatory system.

A: lymph vessels, such as the thoracic duct, return lymphatic fluid to the venous circulation. The lymph fluid has a low O_2 partial pressure.

B: left ventricle blood returned via the pulmonary vein from the lungs, which was oxygenated.

The pulmonary vein is the only vein in the adult that carries oxygenated blood.

D: the inferior vena cava returns deoxygenated blood to the right atrium from the body's lower regions.

22. C is correct.

The afferent arteriole supplies blood to the kidney at the glomerulus. Protein does not typically enter the kidney in the filtrate (fluid destined to become excreted as urine).

The loss of blood plasma into the kidney increases the protein concentration (relative to plasma). The presence of proteins in the filtrate is a clinical indication of kidney abnormalities.

D: vasa recta form a series of straight capillaries of the kidney in the medulla and are parallel to the loop of Henle.

These vessels branch from the efferent arterioles of juxtamedullary nephrons (nephrons closest to the medulla), enter the medulla and surround the loop of Henle.

Each vasa recta has a hairpin turn in the medulla and carries blood at a slow rate, crucial for the support of countercurrent exchange that maintains the concentration gradients established in the renal medulla. The maintenance of this concentration gradient is one of the components responsible for the kidney's ability to produce concentrated urine.

NaCl and urea are reabsorbed into the blood on the *descending* portion of the vasa recta while H_2O is secreted. NaCl and urea are secreted into the interstitium on the *ascending* portion of the vasa recta while H_2O is reabsorbed.

23. D is correct.

Platelets are small pieces of large megakaryocyte cells.

When a wound occurs, platelets are activated to stick to the wound's edge and release their contents that stimulate the clotting reaction.

Platelets release several growth factors, including platelet-derived growth factor (PDGF), a potent chemotactic agent (direct movement from chemical stimuli), and TGF beta, which stimulates the extracellular matrix deposits.

Both growth factors play a significant role in the repair and regeneration of connective tissues.

A: leukocytes (i.e., white blood cells) are immune system cells defending against infectious disease.

There are five types of leukocytes.

Leukocytes are produced and derived from hematopoietic stem cells in the bone marrow. They survive for about three to four days in the human body and are throughout the body, including the blood and lymphatic system.

The name "white blood cell" derives from a blood sample's physical appearance after centrifugation.

White cells are in the *buffy coat* (a thin, typically white layer of nucleated cells between the sedimented red blood cells and the blood plasma).

Leukocytes have common traits, but each is distinct in form and function.

A significant distinction is the presence or absence of granules (i.e., granulocytes or agranulocytes).

Granulocytes (polymorphonuclear leukocytes) are characterized by the presence of differently staining granules in their cytoplasm.

These granules (lysozymes) are membrane-bound enzymes acting primarily to digest endocytosed particles.

There are three granulocytes: *neutrophils*, *basophils,* and *eosinophils,* named according to their staining properties.

Agranulocytes (mononuclear leukocytes) are characterized by the absence of granules in their cytoplasm.

The cells include *lymphocytes*, *monocytes,* and *macrophages*.

B: erythrocytes (i.e., red blood cells) are the most common type of blood cell and the principal means of delivering oxygen (O_2) to the body.

C: lymphocytes are a type of white blood cell in the immune system, a landmark of adaptive immunity.

Lymphocytes are categorized into large or small lymphocytes.

Large granular lymphocytes include natural killer cells (NK cells), while small lymphocytes consist of T cells and B cells.

24. A is correct.

The hepatic portal system directs blood from parts of the gastrointestinal tract to the liver.

The inferior vena cava is a large vein responsible for carrying deoxygenated blood into the heart.

25. B is correct.

The movement of fluids occurs between the capillaries and the tissues.

Blood enters the arterial end of the capillary bed as it passes to the venule end.

In the capillary bed, gases, nutrients, wastes, and fluids exchange with tissue cells in the interstitial space.

Two opposing forces affect the movement of fluids between the capillaries and the interstitial space.

At the arteriole end, hydrostatic pressure (i.e., blood pressure) is higher than the osmotic pressure, which forces fluids out of the capillaries and into the interstitial spaces.

Osmotic pressure is higher than the hydrostatic pressure at the venule end, which forces fluids back into the capillaries and returns the interstitial fluid into the circulatory system.

Starling's hypothesis proposes that not all fluid can be returned to the capillary.

Some fluid travels through the interstitial spaces towards the lymphatic system, where the lymph nodes filter it and then, via the thoracic duct, return to the circulatory system.

26. D is correct.

Hemoglobin exhibits positive cooperative binding (i.e., the sigmoid shape of the O_2 binding curve).

When O_2 binds to the first of four hemoglobin subunits, this initial binding of O_2 changes the shape of Hb and increases the affinity of the three remaining subunits for O_2.

Conversely, as the first O_2 molecule is unloaded, the other O_2 molecules dissociate more easily.

Cooperative binding is necessary for the hemoglobin transport of O_2 to the tissues.

A: hemoglobin reversibly binds carbon dioxide (CO_2) and transports it to the lungs.

In the alveoli of the lungs, CO_2 dissociates from the hemoglobin and is expired (i.e., exhaled) by the body.

B: fetal hemoglobin acquires O_2 from the maternal circulation, so it has a higher O_2 affinity than adult hemoglobin.

C: CO poisoning occurs because this odorless gas has a high hemoglobin affinity and binds tightly to hemoglobin, preventing oxygen from binding to the occupied site of the hemoglobin.

27. A is correct.

Blood volume depends on 1) the amounts of water and sodium ingested and 2) excreted in urine and the gastrointestinal tract, lungs, and skin.

These amounts are highly variable, and it is the function of kidneys to maintain blood volume within a normal range (i.e., homeostasis) by excreting water and sodium into the urine.

When there is an excess of ingested water and sodium, the kidneys respond by excreting more water and sodium in the urine.

28. A is correct.

Blood does not carry cells for injury repair. However, blood is essential for wound healing.

Special blood cells called platelets create clots to stop the bleeding.

Another blood cell called a macrophage acts as a wound protector by fighting infection and assisting in the repair process by producing chemical messengers (i.e., growth factors) helping repair the wound.

Oxygen-rich red blood cells help build new tissue.

Chemical signals prompt the cells to create collagen, which serves as a "scaffolding."

The result is a scar that starts red and dulls eventually.

29. C is correct.

Like O_2, CO binds to hemoglobin but more firmly than O_2, and its binding is almost irreversible.

Carbon monoxide is formed by incomplete combustion of fuels (i.e., hydrocarbons) resulting from natural gas or propane devices (e.g., space heaters, barbecue grills).

A: CO is not irritating, as it is odorless and colorless.

B: CO does not affect the cytochrome chain.

D: CO does not form complexes in blood.

30. A is correct.

Alveoli are thin air sacs and are the sites of gas (e.g., O_2 and CO_2) exchange via passive diffusion in the lungs between the air and blood.

B: pleura is the lungs' outer lining filled with pleural fluid that lubricates the lungs.

C: bronchi are the two main branches of the air intake pathway, with one bronchus for each lung.

D: bronchioles are smaller subdivisions of the bronchi.

31. C is correct.

Cardiac muscle, unlike the other two muscle types, is capable of spontaneous depolarization resulting in contraction.

Smooth and skeletal muscle requires stimulation (e.g., neurotransmitter) to initiate depolarization and contraction.

A: cardiac muscle cells (i.e., cardiomyocytes) are generally single-nucleated but can have up to four centrally located nuclei.

B: cardiac muscle is striated.

D: the autonomic nervous system innervates the cardiac muscle.

The somatic motor system innervates skeletal muscle and controls voluntary actions (e.g., walking, standing).

Notes for active learning

Notes for active learning

Immune System – Detailed Explanations

1. D is correct.

Gamma globulins, also known as immunoglobulins (Ig), are Y-shaped protein antibodies.

Gamma globulins are produced by B cells and used by the immune system to identify and neutralize foreign objects, such as bacteria and viruses.

Antibodies are secreted by white blood cells called plasma cells and recognize a unique part of the foreign target (i.e., antigen).

2. C is correct.

Humoral immunity is the antibody-mediated system (as opposed to cell-mediated immunity) because of macromolecules in the extracellular fluids (e.g., secreted antibodies, complement proteins, specific antimicrobial peptides).

Humoral immunity is so named because it involves substances in the humour (i.e., body fluids).

The immune system divides into a more primitive innate immune system and adaptive immune system (i.e., acquired), each of which contains humoral and cellular components (e.g., phagocytes, antigen-specific cytotoxic T-lymphocytes, cytokines).

The terms *antibody* and *immunoglobulin* are used interchangeably.

Immunoglobulins are glycoproteins that function as antibodies.

Antibodies are synthesized and secreted by plasma cells derived from the immune system's B cells.

Immunoglobulins (antibodies) are in the blood, tissue fluids, and many secretions.

Immunoglobulins are large Y-shaped globular proteins.

In mammals, there are five types of antibodies: IgA, IgD, IgE, IgG, and IgM.

Each immunoglobulin class differs in its biological properties and has evolved to target different antigens.

The acquired immune system uses an antibody to identify and neutralize foreign objects like bacteria and viruses.

Each antibody recognizes a specific antigen unique to its target.

By binding, antibodies cause agglutination and precipitation of antibody-antigen products, prime for phagocytosis by macrophages and other cells, block viral receptors, and stimulate other immune responses (i.e., complement pathway).

B: cytotoxic T cells (also known as cytotoxic T lymphocytes, T-killer cells, cytolytic T cells, CD8, or killer T cells) function in cell-mediated immunity.

Cytotoxic T cells are a type of white blood cell that destroy cancer cells, virally infected cells, or damaged cells.

Most cytotoxic T cells express T-cell receptors (TCRs) that recognize specific antigen molecules capable of stimulating an immune response, often produced by viruses or cancer cells.

3. C is correct.

Antibodies are produced in response to antigens (e.g., viral coat proteins, bacterial cell walls) detected by T cells.

T cells stimulate B cells to become plasma cells and secrete antibodies.

B cells mature into memory cells, or antibody-producing cells, during immune responses.

B cells are stimulated to become memory cells, which remain dormant in the interstitial fluid and lymphatics until the same antigens are detected and provide a rapid humoral response.

A: neurons are not involved in the immune response.

B: T cells are the immune response organizers and can become cytotoxic cells (i.e., kill invading cells), helper T cells (i.e., recruit other T and B cells), and suppressor T cells (i.e., inactivate the immune response when the antigen has been cleared).

D: macrophages are phagocytic cells that engulf and digest bacterial cells and foreign material.

4. B is correct.

Phagocytes are cells that function to eliminate foreign invaders (e.g., particles, microorganisms, dying cells) in the body by phagocyting (i.e., digesting) them.

Macrophages and neutrophils play an essential role in the inflammatory response by releasing proteins and other substances that control infection and damage the host tissue.

5. B is correct.

The innate immune system responds to foreign invaders with white blood cells (i.e., granulocytes) and inflammation.

Granulocytes include neutrophils, eosinophils, and basophils because of granules' presence in their cytoplasm.

They are referred to as polymorphonuclear cells (PMNs) due to their distinctive lobed nuclei.

Neutrophil granules contain toxic substances that kill or inhibit the growth of bacteria and fungi.

Like macrophages, neutrophils attack pathogens by activating a respiratory burst of potent oxidizing agents (e.g., hydrogen peroxide, free oxygen radicals, and hypochlorite).

Neutrophils are the most abundant type of phagocyte (i.e., 50 to 60% of the total circulating leukocytes) and are usually the first cells to arrive at the site of an infection.

Innate immunity does not involve humoral immunity (B cell) or cell-mediated immunity (T cell).

6. B is correct.

$CD4^+$ cells (also known as T helper cells) are white blood cells that are an essential part of the human immune system.

They are called helper cells because one of their primary roles is to send signals to other types of immune cells, including CD8 killer cells that destroy the infection or virus.

If $CD4^+$ cells become depleted (for example, in an untreated HIV infection or following immune suppression before a transplant), the body is left vulnerable to a wide range of infections that it would otherwise be able to fight.

$CD4^+$ cells are activators of cell-mediated and humoral responses.

AIDS is a form of *severe combined immunodeficiency* (SCID) since the loss of an adequate $CD4^+$ suppresses humoral and cell-mediated responses.

The immune system can be subdivided into non-specific (innate) and adaptive (i.e., acquired) immunity.

The innate immune system consists of physical barriers (e.g., skin and mucous membranes) and cells that non-specifically remove invaders (e.g., mast cells, macrophages, neutrophils).

The adaptive immune system synthesizes cells directed against specific pathogens and responds rapidly if a similar pathogen is encountered.

The innate system responds with the same mechanisms (and efficiency) to every infection.

7. D is correct.

Antigens are foreign substances (e.g., bacteria, virus) that enter the body and induce an immune response, specifically the production of antibodies to destroy the antigen.

8. B is correct.

The effectiveness of a vaccine depends on the organism's immune system.

Plasma cells (a differentiated form of a B cell) secrete antibodies (i.e., immunoglobulin).

The acquired immune system consists of humoral (i.e., antibodies) and cell-mediated (i.e., T cells) responses.

The humoral response is mediated by B cells synthesized and matured in the bone marrow.

When stimulated by antigens, B cells differentiate into several cells, including memory B cells, and plasma B cells secrete antibodies.

A: cell-mediated immunity involves T cells that originate in the bone marrow (as B cells do) but then migrate to and mature in the thymus.

T-helper cells regulate the activity of other T and B cells.

C: macrophages are non-specific immune cells released by the bone marrow and are involved in phagocytosis but do not synthesize or release immunoglobulin (i.e., antibodies).

D: the thymus is the structure where T cells mature.

T cells start as hematopoietic precursors and migrate from the bone marrow to the thymus.

T cells undergo maturation, which involves ensuring that the cells react against antigens (i.e., *positive selection*) and do not react against antigens on body tissues (i.e., *negative selection*).

Once mature, T cells leave the thymus to perform their vital functions in the immune system.

9. D is correct.

T cells are categorized as helper, cytotoxic (i.e., killer), memory, regulatory (i.e., suppressor), and natural killer T cells.

10. B is correct.

B-lymphocytes (B-cells) are a type of white blood cell that creates antibodies.

B-lymphocytes develop from stem cells in the bone marrow.

11. D is correct.

Tissue repair after an injury is not part of the inflammatory response—the restoration of tissue structure and function via two separate processes: regeneration and replacement.

For regeneration, the cell type destroyed must be replicated.

Replacement is when tissue is replaced with scar tissue.

12. C is correct.

When exposed to a specific antigen, B lymphocytes (B cells) differentiate and divide into memory or plasma cells.

The plasma cells produce antibodies that recognize and bind the specific antigen that activated the precursor B lymphocytes.

A: cytotoxic T cells originate from T cells (T lymphocytes), not B lymphocytes, and bind to and destroy the antigen directly.

B: lymphokines originate from T cells (T lymphocytes), not B lymphocytes, and are secreted by helper T cells that activate other T cells, B cells, and macrophages.

D: macrophages are highly specialized for removing dying/dead cells and cellular debris.

Macrophages do not originate from lymphocytes.

When a monocyte enters damaged tissue through the endothelium of a blood vessel (i.e., leukocyte extravasation), it undergoes a series of changes to become a macrophage.

Monocytes are attracted to a damaged site by chemotaxis (i.e., chemical signals) triggered by stimuli such as damaged cells, pathogens, and cytokines released by macrophages already at the site.

13. C is correct.

Memory B cells are formed after primary infection and differentiate into plasma cells (i.e., B lymphocytes) when exposed to a specific antigen.

A: phagocytosis is an example of nonspecific (innate) immunity.

Phagocytes include monocytes (large white blood cells), macrophages (which stimulate lymphocytes and other immune cells to respond to pathogens), neutrophils (one of the first-responders of inflammatory cells to migrate towards the site of inflammation), mast cells (involved in wound healing and defense against pathogens – prominent in allergy and anaphylaxis) and dendritic cells (antigen-presenting cells that function as a bridge between the innate and adaptive immune systems).

B: stomach acid is an example of innate (i.e., nonspecific) immunity.

D: skin is an example of innate (i.e., nonspecific) immunity.

14. C is correct.

Interferons are signaling proteins developed in response to pathogens (i.e., viruses, bacteria, parasites, tumor cells).

Interferons are named for their ability to interfere with replicating viral RNA and DNA.

15. D is correct.

Within the first few minutes after the injury, platelets (i.e., thrombocytes) aggregate at the injury site to form a fibrin clot that reduces active bleeding.

The speed of wound healing depends on bloodstream levels of platelets, fibrin, and hormones, such as oxytocin.

During the inflammation phase, bacteria and cell debris are removed via phagocytosis by white blood cells.

The proliferation phase is characterized by the release of blood factors into the wound that causes the migration and division of cells, angiogenesis (i.e., new vascular tissue growth), collagen deposition, granulation tissue formation, and wound contraction.

Fibrin is a fibrous protein involved in the clotting of blood. It is formed from fibrinogen by the protease thrombin.

Fibrin form a "mesh" that plugs or clots (in conjunction with platelets) over a wound site.

Fibrin is involved in signal transduction, blood coagulation, and platelet activation.

16. A is correct.

AB blood has neither anti-A nor anti-B antibodies, so there is no agglutination reaction when any type of blood is transfused into an AB recipient.

17. A is correct.

O type blood contains neither A nor B antigens, so there is no agglutination reaction when O type blood is infused into a patient with any other blood type.

18. D is correct.

Plasma cells are white blood cells that secrete antibodies that target and bind to antigens (foreign substances).

19. D is correct.

B cells (also called B lymphocytes) are a type of lymphocyte.

B cells participate in the humoral immunity of the adaptive immune system. B cells mature in the bone marrow.

Activated B cells go through a two-step differentiation process producing short-lived plasmablasts (for immediate immune response) and long-lived plasma cells and memory B cells (for persistent protection).

Plasma cells release into the blood and lymph antibody molecules closely modeled after the precursor B cell receptors.

When released, they bind to the target antigen (i.e., foreign substance) and initiate its destruction.

20. B is correct.

Lymph nodes are linked by lymph vessels and contain phagocytic cells (leukocytes) that filter the lymph, remove and destroy foreign particles/pathogens.

The lymphatic fluid absorbs fat and fat-soluble vitamins.

A: the liver breaks down hemoglobin and uses its components to produce bile salts.

C: amino acids are not absorbed from the digestive tract by the lymphatic system or lymph nodes.

D: glucagon is produced by the pancreas and increases the blood concentration of glucose.

21. A is correct.

The lymphatic system transports excess fluid and proteins from the interstitial space into the circulatory system.

The lymphatic system does not regulate erythropoiesis.

Erythropoiesis is the process of red blood cell production and occurs in the bone marrow. Erythropoiesis is stimulated by decreased oxygen detected by the kidneys, which secrete the hormone erythropoietin.

B: macrophages in lymph nodes engulf bacteria and other foreign particles.

C: the lymphatic system functions to remove proteins from interstitial spaces.

D: the lymphatic system is essential in absorbing nutrients (particularly lipids) from the small intestine via lacteals (i.e., specialized lymph vessels).

22. B is correct.

Antibodies bind to antigens through interactions between the antibody's variable region and the antigen.

The fragment antigen-binding (Fab fragment) is a region on an antibody that binds to antigens.

It is composed of one constant and one variable domain of each heavy and light chain at the monomer's amino-terminal end. The two variable domains bind the epitope on their specific antigens.

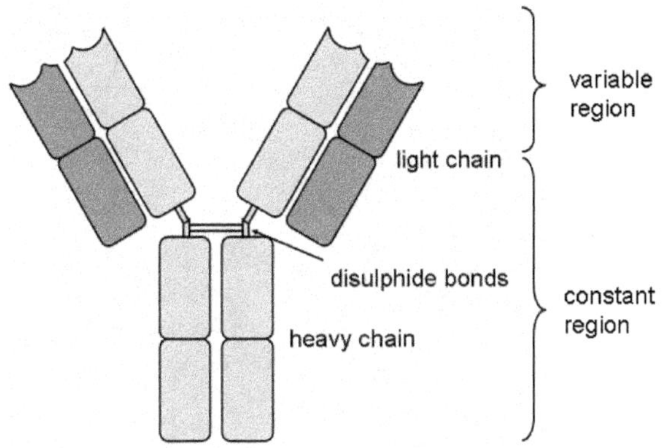

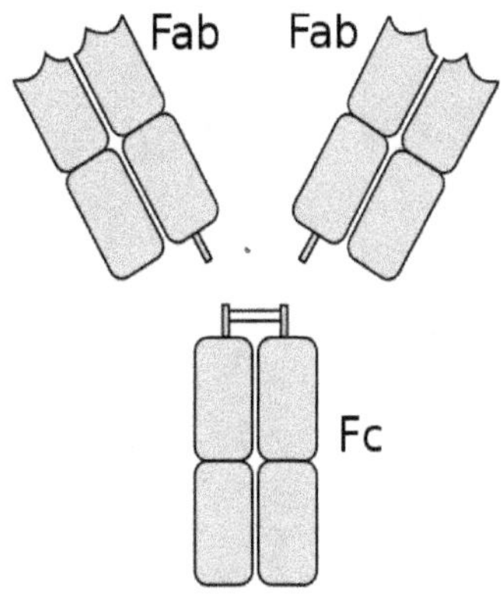

A: antibodies do not assist in the phagocytosis of cells.

D: antibodies are produced by plasma cells derived from stem cells in the bone marrow.

23. B is correct.

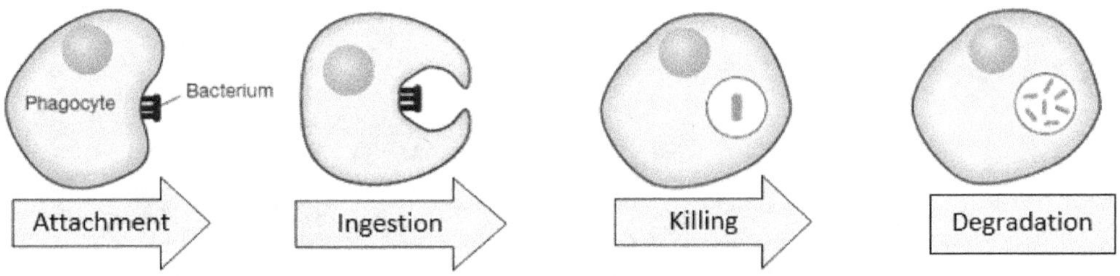

Stages of phagocytosis from attachment to degradation

24. B is correct.

Humans have approximately 500-600 lymph nodes distributed throughout the body, with clusters in the underarms, groin, neck, chest, and abdomen.

Enlargement of the lymph nodes is a common indicator of infection, malignancy, or autoimmune disease.

C: inflammatory responses draw fluid into the affected area.

Notes for active learning

Central Nervous System – Detailed Explanations

1. B is correct.

Frequency of action potentials is related to the *intensity* of the stimulus.

Strength of the stimulus is coded into the *frequency* of the action potentials generated.

The stronger the stimulus, the higher the frequency of action potentials.

Thus, the nervous system is *frequency-modulated* and *not amplitude-modulated.*

2. A is correct.

Blood-brain barrier helps block harmful substances (e.g., bacteria, toxins, metabolic waste products) from entering the brain.

The blood-brain barrier does allow the passage of oxygen, glucose, amino acids, alcohol, and anesthetics.

3. C is correct.

Medulla oblongata controls many vital functions, such as breathing, heart rate, and gastrointestinal activity.

A: *cerebrum* (i.e., cerebral cortex) processes and integrates sensory input and motor responses and participates in memory and creativity.

B: *cerebellum* is vital in coordinating muscles and aids in balance by receiving input from the inner ear, hand-eye coordination, and the timing of rapid movements. The primary function of the cerebellum is *coordinating unconscious movement.*

D: *hypothalamus* controls hunger, thirst, sleep, water balance, blood pressure, temperature regulation, and libido. It is essential in modulating the endocrine system.

4. C is correct.

In anatomy, the term *decussation* (i.e., crossing) means the same thing as *chiasma* (or chiasm).

Chiasma is where two chromatids are interwoven in genetics, as observed between the S phase and anaphase.

5. A is correct.

Cerebellum, along with the pons and medulla oblongata, is in the hindbrain. Higher brain sensory neurons and motor neurons pass through the hindbrain. Primary function of the cerebellum is coordinating unconscious movement (e.g., hand-eye coordination, posture, balance).

Cerebellum damage would likely result in *loss of muscle coordination.*

Cerebrum divides into left and right hemispheres, subdivided into *four lobes.*

Cerebrum coordinates most voluntary activities, sensations, and *higher functions* (including speech and cognition). The spinal cord also controls the extremities.

6. C is correct.

Medulla oblongata monitors CO_2 levels and pH of the blood. It regulates breathing, temperature, heart rate, and reflex activities such as sneezing, coughing, and swallowing.

Reticular activating system (RAS) is a network of neurons in the brainstem that receives input from all sensory systems and sends non-specific information to the brain.

Hypothalamus controls hunger, thirst, sleep, water balance, blood pressure, temperature regulation, and libido. It has a vital role in modulating the endocrine system.

Cerebral cortex controls vision, hearing, smell, voluntary movement, and memory.

7. D is correct.

Cerebral cortex is the outer layer of the brain, made of gray matter.

White matter is a different type of nerve fiber in the brain's inner layer.

8. B is correct.

Cerebellum is part of the hindbrain (i.e., the posterior part of the brain) and consists of the pons and medulla oblongata. Cerebellum receives sensory information from the visual and auditory systems and information about the orientation of joints and muscles.

Cerebellum's primary function is hand-eye coordination. It receives information about the motor signals being initiated by the cerebrum and integrates inputs to produce balance and unconscious coordinated movement.

Damage to the cerebellum could compromise these functions, while destruction would eliminate all and seriously *impair coordinated movement*.

A: *thermoregulation* is a function of the hypothalamus (a part of the cerebrum).

C: *sense of smell* (i.e., olfaction) is a cerebrum function.

D: *urine formation* is the primary function of the kidneys, along with some hormonal regulation.

9. B is correct.

Pressure waves (i.e., sound) are converted into *neural signals* by hair cells in the organ of Corti of the cochlea.

Organ of Corti (in mammals) is part of the cochlea of the inner ear with hair cells (i.e., auditory sensory cells). Organ of Corti is the structure that *transduces pressure waves into electrical signals* as action potentials.

10. C is correct.

Brain stem is the posterior part of the brain that joins it with the spinal cord. It consists of pons, medulla, and midbrain.

Pons contains neural pathways that carry signals from and to the brain.

Medulla is associated with breathing, heart rate, and blood pressure functions.

Midbrain oversees vision, hearing, motor control, alertness, and temperature regulation.

11. B is correct.

Hypothalamus controls hunger, thirst, sleep, water balance, blood pressure, temperature regulation, and libido.

Hypothalamus is essential in modulating the endocrine system.

12. D is correct.

Cerebrospinal fluid is a clear fluid in the brain and spinal cord. It acts as a buffer for the brain and provides mechanical and immunological protection.

Red blood cells are usually *not* in cerebrospinal fluid.

13. A is correct.

Temporal lobe of the brain is where the *primary auditory cortex* is located.

14. D is correct.

Dorsal root ganglia of the spinal cord contain cell bodies of sensory neurons (or *first-order neurons*), which bring information from the periphery to the spinal cord.

15. A is correct.

White matter consists of bundles of myelinated nerve fibers.

Myelin is a white fatty substance enclosing nerve fibers and increasing the transmission speed of neurons.

16. A is correct.

Broca's area is in the frontal lobe of the brain (in the dominant hemisphere – usually the left) and is responsible for speech production.

Damage to Broca's area is commonly associated with speech where the content of the information is correct, but *sentence structure* and *fluidity* are impaired.

17. C is correct.

Organ of Corti and the *semicircular canals* contain hair cells with micro cilia (i.e., small hair) projecting from the apical surface of the cell into the surrounding fluid.

Movement of fluid around the hair cells in the *organ of Corti detects sound.*

Fluid movement around the hair cells in the *semicircular canals* detects changes in body orientation.

Skin hair is different and is formed by keratinized dead epithelial cells.

18. C is correct.

Prefrontal cortex lies at the front of the brain and is responsible for cognitive behavior, personality, decision making, and social behavior.

Limbic association area helps form memories and guides emotional responses.

Primary somatosensory cortex processes sensations such as touch, pain, and temperature.

Posterior association area is where visual, auditory, and somatosensory association areas meet, giving spatial awareness.

19. C is correct.

Afferent nerves are sensory neurons that carry impulses from sensory stimuli toward the central nervous system.

Efferent nerves carry neural impulses from the central nervous system toward muscles.

20. A is correct.

Regeneration within the central nervous system (CNS) is prevented due to *growth-inhibiting proteins* of oligodendrocytes.

Notes for active learning

Notes for active learning

Reproductive System – Detailed Explanations

1. D is correct.

Menstrual cycle is regulated by the hypothalamus, anterior pituitary, and ovaries.

Follicle-stimulating hormone (FSH) is released by the anterior pituitary, stimulating the maturation of an ova.

Progesterone, synthesized by the ovary and corpus luteum, regulates the development and shedding of the endometrial lining of the uterus.

Adrenal medulla synthesizes epinephrine (adrenaline) and norepinephrine, which assist the sympathetic nervous system in stimulating the *fight or flight* response but are *not* involved in the menstrual cycle.

2. B is correct.

Gametes form via *meiosis* and are double-stranded haploids.

A single chromosome consists of two hydrogen-bonded complementary antiparallel DNA strands.

3. A is correct.

Oogenesis produces one viable egg and (up to) three polar bodies resulting from the cytoplasm's unequal distribution during meiosis.

Gametes (e.g., egg and sperm) become haploid (1N) through reductive division (i.e., meiosis), in which a diploid cell (2N) gives rise to four haploid sperm or one haploid egg and (up to) three polar bodies.

B: *interstitial cells* are stimulated by luteinizing hormone (LH) to produce testosterone, stimulating sperm development within the seminiferous tubules, along with FSH.

C: *follicle-stimulating hormone* (FSH) stimulates eggs to develop in follicles within the ovaries.

D: *follicle-stimulating hormone* (FSH) is involved in gamete production for males and females.

4. C is correct.

Let C designate wild-type and c designate the colorblind allele.

Mother is Cc, and the father is CY (a normal allele with a single copy of the X gene).

From mating, the mother's gamete (as a carrier due to her dad) is C or c, with a 50% probability of the gamete inheriting the C or c allele.

Assuming a boy (i.e., the father transmits Y and not X), the father's allele of Y (boy) is 100%, and the probability of the mother passing a c (colorblind) is 50%.

Probability of a son being colorblind is 50% or ½.

If the question had asked, "what is the probability that they will have a colorblind child?" the analysis changes to determine the probability for all children (not just boys in the original question).

continued…

Gametes produced by the mother are C and c with a 50% probability each.

The affected child is a boy.

What is the probability that the father passes the X or Y gene to the offspring? Here, the probability is 50%.

Individual event probabilities are multiplied to determine the overall probability:

$\frac{1}{2} \times \frac{1}{2} = \frac{1}{4}$

5. C is correct.

Primary oocytes are arrested in meiotic prophase I from birth until ovulation within the ovaries.

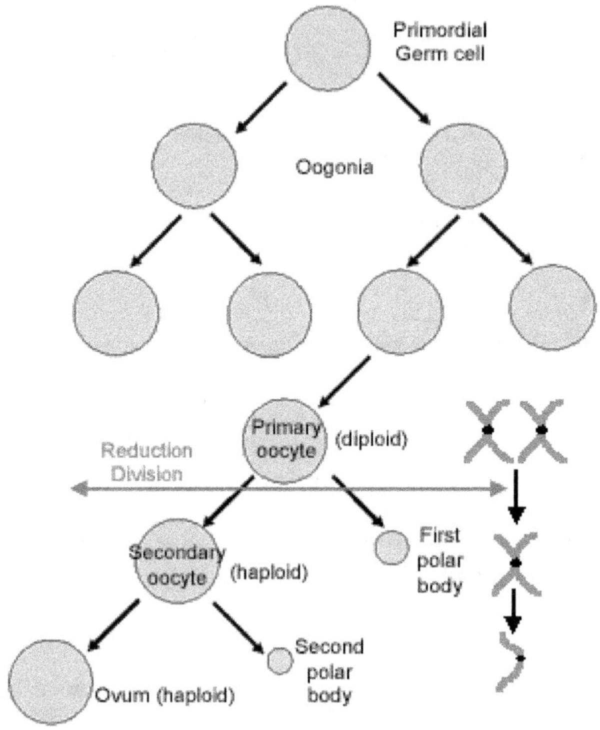

Diploid primordial germ cell undergoes two rounds of meiosis to form haploid ovum and polar bodies

6. C is correct.

Seminiferous tubules are in the testes and are the site of sperm production.

Spermatozoa are the mature male gametes in many sexually reproducing organisms.

Spermatogenesis is how spermatozoa are produced from male primordial germ cells by mitosis and meiosis.

Initial cells in this pathway are spermatogonia, which yield 1° spermatocytes by mitosis.

1° spermatocyte divides via meiosis into two 2° spermatocytes.

Meiosis converts 1° spermatocyte (2N) into four spermatids (1N).

continued…

2° spermatocytes complete meiosis by dividing into 2 spermatids as mature spermatozoa (i.e., sperm cells).

1° spermatocyte gives rise to two 2° spermatocytes that, by further meiosis, produce four spermatozoa.

Seminiferous tubules are in the testes and the location of meiosis and subsequent creation of gametes (e.g., spermatozoa for males or ova for females).

Epithelium of the tubule consists of *Sertoli cells* whose primary function is to nourish the developing sperm cells through the stages of spermatogenesis.

Sertoli cells act as phagocytes, consuming the residual cytoplasm during spermatogenesis.

Spermatogenic cells differentiate between the Sertoli cells, which differentiate through meiosis into sperm cells.

7. C is correct.

A human cell after the first meiotic division is 1N with 2 chromatids.

8. D is correct.

Anterior pituitary secretes *follicle-stimulating hormone* (FSH) and *luteinizing hormone* (LH) to influence the maturation of the follicle.

Follicle-stimulating hormone (FSH) stimulates the maturation of gametes (e.g., ova and spermatids).

Follicle-stimulating hormone (FSH) stimulates estrogen production, which aids in maturing the primary follicle.

Luteinizing (LH) stimulates ovulation and development of the corpus luteum; the mature corpus luteum secretes progesterone, causing the uterine lining to thicken and become vascular in preparation for implantation of the fertilized egg (i.e., zygote).

9. A is correct.

Sperm synthesis begins within the seminiferous tubules of the testes.

Sperm undergoes maturation and storage in the *epididymis*.

During ejaculation, sperm are released from the *vas deferens*.

Seminal vesicles are a pair of male accessory glands that produce about 50-60% of the liquid of the semen.

Mnemonic: SEVEN UP — sperm path during ejaculation:

Seminiferous tubules > **E**pididymis > **V**as deferens > **E**jaculatory duct > **N**othing > **U**rethra > **P**enis

10. D is correct.

Genetic recombination is when two DNA strand molecules exchange genetic information (i.e., a base composition within the nucleotides), resulting in new combinations of alleles (i.e., alternative forms of genes).

In eukaryotes, the natural process of genetic recombination during meiosis (i.e., the formation of gametes – eggs and sperm) results in genetic information passed to progeny.

continued…

Genetic recombination in eukaryotes involves pairing homologous chromosomes (i.e., a set of maternal and paternal chromosomes), which may involve nucleotide exchange between the chromosomes.

Information exchange may occur without physical exchange when a section of genetic material is duplicated without a change in the donating chromosome or by breaking and rejoining the DNA strands (i.e., forming new molecules of DNA).

Mitosis may involve recombination, where two sister chromosomes form after DNA replication. New combinations of alleles are often not produced because the sister chromosomes are usually identical.

In meiosis and mitosis, recombination occurs between similar DNA molecules (homologous chromosomes or sister chromatids, respectively).

In meiosis, non-sister (i.e., same parent) homologous chromosomes pair with each other, and recombination often occurs between non-sister homologs.

For somatic cells (i.e., undergo mitosis) and gametes (i.e., undergo meiosis), recombination between homologous chromosomes or sister chromatids is a common DNA repair mechanism.

11. C is correct.

Primary oocytes (2N cells) are arrested in prophase I of meiosis I until puberty. There is additional development into secondary oocytes that occurs prior to ovulation.

After ovulation, the oocyte is arrested in metaphase of *meiosis II until fertilization.*

12. C is correct.

Hypothalamus secretes releasing factors (i.e., tropic hormones), causing the anterior pituitary to release LH and FSH.

Follicle-stimulating hormone (FSH) stimulates ovaries to produce mature *ovarian follicles.*

During the follicular stage, the ovary produces estrogen.

As estrogen is released, FSH levels drop, and LH levels increase, which triggers the follicle to release the ovum (i.e., ovulation).

Luteinizing hormone (LH) affects the corpus luteum (i.e., formerly the follicle) that secretes progesterone.

Estrogen is first secreted by ovaries (under the influence of FSH) during the follicular stage (day 1 to 14), decreasing FSH secretions.

Decrease in FSH, along with an increase in LH, causes rupture of the follicle (i.e., ovulation) and formation of the *corpus luteum.*

As the corpus luteum matures, *progesterone levels increase.*

13. B is correct.

Pseudoautosomal regions are named because any genes within them are inherited, like autosomal genes. Pseudoautosomal regions allow males to pair and segregate X and Y chromosomes during meiosis.

Males have two copies of genes: one in the pseudoautosomal region of their Y chromosome and the other in their X chromosome's corresponding portion.

Typical females possess two copies of pseudoautosomal genes, as each X chromosome contains pseudoautosomal regions.

Crossing over (during prophase I) between the X and Y chromosomes is usually restricted to the pseudoautosomal regions.

Pseudoautosomal genes exhibit an *autosomal*, rather than sex-linked, inheritance pattern.

Females can inherit an allele initially on the Y chromosome of their father, and males can inherit an allele initially on the X chromosome of their father.

14. B is correct.

Primary spermatocyte completes the synthesis (S) phase of interphase, not the first meiotic division, and remains diploid (2N) with 46 chromosomes (i.e., 23 pairs).

15. B is correct.

Chromosomes (not chromatids) *segregate* during mitosis to produce identical somatic (body) 2N (diploid) cells from the parental cell. Meiosis is the process for germline cells (i.e., egg, sperm).

Mutations are inheritable changes in the cell's genetic (DNA) material; recombination occurs during prophase I (at the chiasma of the tetrad) of meiosis, and homologous chromosomes segregate during meiosis I.

During meiosis I, homologous chromosomes are separated. Subsequently, during meiosis II, the sister chromatids separate to produce four 1N (haploid) products, each with half the number of chromosomes as the original cell.

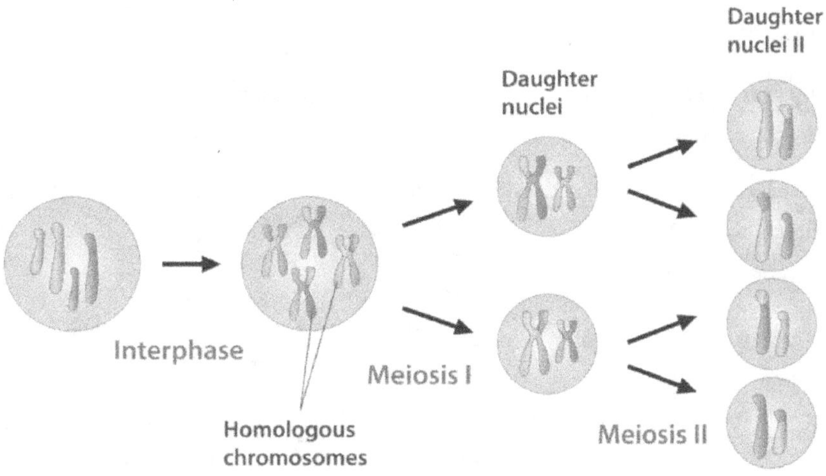

Chromosomes replicate in S phase of mitosis to produce homologous chromosomes. Meiosis I daughter cells are diploid (2N), while the gametes formed from meiosis II (reductive phase) are haploid (1N).

16. D is correct.

At day 14, LH surges, causing a mature follicle to burst and release the ovum from the ovary (i.e., ovulation).

Following ovulation, luteinizing hormone (LH) induces the ruptured follicle to develop into the corpus luteum, secreting progesterone and estrogen.

Prolactin stimulates milk production after birth.

Ovaries secrete estrogen and progesterone, while the *anterior pituitary secretes LH.*

Progesterone and estrogen *inhibit GnRH release*, inhibiting the release of FSH and LH, preventing additional follicles from maturing.

Progesterone stimulates development and maintenance of endometrium in preparation for embryo implantation.

17. B is correct.

Luteinizing hormone (LH) stimulates the release of testosterone in males by Leydig cells.

Testosterone is necessary for proper development of the testes, penis, and seminal vesicles.

Testosterone surges between the first and fourth months of life, and testosterone inadequacy is a primary cause of *cryptorchidism*.

During puberty, testosterone is needed for *secondary male sex characteristics* (e.g., growth of body hair, broadening of the shoulders, enlarging of the larynx, deepening of the voice, and increased secretions of oil and sweat glands)

Cortisol is a stress hormone that elevates blood glucose levels.

18. C is correct.

Leydig cells are *androgens* (i.e., male hormones such as 19-carbon steroids) that secrete testosterone, androstenedione, and dehydroepiandrosterone (DHEA) when stimulated by luteinizing hormone (LH) released by the pituitary gland.

Leydig cells are adjacent to *seminiferous tubules* in testicles.

Luteinizing hormone (LH) increases conversion of cholesterol to pregnenolone leading to testosterone synthesis and secretion by Leydig cells.

Prolactin (PRL) increases the response of Leydig cells to LH by increasing the number of LH receptors expressed on Leydig cells.

19. C is correct.

In females, secondary oocytes are haploid (i.e., single copies of 23 chromosomes, each with a pair of chromatids). They do not complete meiosis II (i.e., haploid with a single chromatid) until fertilized by a sperm.

Each month during puberty, one primary oocyte (i.e., diploid, each with a pair of chromatids) completes meiosis I to produce a secondary oocyte (1N) and a polar body (1N).

The 1N secondary oocyte (i.e., 23 chromosomes each with a pair of chromatids) is expelled as an ovum from the follicle during ovulation, but meiosis II does not occur until fertilization.

Menarche is marked by the onset of menstruation and signals the possibility of fertility.

Menstruation is the sloughing of the *endometrial lining* of the uterus during the monthly hormonal cycle.

Menopause is when menstruation ceases.

20. D is correct.

Klinefelter syndrome describes the symptoms from additional X genetic material in males.

Turner syndrome is the condition in females with a single X (monosomy X) chromosome.

XYY syndrome is a genetic condition in which a human male has an extra male (Y) chromosome, giving 47 chromosomes instead of 46.

XYY is not inherited but usually occurs during the formation of sperm cells.

Nondisjunction error during anaphase II (meiosis II) results in sperms with an extra Y chromosome.

If an atypical sperm contributes to genetic makeup, the child has an extra Y chromosome in each somatic cell.

Triple X syndrome is not inherited but usually occurs during the formation of gametes (e.g., ovum and sperm) because nondisjunction in cell division results in reproductive cells with additional chromosomes.

During cell division, errors (non-disjunction) can result in gametes with additional chromosomes.

An egg or a sperm may gain an extra X chromosome due to non-disjunction, and if one gamete contributes to the zygote, the child will have an extra X chromosome in each cell.

21. B is correct.

Spermatozoon development produces motile, mature sperm, fusing with an ovum to form the zygote.

Eukaryotic *cilia* are structurally identical to eukaryotic flagella, but distinctions are based on function and length.

Microtubules form the sperm's flagella, which is necessary for movement through the cervix, uterus, and along the fallopian tubes (i.e., oviducts).

Acrosomal enzymes digest the outer zona pellucida (a glycoprotein membrane surrounding the plasma membrane of an oocyte) to permit the fusion of sperm and egg.

continued…

Sperm midpieces (diagram below) have many mitochondria used for ATP production for the sperm's movement (via the flagellum) through the female cervix, uterus, and Fallopian tubes.

D: *testosterone* is necessary for *spermatogenesis*.

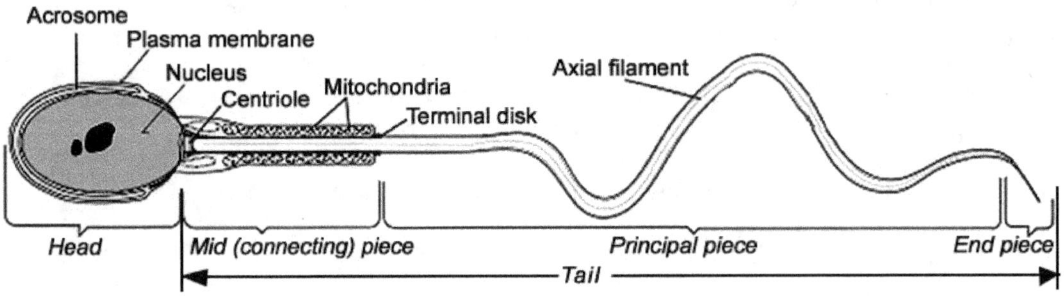

Human spermatozoon

22. A is correct.

Seminiferous tubules are in the testes where meiosis occurs to create gametes (e.g., spermatozoa).

Epithelium of *seminiferous tubules* consists of *Sertoli* (i.e., *nurse cells*) cells that nourish developing sperm and act as phagocytes by consuming residual cytoplasm during spermatogenesis.

Between the Sertoli cells are spermatogenic cells, which differentiate through meiosis into sperm cells.

23. C is correct.

In males, diploid spermatogonia (2N) cells undergo meiosis I to produce 1° diploid spermatocyte (2N), which undergoes meiosis II to yield 2° haploid spermatocytes (1N).

2° spermatocytes undergo meiosis II to produce four spermatids (1N) that mature into spermatozoa (i.e., sperm).

In females, 1° oocyte is (2N), whereby 2° oocyte undergoes the second meiotic division to produce two (1N) cells – a mature oocyte (i.e., ovum – female gamete) and another polar body.

Primary oocyte is diploid (2N).

During fertilization, a (1N) ovum and a (1N) sperm fuse to produce a (2N) zygote.

Spermatogonium is a diploid (2N) cell.

continued…

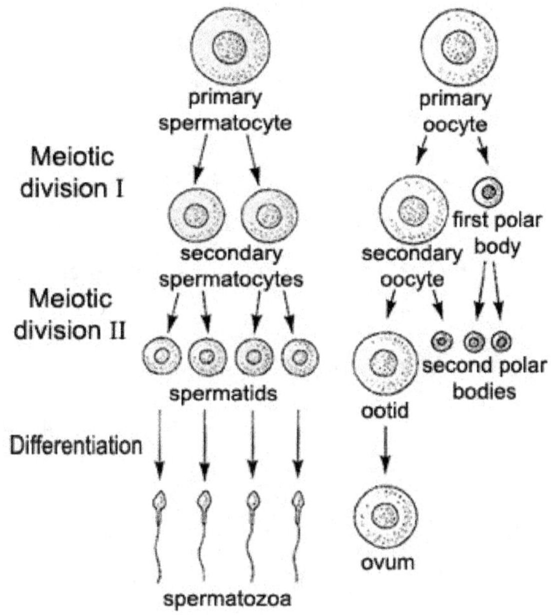

Spermatogenesis Oogenesis

Diploid primary oocytes and oocytes undergo two rounds of meiosis forming haploid spermatozoa and ovum

24. D is correct.

Probability that a child will be male (or female) is ½.

Each event is *independent*; the first (or second or third) child does *not* affect future events.

Probability for 4 children is:

$$\frac{1}{2} \times \frac{1}{2} \times \frac{1}{2} \times \frac{1}{2} = 1/16$$

Same probability if the first and third child are female, with the second and fourth as male (or combination).

25. C is correct.

Chlamydia and other infections may scar the reproductive tract, preventing the ova from reaching the uterus and resulting in infertility.

26. C is correct.

Seminal vesicles secrete a significant proportion of the fluid that ultimately becomes semen.

50-70% of the seminal fluid originates from the seminal vesicles but is not expelled in the first ejaculate fractions dominated by spermatozoa and zinc-rich prostatic fluid.

27. D is correct.

Unequal division of cytoplasm occurs during the meiotic process of oogenesis (i.e., production of an egg cell). Meiotic divisions occur in two stages.

First stage produces precursor (2N) cells when a daughter cell (precursor egg) receives most of the cytoplasm; the other is devoid of sufficient cytoplasm (but genetically identical) and a nonfunctioning polar body.

In the *second division of oogenesis*, the large (2N) daughter cell divides again, and again one (1N) daughter cell receives the most cytoplasm (i.e., putative *ovum*).

The other daughter eggs become small nonfunctional (1N) *polar bodies*.

Polar bodies (from meiosis I) may divide (via meiosis II) to form two nonfunctional (1N) polar bodies.

The result is a potential of four (1N) cells.

However, only one—the one with a greater amount of cytoplasm during each meiotic division—becomes a functional egg (i.e., ovum) cell along with (up to) three (1N) polar bodies.

Bacterial cells divide for reproduction, and cytoplasmic division is equal.

Mitosis of kidney cells distributes cytoplasm equally.

Spermatogenesis is when one (2N) precursor cell forms four functional (1N) sperm cells because both meiosis divisions are equal and contain an equal amount of cytoplasm.

28. B is correct.

Dyneins are motor proteins (or *molecular motors*) using ATP to perform movement as *retrograde transport*.

Dynein transports cellular contents by *walking* along cytoskeletal microtubules towards the *minus-end* of the microtubules, usually oriented towards the cell center (i.e., *minus-end directed motors.*)

Kinesins are motor proteins that move toward the *plus-end* of the microtubules (i.e., *plus-end directed motors*).

Ovulation is the phase of the female menstrual cycle. A partially mature ovum that has yet to complete meiosis II is released from the ovarian follicles into the Fallopian tube (i.e., oviduct).

After ovulation, the egg can be fertilized by sperm during the *luteal phase*.

Ovulation is determined by circulating hormone levels and is not affected by a defect in dynein motor proteins.

Lungs require cilia to remove bacteria and other particulates.

Kartagener's syndrome males would be *infertile due to sperm immobility*.

Ova would not typically enter the Fallopian tubes (i.e., oviduct) because of the lack of cilia, which would cause an increased risk of ectopic pregnancy.

29. A is correct.

Estrous cycle (i.e., sexual desire) comprises the recurring physiologic changes induced by reproductive hormones in most mammalian females.

Estrous cycles start after sexual maturity in females and are interrupted by anestrous phases or pregnancies.

Menstrual cycle (diagram below) changes the uterus and ovary for sexual reproduction. The menstrual cycle is essential for producing eggs and preparing the uterus for pregnancy.

In humans, the length of a menstrual cycle varies significantly among women (ranging from 21 to 35 days), with 28 days designated as the average length.

Each cycle has three phases based on the ovary (ovarian cycle) or the uterus (uterine cycle).

Ovarian cycle consists of the *follicular, ovulation, and luteal phases*, whereas the uterine cycle has *menstruation, proliferative,* and *secretory phases.*

Endocrine system controls both cycles, and the regular hormonal changes can be interfered with using hormonal contraception to prevent reproduction.

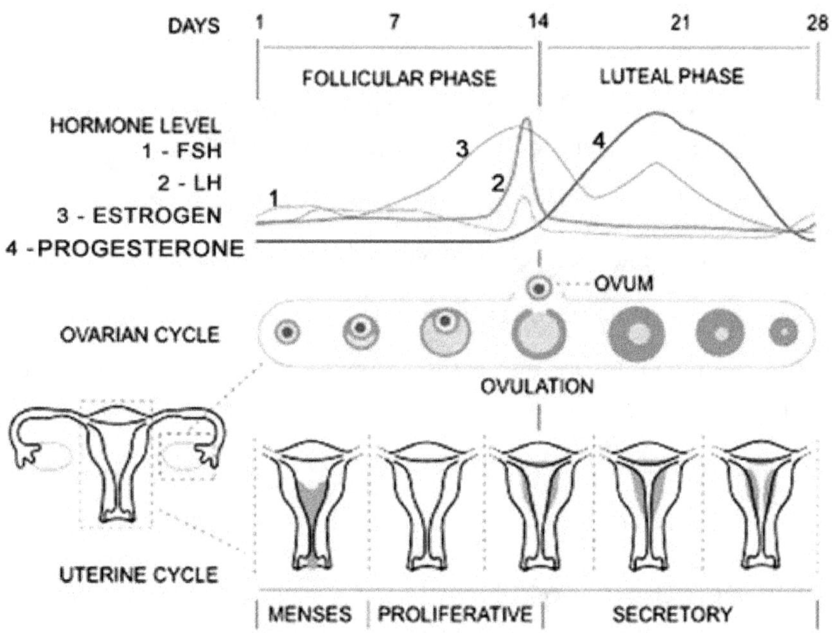

Menstrual cycle

30. D is correct.

DNA replication occurs during the synthesis (S) phase of interphase (i.e., G1, S, G2) to form *sister chromatids* joined by a centromere.

Replication occurs during the cell cycle *S* phase (i.e., interphase).

Transcription (in the nucleus) and *translation* (in the cytoplasm) occur during the S phase.

31. C is correct.

Progesterone is a steroid hormone involved in the female menstrual cycle, pregnancy (supports *gestation*), and embryogenesis.

Progesterone levels are relatively low in women during the pre-ovulatory phase of the menstrual cycle, rise after ovulation, and are elevated during the luteal phase.

During pregnancy, human chorionic gonadotropin (HCG) is released, maintaining the corpus luteum and allowing it to maintain progesterone levels.

At 12 weeks, the placenta begins to produce progesterone in place of the corpus luteum – this process is the *luteal-placental shift*.

After the luteal-placental shift, progesterone levels start to rise further.

After delivery of the placenta and during lactation, progesterone levels are very low.

Progesterone levels are relatively low in children and postmenopausal women.

Adult males have levels similar to those in women during the follicular phase of the menstrual cycle.

32. A is correct.

During meiosis, the gamete (e.g., ovum and sperm) reduces its genetic component from diploid (2N) to haploid (1N) with half the typical chromosome number for somatic (i.e., body cells).

When a haploid egg and sperm unite, they form a diploid zygote.

Ova contains an X chromosome, while sperm contains an X or a Y chromosome.

Gametes form in meiosis (i.e., two reduction divisions) without intervening chromosome replication.

During prophase I of meiosis I, *tetrads* form, and *sister chromatids* (i.e., a chromosome replicated in a prior S phase) undergo homologous recombination as *crossing over*.

Crossing over increases genetic variance within the progeny and is a driving force in the evolution of a species.

33. B is correct.

Fertilized egg is a *zygote* undergoing rapid cell divisions without significant growth, producing a cluster of cells of the same total size as the original zygote.

Cells derived from zygote cleavage are *blastomeres* and form a solid mass known as a *morula*.

Cleavage ends with the formation of the blastula.

In different species, mainly depending on the amount of yolk in the egg, the cleavage can be *holoblastic* (i.e., total or entire cleavage) or *meroblastic* (i.e., partial cleavage).

Egg pole with the highest yolk concentration is the *vegetal pole*, while the opposite end is the *animal pole*.

Humans undergo *holoblastic cleavage*.

34. C is correct.

Estrogen is a steroid hormone necessary for typical female maturation.

Estrogen stimulates the development of the female reproductive tract and contributes to *secondary sexual characteristics* and l*ibido*.

Estrogen is secreted by the *follicle* during the menstrual cycle and is responsible for the thickening of the endometrium in preparation for the implantation of the fertilized egg.

A: *luteinizing hormone* (LH) induces the ruptured follicle to develop into the *corpus luteum*.

B: *luteinizing hormone* stimulates testosterone synthesis in males.

D: *follicle-stimulating hormone* (FSH) is released by the anterior pituitary and promotes the development of the follicle, which matures and begins secreting estrogen.

35. C is correct.

X-linked dominance is a mode of inheritance whereby a dominant gene is on the X chromosome.

X-linked dominant is less common than X-linked recessive.

For X-linked dominant inheritance, one allele is sufficient to cause the disorder when inherited from a parent who has the disorder.

X-linked dominant traits do not necessarily affect males more than females (unlike X-linked recessive traits).

An affected father has all affected daughters but no affected sons (unless the mother is also affected).

36. A is correct.

Increased estrogen secretions initiate the luteal surge that increases LH secretion → ovulation.

37. B is correct.

Meiosis is cell division in sexually reproducing eukaryotes (animals, plants, and fungi), whereby the chromosome number is reduced by half, resulting in four genetically distinct haploid daughters.

DNA replication is followed by mitosis and *two rounds of cell division* in meiosis to produce four (1N) cells.

Two rounds of meiotic divisions are Meiosis I and Meiosis II.

Meiosis I is a *reductive division*, as the cells are reduced from diploid (2N) to haploid (1N).

Meiosis II (only G phase separates I and II) is an *equational division*, as the cells begin and end as haploids.

38. C is correct.

During prophase I, the *chromatin condenses* into chromosomes, the centrioles migrate to the poles, the spindle fibers begin to form, and the nucleoli and nuclear membrane disappear.

Homologous chromosomes physically pair and intertwine in the process of *synapsis*.

Prophase I: *chromatids* (i.e., a strand of the replicated chromosome) of *homologous chromosomes* break and exchange equivalent pieces of DNA via *crossing over*.

Metaphase I: *homologous pairs align* at the equatorial plane, and each pair attaches to a separate spindle fiber by its *kinetochore* (i.e., protein collar around the centromere of the chromosome).

Anaphase I: *homologous chromosomes separate* and are pulled by the spindle fibers to opposite cell poles.

Disjunction (i.e., separation of homologous chromosomes) is essential for segregation, as described by Mendel.

Telophase I: *nuclear membrane forms* around each new nucleus, and chromosomes consist of sister chromatids joined at the centromere.

The cell divides into two daughter cells, each receiving a haploid nucleus (1N) of chromosomes.

39. B is correct.

Sperm mature and are stored until ejaculation in the *epididymis*.

During ejaculation, sperm flow from the lower portion of the epididymis. Products from the prostate gland have not activated them, and they cannot swim.

Sperm is *transported via peristaltic action* of muscle layers within the *vas deferens* and forms *semen* mixed with diluting fluids of the seminal vesicles and other accessory glands before ejaculation.

Seminiferous tubules secrete testosterone. The primary functions of the testes are to produce sperm (i.e., spermatogenesis) and produce androgens (e.g., testosterone).

Leydig cells are localized between seminiferous tubules and produce and secrete testosterone and other androgens (e.g., DHT, DHEA) necessary for sexual development and puberty, secondary sexual characteristics (e.g., facial hair, sexual behavior, and libido), supporting spermatogenesis and erectile function.

Luteinizing hormone (LH) results in testosterone release.

Testosterone and follicle-stimulating hormone (FSH) are needed to support spermatogenesis.

40. D is correct.

Crossing over occurs during prophase I of meiosis.

Recombination occurs at the chiasma (physical contact) of the tetrad (two sets of sister chromatids).

Homologous chromosomes segregate during meiosis I.

41. A is correct.

In prophase I, *tetrads form*, genetic recombination occurs, and the spindle apparatus forms.

Chromosomes migrate to the poles of the cell during anaphase from the *splitting of the centromere*.

42. A is correct.

Spermatogenesis and *oogenesis* are *gametogenesis*.

Haploid (1N) gametes (i.e., ova and sperm) are produced by diploid (2N) cell reductive divisions (i.e., meiosis).

Spermatogenesis occurs in the gonads, whereby the cytoplasm equally divides during meiosis with the production of four viable sperm 1N cells.

Oogenesis occurs in the gonads, whereby the cytoplasm divides unequally. One 1N ovum (e.g., egg) receives the bulk of the cytoplasm and (up to) three additional 1N polar bodies.

Polar bodies contain a 1N genome (like sperm and egg) but lack sufficient cytoplasm for a viable gamete.

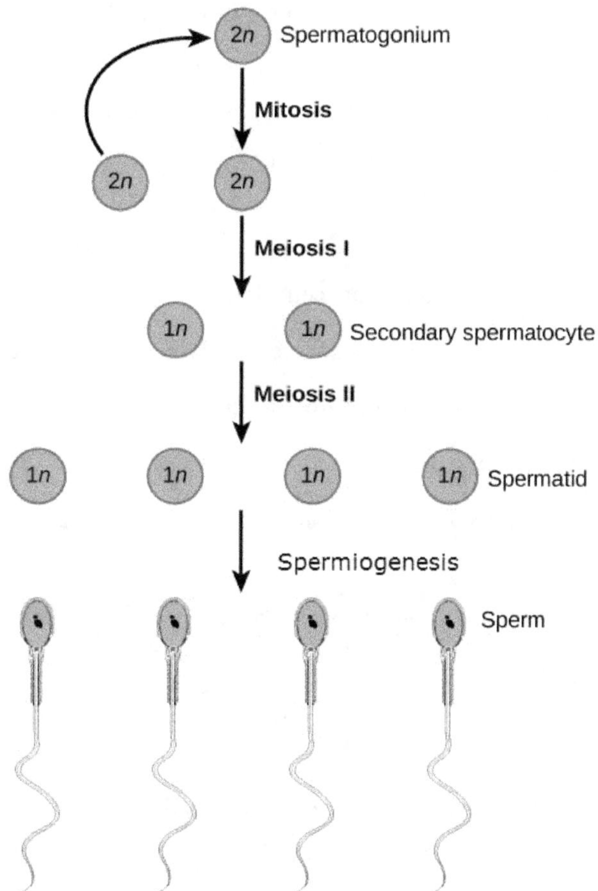

Diploid primary oocytes and oocytes undergo two rounds of meiosis forming haploid spermatozoa and ovum

43. B is correct.

Inner lining of the Fallopian tubes is covered with cilia, helping the egg move towards the uterus, where it implants if fertilized.

Fertilization usually happens when a sperm fuses with an egg in the Fallopian tube.

C: cilia lining the respiratory tract perform this function.

D: *Fallopian tubes* are isolated from the external environment, so pH fluctuations are not an issue.

44. B is correct.

Barr body is an *inactive X chromosome* in a female somatic cell rendered inactive by *lyonization* for species in which sex is determined by the Y chromosomes (i.e., humans and other species).

Lyon hypothesis states that cells with multiple X chromosomes inactivate all but one X randomly during mammalian embryogenesis.

In humans with more than one X chromosome, the number of Barr bodies visible at interphase is one less than the total number of X chromosomes.

For example, a man with Klinefelter syndrome of 47, XXY karyotype has a single Barr body.

A woman with a 47, XXX karyotype has two Barr bodies.

45. A is correct.

Polar bodies form in females as 1N, nonfunctional cells during meiosis.

Meiosis is a two-stage process whereby a 2N cell undergoes a reduction division to form two 1N cells.

Each 1N cell forms two 1N cells for four 1N cells from a parental 2N germ cell in the *second meiotic* division.

For sperm cells, four functional, unique 1N gamete sperm cells form.

For egg cells, the first meiotic division involves an *unequal division of cytoplasm* that results in one large cell and one small cell (i.e., the first polar body).

The large cell divides (second meiotic division) to generate *one functional egg* (1N ovum containing most of the cytoplasm) and another polar body.

First polar body divides again to form two other polar bodies. One large ovum (i.e., functional egg cell) and three polar bodies form during a female meiotic division.

Polar bodies form in mitosis, which is how somatic (i.e., body cells) cell divisions occur; equal cells do not form gametes (i.e., germline cells).

46. B is correct.

In males, *testosterone* is primarily synthesized in *Leydig cells* in the testis.

Testosterone synthesis is regulated by the *hypothalamic–pituitary–testicular axis.*

When testosterone levels are low, the hypothalamus releases *gonadotropin-releasing hormone* (GnRH), which stimulates the pituitary gland to release FSH and LH.

FSH and LH stimulate the testis to produce testosterone.

Testosterone acts on the hypothalamus and pituitary with high levels inhibiting GnRH and FSH/LH release through a negative feedback loop.

47. A is correct.

Turner syndrome (i.e., 45, X) describes several conditions in females, of which monosomy X (i.e., the absence of the entire sex (X) chromosome, the Barr body) is the most common.

Turner syndrome is a chromosomal abnormality when all or part of a sex chromosome is absent or abnormal.

48. B is correct.

If fertilization of the ovum does not occur, the *corpus luteum stops secreting progesterone and degenerates.*

Menstruation phase follows decreased progesterone secretion but does not result from increased estrogen levels.

Increased estrogen secretion (not LH) causes the luteal surge, which occurs earlier in the cycle.

Thickening of the endometrial lining occurs while estrogen and progesterone levels are high.

49. A is correct.

Vasectomy procedures prevent the movement of sperm along the vas deferens.

Notes for active learning

Fetal Development – Detailed Explanations

1. A is correct.

Respiratory exchanges during fetal life occur through the *placenta*.

Placenta connects the developing fetus to the uterine wall and allows nutrient uptake, waste elimination, and gas exchange via the mother's blood supply.

Umbilical cord connects the developing embryo (or fetus) and the placenta. During prenatal development, the umbilical cord is physiologically and genetically part of the fetus and contains two arteries (the umbilical arteries) and one vein (the umbilical vein).

Umbilical vein supplies the fetus with oxygenated, nutrient-rich blood from the placenta. Conversely, the fetal heart pumps deoxygenated, nutrient-depleted blood through the umbilical arteries back to the placenta.

Fetal circulatory systems change at birth as the newborn uses its lungs. After the infant's first breath, the newborn's cardiovascular system constricts the ductus arteriosus (i.e., connects the pulmonary artery to the aorta) and converts it to the ligamentum arteriosum. Resistance in the pulmonary blood vessels decreases, increasing blood flow to the lungs.

At birth, umbilical blood flow ceases, and blood pressure in the inferior vena cava decreases, causing a decrease in pressure in the *right atrium*. The *left atrial pressure* increases due to increased blood flow from the lungs. Increased left atrial pressure, coupled with decreased right atrial pressure, causes *closure* of the *foramen ovale*.

Ductus venosus, which shunts blood from the left umbilical vein directly to the inferior vena cava to allow oxygenated blood from the placenta to bypass the liver, completely closes within three months after birth.

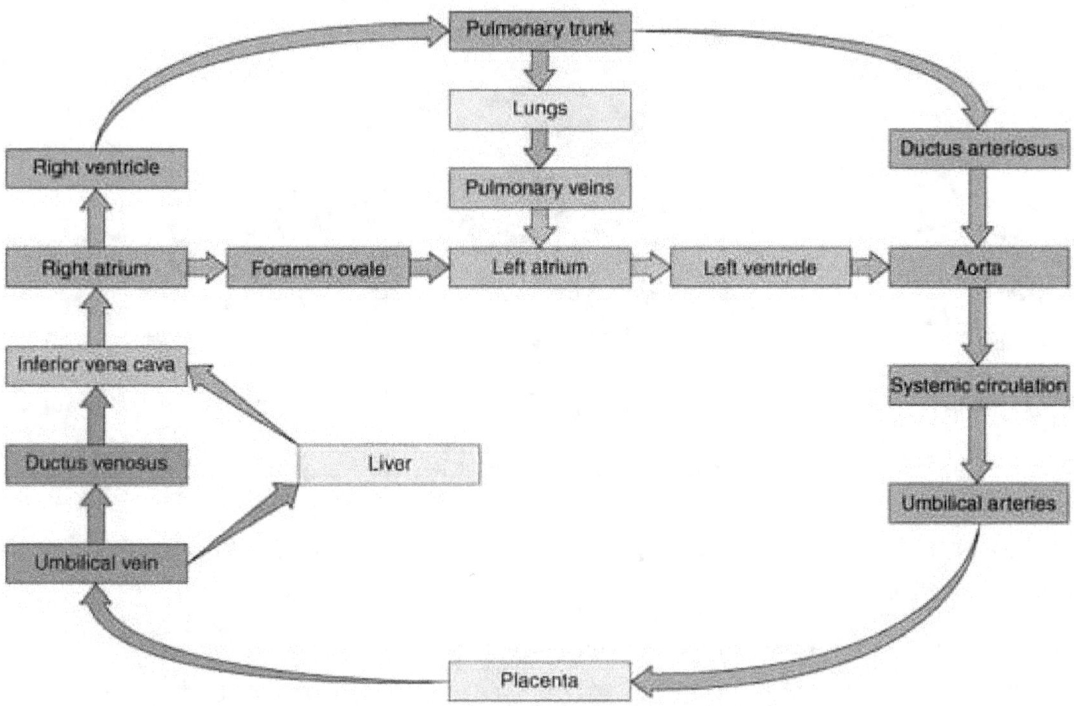

Adult and fetal circulatory with ductus arteriosus and ductus venosus providing shunt pathways

2. D is correct.

Indeterminate cleavage results in cells that maintain the ability to develop into a complete organism.

A: *zygote implantation* into the uterus causes cell migration transforming the blastula from a single-cell layer into a three-layered gastrula (i.e., ectoderm, mesoderm, and endoderm).

B: *blastulation* begins when the morula develops the blastocoel as a fluid-filled cavity that (by the fourth day) becomes the blastula as a hollow sphere of cells.

C: *determinate cleavage* results in cells whose differentiation potentials are determined early in development.

3. B is correct.

Mutations in *Drosophila*, transforming one body segment into a different one, are in homeotic genes.

Homeotic genes encode the related homeodomain proteins involved in developmental patterns and sequences.

Homeotic genes influence the development of specific structures in plants and animals, such as the Hox and ParaHox genes, which are essential for segmentation.

Homeotic genes determine where, when, and how body segments develop.

For example, alterations in homeotic genes in laboratory flies cause changes in patterns of body parts, sometimes producing dramatic effects, such as legs growing in place of antennae or an extra set of wings.

Hox genes are *homeotic transcription factors* are crucial for controlling the body plan along the *anterior-posterior axis* (i.e., craniocaudal axis) and specify the segment identity of tissues within the embryo.

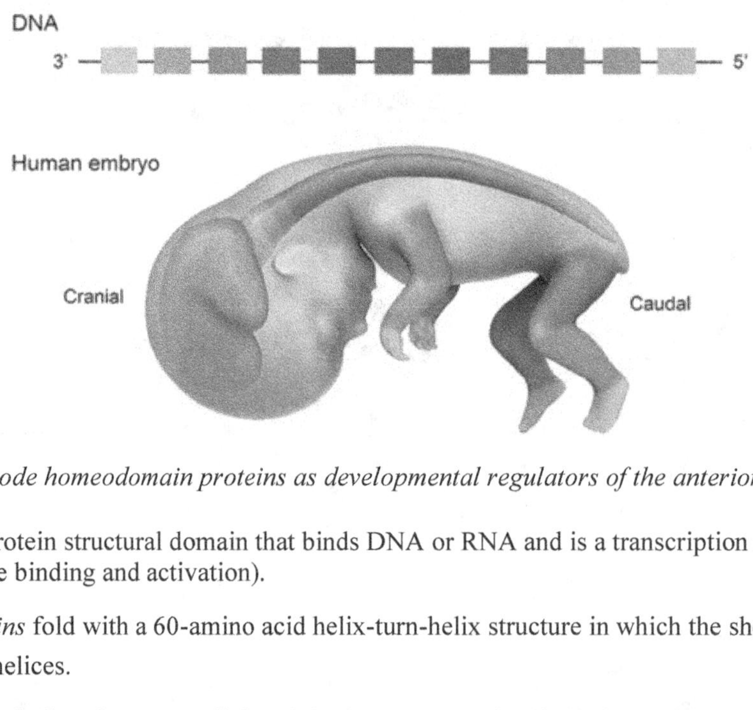

HOX genes encode homeodomain proteins as developmental regulators of the anterior-posterior axis

Homeodomain is a protein structural domain that binds DNA or RNA and is a transcription factor (i.e., facilitates polymerase binding and activation).

Homeodomain proteins fold with a 60-amino acid helix-turn-helix structure in which the short loop regions connect three alpha-helices.

N-terminus of the two helices is *antiparallel,* and the longer C-terminal helix is roughly perpendicular to the axes established by the first two.

Third helix functions as a *transcription factor* and interacts directly with DNA.

4. B is correct.

Ectoderm gives rise to hair, nails, skin, brain, and nervous system. Ectoderm forms "*outer linings*," including the epidermis (i.e., outermost skin layer) and hair. Ectoderm is the precursor to *mammary glands* and *central* (CNS) and *peripheral nervous systems* (PNS).

Connective tissue is derived from *mesoderm.*

Rib cartilage is derived from *mesoderm.*

Epithelium of the digestive system is derived from *endoderm.*

5. D is correct.

Chorion is a membrane between the developing fetus and mother in humans and most mammals.

Chorion consists of two layers: an outer layer formed by the *trophoblast* and an inner layer formed by the *somatic mesoderm* (in contact with the amnion).

6. A is correct.

The embryo does *not* support the maintenance of the corpus luteum when an embryo lacks synthesis of human chorionic gonadotropin (hCG).

7. C is correct.

Mesoderm develops into the circulatory system, the musculoskeletal system, the excretory system, the outer coverings of internal organs, the gonads, and various types of muscle tissue.

A: *ectoderm* develops into the brain and nervous system, hair and nails, lens of the eye, inner ear, sweat glands, the lining of the nose and mouth, and skin epidermis.

B: *epidermis* is not an embryonic germ layer but the layer of the skin that covers the dermis.

D: *endoderm* develops into the epithelial lining of the digestive tract, respiratory tracts, lining of the liver, bladder, pancreas, thyroid, and alveoli of the lungs.

8. B is correct.

Gastrulation is a phase early in the embryonic development of animals, during which the single-layered blastula is reorganized into a trilaminar (*three-layered*) structure of the gastrula. Gastrulation occurs after cleavage, the formation of the blastula and primitive streak, followed by organogenesis when individual organs develop within the newly formed germ layers.

Following gastrulation, cells are organized into sheets of connected cells (as in epithelial) or mesh of isolated cells (i.e., *mesenchyme*). Each germ layer gives rise to specific tissues and organs in the developing embryo.

The three germ layers are the *ectoderm*, *mesoderm,* and *endoderm.*

Ectoderm gives rise to the epidermis and other tissues that will form the nervous system.

Mesoderm is between the ectoderm and the endoderm. It gives rise to somites that form muscle, cartilage of the ribs and vertebrae, dermis, notochord, blood and blood vessels, bone, and connective tissue.

Endoderm gives rise to the epithelium of the respiratory system, digestive system, and organs associated with the digestive system (e.g., liver and pancreas).

Gastrulation occurs when a blastula of one layer folds inward and enlarges to create the *three primary germ layers* (i.e., endoderm, mesoderm, and ectoderm).

Archenteron gives rise to

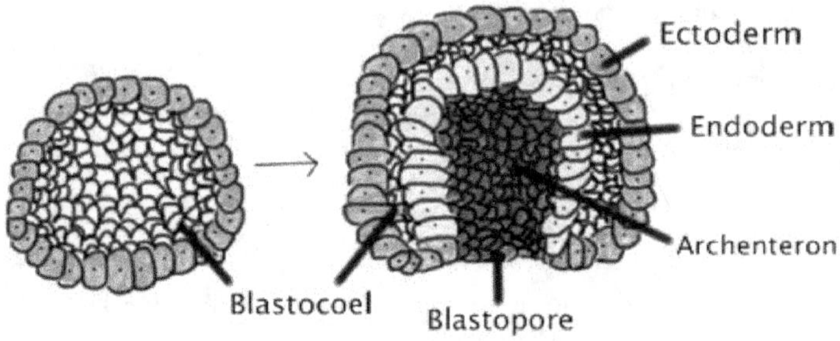

Blastula (left) becomes the gastrula (right) with the formation of the three primary germ layers

9. A is correct.

During the first *eight weeks* of development, myelination of the spinal cord does not occur.

10. A is correct.

Ectoderm germ layers give rise to the skin epidermis and the nervous system.

Endoderm germ layer gives rise to the lining of the digestive system, its associated glands, organs (e.g., liver and pancreas), and the lungs.

Mesoderm gives rise to most organs and systems, including excretory, reproductive, muscular, skeletal, and circulatory systems. If tissue is not specifically endoderm or ectoderm, it is likely to *be mesoderm-derived.*

The incorrect choices are *mesoderm-derived* tissues or structures.

11. C is correct.

Fate mapping traces the embryonic origin of tissues in the adult by establishing the correspondence between individual cells (or groups of cells) at one stage of development and their progeny at later stages.

When carried out at single-cell resolution, this process is *cell lineage tracing*.

A: *phylogenetic tree* (i.e., evolutionary tree) is a branching diagram showing the inferred evolutionary relationships (phylogeny) among biological species or other entities based upon similarities and differences in their physical or genetic characteristics.

The taxa joined in the tree are implied to have descended from a *common ancestor*.

B: *pedigree diagram* shows the occurrence and appearance (i.e., phenotypes) of a gene (or phenotype) in ancestors.

A pedigree presents family information in an easily readable chart.

Pedigrees use standardized symbols: squares represent males and circles females. Pedigree construction is a family history, and details about an earlier generation may be uncertain as memories fade.

For an unknown sex, a diamond is used. Phenotype in question is represented by a filled-in (darker) symbol.

Heterozygotes are indicated by a shaded dot inside a symbol or a half-filled symbol.

D: *linkage map* is a genetic map of an experimental population that shows the position of its known genes or genetic markers relative to each other based on recombination frequency rather than a specific physical distance along each chromosome.

Genetic linkage is the tendency of genes proximal on a chromosome to be inherited together during meiosis.

Genes whose loci (i.e., position) are closer to each other are less likely to be separated onto different chromatids during chromosomal crossover (i.e., during prophase I) and are genetically *linked.*

12. A is correct.

Hyaline cartilage tissue is the precursor of long bones in the embryo.

13. D is correct.

Gene expression changes as development proceeds; different proteins are encoded by expressed genes.

During development, cells change in their ability to respond to signals from other tissues and in their ability to induce changes in other cells. These changes during development are inherited by daughter cells.

A: *microtubules* are involved in mitosis but do not change during development.

C: *somatic cells* have the same genes (exceptions include gametes and B and T cells).

14. C is correct.

Acrosome is at the tip of the sperm and contains specialized secretory molecules.

Acrosomal reaction is due to a signaling cascade involving the glycoproteins on the egg's surface.

Acrosin digests the zona pellucida and membrane of the oocyte.

Part of the sperm cell membrane fuses with the egg cell's membrane, and the contents of the head enter the egg fused with the plasma membrane. This allows the sperm to release its degradation enzymes, penetrate the egg's tough coating, and bind and fuse.

A *pronucleus* is the nucleus of a sperm or an egg cell after the sperm enters the ovum but before they fuse.

15. D is correct.

Three primary germ layers form during gastrulation in embryogenesis: *ectoderm*, *mesoderm,* and *endoderm*.

Ectoderm is the external germ layer giving rise to the skin, fingernails, and nervous system (including the eye).

Mesoderm is the middle germ layer and gives rise to most organ systems, including the musculoskeletal, cardiovascular, reproductive, and excretory systems.

Endoderm is the innermost germ layer and gives rise to the gallbladder, liver, pancreas, and epithelial lining of luminal structures and accessory digestive organs.

16. D is correct.

Ductus arteriosus is a blood vessel connecting the pulmonary artery to the proximal descending aorta in the developing fetus. It allows most of the blood from the right ventricle to bypass the fetus's fluid-filled non-functioning lungs. Upon closure at birth, the *ductus arteriosus* becomes the *ligamentum arteriosum.*

Ligamentum teres (i.e., round ligament) refers to several structures:

ligamentum teres uteri (i.e., round ligament of the uterus),

ligamentum teres hepatis (round ligament of the liver), and

ligamentum teres femoris (i.e., ligament of head of the femur).

17. C is correct.

During gastrulation, invagination of the blastula forms the mesoderm as the third primary germ layer.

18. C is correct.

Homeotic genes encode homeodomain proteins involved in developmental patterns and sequences.

Homeodomain is a protein structural domain that binds DNA or RNA and is often in transcription factors.

Homeodomain protein folding consists of a 60-amino acid helix-turn-helix structure in which short loop regions connect three alpha-helices. The N-terminus of the two helices is antiparallel, and the longer C-terminal helix is roughly perpendicular to the axes established by the first two.

The third helix functions as a transcription factor and interacts directly with DNA.

19. D is correct.

Neurulation and *organogenesis* follow gastrulation.

A: mitosis continues throughout development, but cleavage is a specific term reserved for the first few cell divisions when the zygote becomes the morula.

During cleavage, no growth occurs, and the morula is the same approximate size as the zygote.

B: *blastula formation* precedes gastrulation.

C: *blastocoel* is the fluid-filled central region of a blastula and forms early after fertilization when the zygote divides into many cells.

20. A is correct.

Trophoblast cells give rise to the outer layer of a blastocyst, provide nutrients to the embryo and develop into a large part of the placenta.

Trophoblasts form during the first stage of pregnancy and are the first cells to differentiate from the fertilized egg.

21. B is correct.

Placenta connects the developing fetus to the uterine wall and allows nutrient uptake, waste elimination, and gas exchange via the mother's blood supply.

Umbilical cord connects the developing embryo (or fetus) and the placenta.

During prenatal development, the *umbilical cord* is physiologically and genetically part of the fetus and contains two arteries (the *umbilical arteries*) and one vein (the *umbilical vein*).

Umbilical veins supply the fetus with oxygenated, nutrient-rich blood from the placenta.

The *fetal heart* pumps deoxygenated, nutrient-depleted blood through the umbilical *arteries* to the placenta.

22. C is correct.

For vertebrates, induction is how a group of cells cause *differentiation* in another group of cells.

For example, cells that form the notochord induce the formation of the neural tube.

An example of induction in vertebrate development is eye formation, where the optic vesicles induce the ectoderm to thicken and form the lens placode (i.e., thickened portion of ectoderm becoming lens), which induces the optic vesicle to form the optic cup, which induces the lens placode to form the cornea.

A: *neural tube* does develop into the nervous system but is not induced by another group of cells or tissue.

B: *thyroxin stimulating hormone* (TSH) does stimulate the secretion of thyroxine but is not induction.

D: *neurons synapse* with other neurons via neurotransmitters (i.e., chemical messenger), but they do not induce changes in other tissues as does induction.

23. D is correct.

Acrosomal reaction by the sperm is hydrolytic enzymes degrading the plasma membrane.

Acrosome is at the tip of the sperm and contains specialized secretory molecules.

Acrosomal reaction is due to a signaling cascade involving the glycoproteins on the egg's surface.

Acrosin digests the zona pellucida and membrane of the oocyte, and the sperm releases its degradation enzymes to penetrate the egg's tough coating and allow the sperm to bind and fuse with the egg.

24. C is correct.

A: *endoderm* is the innermost germ layer and gives rise to the inner lining of the respiratory and digestive tracts and associated organs.

B: *blastula* refers to a hollow ball of embryonic cells arising from the morula. The blastula is *not* a germ layer.

D: *ectoderm* is the outermost germ layer and gives rise to the hair, nails, eyes, skin, and central nervous system.

25. B is correct.

Inner cell mass is the cells inside the primordial embryo that forms before implantation and gives rise to the definitive structures of the fetus.

Primitive streak is a structure that forms in the blastula during the early stages of embryonic development.

The presence of the primitive streak establishes bilateral symmetry, determines the site of gastrulation, and initiates germ layer formation.

26. D is correct.

The *fetal circulatory system* changes as newborns begin using their lungs at birth.

Resistance in the pulmonary blood vessels decreases, causing an increase in blood flow to the lungs.

Umbilical blood flow ceases at birth, and blood pressure in the inferior vena cava decreases, which causes a decrease in pressure in the right atrium.

In contrast, the left atrial pressure increases due to increased blood flow from the lungs. Increased left atrial pressure, coupled with decreased right atrial pressure, causes the closure of the foramen ovale.

Ductus arteriosus (i.e., connects the pulmonary artery to the aorta) constricts and subsequently is sealed.

Ductus venosus, which shunts blood from the left umbilical vein directly to the inferior vena cava to allow oxygenated blood from the placenta to bypass the liver, completely closes within three months after birth.

The fetus produces adult hemoglobin a few weeks before birth (2 α and 2 β chains; alpha and beta) though lower amounts of fetal hemoglobin continue until the production completely stops.

After the first year, low fetal hemoglobin levels (2 α and 2 γ chains; alpha and gamma) are in the infant's blood.

27. B is correct.

Homeotic genes encode the related homeodomain proteins involved in developmental patterns and sequences.

The *homeobox* is a stretch of DNA about 180 nucleotides long that encodes a homeodomain (i.e., protein) in vertebrates and invertebrates.

Exons (expressed sequences) are retained during RNA processing of the primary transcript (hnRNA) into mRNA for translation into proteins.

28. A is correct.

Somatic cells have the same genome, but individual cells express different genes.

Differences in expression are *spatial* (i.e., cell type) or *temporal* (i.e., stages of development).

I: *ectoderm tissue* arises after gastrulation as a primary germ layer. Each germ layer retains its *determination.*

II: *somatic cells* have identical genomes compared to their parents.

Gametes (via segregation and recombination during prophase I) have unique 1N genomes compared to their parents.

29. C is correct.

Yolk sac in humans gives rise to *blood cells* and *gamete-forming cells.*

Chorion is a double-layered membrane of trophoblast and extra-embryonic mesoderm.

Chorion gives rise to the fetal part of the placenta.

Luteal placental shift (7-9 weeks) is when the placenta develops enough to produce hormones to sustain the pregnancy. Before this shift, the *placenta secretes progesterone* instead of the *corpus luteum.*

In humans, waste goes to the placenta and is received by the mother.

Non-placental organisms use allantois to collect waste.

30. A is correct.

Trophoblast is primarily responsible for forming placental tissue. They give rise to the outer layer of a blastocyst, provide nutrients to the embryo, and develop into a large part of the placenta.

Trophoblasts form during the first stage of pregnancy and the first cells to differentiate from the fertilized egg.

31. C is correct.

Spina bifida is a bone abnormality that arises from the embryonic mesoderm germ layer.

A lesion to the mesoderm would likely affect the development of other structures based on different connective tissue (e.g., blood, blood vessels, muscles, and connective tissue of organs).

Thus, this lesion affects the development of blood vessels and muscles.

I: *intestinal epithelium* develops from the *endoderm.*

II: skin, hair, and the nervous system develop from the *ectoderm.*

32. B is correct.

Homeotic genes encode the related homeodomain proteins involved in developmental patterns and sequences.

Homeotic genes are involved in developmental patterns and sequences. They determine where, when, and how body segments develop.

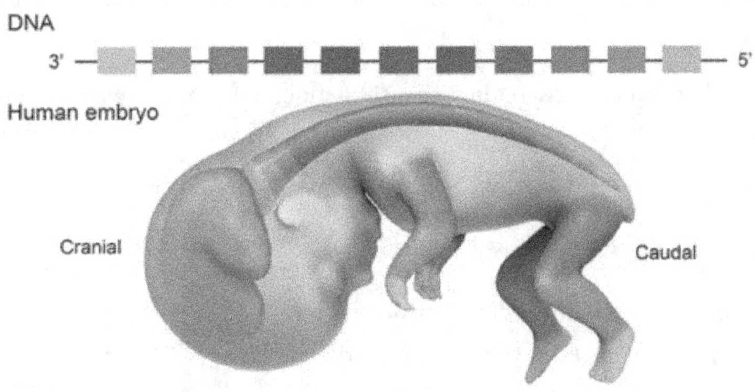

HOX genes encode homeodomain proteins as developmental regulators of the anterior-posterior axis. The linear sequence of genes on the chromosome corresponds to the cranial-caudal orientation

Alterations in these genes cause changes in body parts and structure patterns, sometimes resulting in dramatic effects.

Homeodomain is a protein structural domain that binds DNA or RNA and is a transcription factor (i.e., facilitates polymerase binding and activation).

33. D is correct.

Mesoderm forms muscles, blood, bone, reproductive organs, and kidneys.

34. C is correct.

The *primitive streak* is a transitional structure formed at the onset of gastrulation.

Inner cell mass is converted into the trilaminar embryonic disc comprised of the three germ layers: ectoderm, mesoderm, and endoderm.

Primitive streak formation is not part of implantation but an early process of embryonic development.

35. A is correct.

During labor, the *oxytocin* hormone stimulates contractions of uterine smooth muscle.

36. D is correct.

Ectoderm cells are determined but not differentiated because they have the potential to develop into more than one type of tissue, but not any type.

Experimentally, ectoderm development was influenced (i.e., induced) by changing locations within the developing embryo.

The underlying *mesoderm* differentiated because it released the molecular inducers as signals to the overlying ectoderm (undifferentiated).

A: *ectoderm* cell *location* is essential because their development (i.e., differentiation into specific tissue/structures) is induced by the underlying mesoderm cells that send chemical substances (i.e., inducers) specific to the position of the mesoderm.

Mesoderm, not ectoderm, determines cellular differentiation at this stage of development.

B: *differentiated fate* of the cells has not yet been determined because the transplanted cells would develop into wing feathers instead of claws.

C: ectoderm cells cannot develop into any tissue because the location influences the development of the transplanted ectoderm cells.

37. D is correct.

Recognition of the sperm by the *vitelline envelope* (i.e., a membrane between the outer zona pellucid and inner plasma membrane) triggers the cortical reaction, transforming it into the hard *fertilization membrane* as a physical barrier to other spermatozoa (i.e., *slow block to polyspermy*).

Cortical reaction within the egg is analogous to the acrosomal reaction within the sperm.

38. C is correct.

Polyspermy in humans results in a nonviable zygote.

39. C is correct.

Mesoderm germ layer gives rise to many tissues.

A: *intestinal mucosa* is derived from the endoderm

B: *nerve* is derived from the ectoderm.

D: *lung epithelium* is derived from the endoderm.

40. C is correct.

Genomic imprinting is *epigenetic* (i.e., heritable change) when specific genes are expressed in a parent-of-origin-specific manner. Genomic imprinting is an epigenetic process involving *DNA methylation* and *histone modulation* to achieve single allelic gene expression without altering the original genetic sequence.

41. C is correct.

An embryo requires a *greater protein translation rate* using ribosomes to read mRNA transcripts.

42. C is correct.

Zygote (i.e., fertilized egg) is the first structure to form during fertilization in humans.

43. A is correct.

Capacitation is the final step in the maturation of spermatozoa. It is required for them to be competent to fertilize an oocyte.

Capacitation destabilizes the acrosomal sperm head membrane, allowing greater binding between sperm and oocyte by removing steroids (e.g., cholesterol) and non-covalently bound glycoproteins, which increases membrane fluidity and Ca^{2+}permeability. Ca^{2+} influx produces intracellular cAMP levels and increasesd sperm motility.

44. C is correct.

Endoderm develops into the *epithelial linings* of the digestive and respiratory tracts, parts of the liver, thyroid, pancreas, and bladder lining.

45. C is correct.

Homeotic genes influence the development of specific structures in plants and animals, such as the Hox and ParaHox genes, which are essential for segmentation.

Homeotic genes determine where, when, and how body segments develop in organisms.

Alterations in these genes cause changes in patterns of body parts, sometimes causing dramatic effects, such as legs growing in place of antennae or an extra set of wings.

B: *loss-of-function* mutations result in the gene product with less or no function. When the allele has a complete loss of function (i.e., null allele), it is *amorphic* (i.e., complete loss of gene function) mutation.

Phenotypes associated with such mutations are most often recessive.

Exceptions are haploid or haploinsufficiency (i.e., a diploid organism has a single copy of the functional gene) when the reduced dosage of a normal gene product is not enough for a typical phenotype.

46. A is correct.

The cells can assume several different fates and are not yet terminally differentiated.

B: gastrula cells influenced by their surroundings are *competent.*

C: *ectoderm layer* gives rise to the eye, among other structures.

D: cells can become other ectoderm tissue (e.g., gills).

47. B is correct.

Proteases and *acrosin* enzymes degrade the protective barriers around the egg and allow the sperm to penetrate.

Acrosomal reaction by the sperm is hydrolytic enzymes degrading the plasma membrane.

Acrosome is at the tip of the sperm and contains specialized secretory molecules.

Acrosomal reaction is due to a signaling cascade involving the glycoproteins on the egg's surface.

Acrosin digests the zona pellucida and membrane of the oocyte, and the sperm releases its degradation enzymes to penetrate the egg's tough coating and allow the sperm to bind and fuse with the egg.

48. A is correct.

The human blastocyst implants in the uterine wall about a week after fertilization.

49. B is correct.

The *ectoderm* develops into the nervous system, the epidermis, the eye lens, and the inner ear.

I: *endoderm* develops into the lining of the digestive tract, lungs, liver, and pancreas.

III: *mesoderm* develops into the connective tissue, muscles, skeleton, circulatory system, gonads, and kidneys.

50. A is correct.

After the infant's first breath, the newborn's cardiovascular system constricts the *ductus arteriosus* (i.e., connects the pulmonary artery to the aorta) and converts it to the *ligamentum arteriosum.*

Fetal circulatory systems change at birth as the newborn uses its lungs. Resistance in the pulmonary blood vessels decreases, increasing blood flow to the lungs.

At birth, umbilical blood flow ceases, and blood pressure in the inferior vena cava decreases, which causes a decrease in pressure in the right atrium.

continued...

In contrast, the left atrial pressure increases due to increased blood flow from the lungs. Increased left atrial pressure, coupled with decreased right atrial pressure, causes *closure* of the *foramen ovale*.

Ductus venosus, which shunts blood from the left umbilical vein directly to the inferior vena cava to allow oxygenated blood from the placenta to bypass the liver, completely closes within three months after birth.

Diagram below

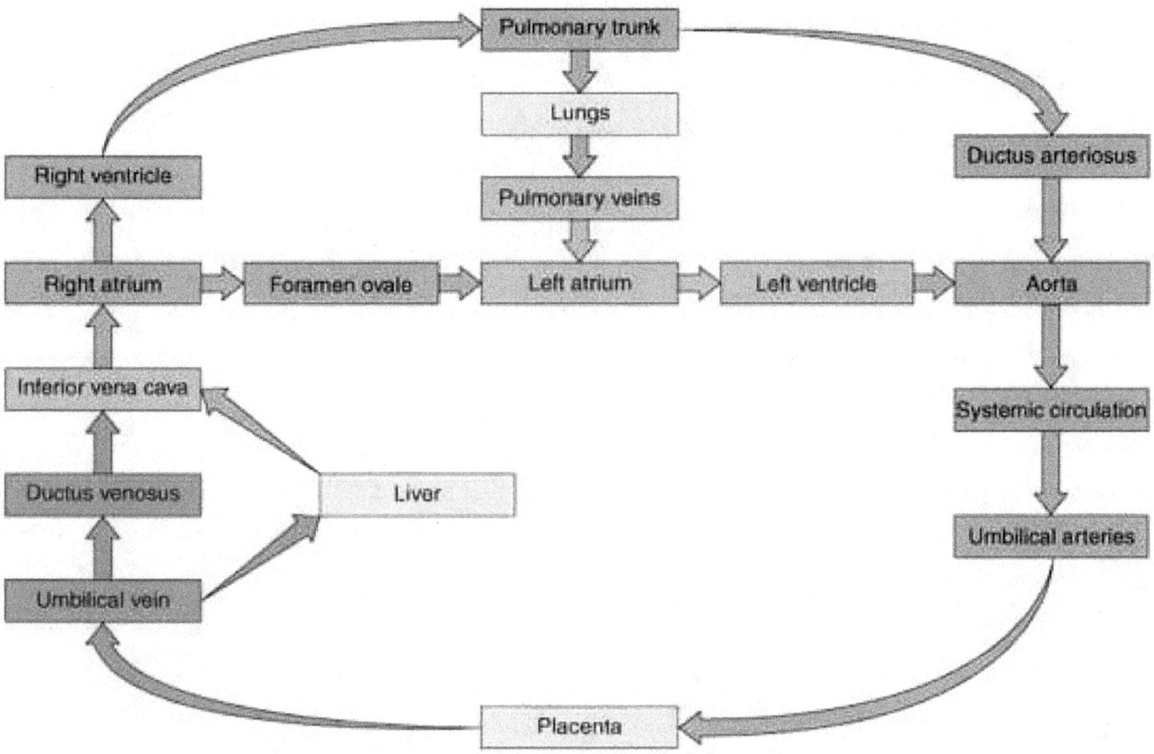

Adult and fetal circulatory with ductus arteriosus and ductus venosus providing shunt pathways

51. C is correct.

Genomic imprinting is an epigenetic (i.e., heritable changes in gene activity) process involving DNA methylation and histone remodeling for monoallelic (i.e., single allele) gene expression without altering the genetic sequence within the genome.

Epigenetic changes are established in the germline (i.e., cells that give rise to gametes of egg/sperm) and can be maintained through mitotic divisions.

52. A is correct.

Amniotic fluid, surrounded by the amnion, is a liquid environment around the egg that protects it from shock.

Embryonic membranes include:

1) *chorion* lines the inside of the shell and permits gas exchange;

2) allantois is a saclike structure developed from the digestive tract and functions in respiration, excretion, and gas exchange with the external environment;

3) amnion encloses *amniotic fluid* as a watery environment for embryogenesis and shock protection;

4) yolk sac encloses the yolk and transfers food to the developing embryo.

53. A is correct.

Mesoderm gives rise to the entire circulatory system, muscle, and most other tissue between the gut and skin (excluding the nervous system).

B: *ectoderm* gives rise to skin, the nervous system, retina, and lens.

C: *differentiated endoderm* has a restricted fate that does not include heart tissue.

D: *endoderm* gives rise to the inner lining of the gut.

54. C is correct.

At the end of the *first trimester*, an ultrasound can determine the sex of the fetus.

55. C is correct.

The egg is viable and able to be fertilized 12-24 hours after ovulation.

56. B is correct.

Blastulation begins when the *morula* develops a fluid-filled cavity (i.e., blastocoel). By the fourth day, it becomes a hollow sphere of cells (i.e., blastula).

A: *gastrula* is the embryonic stage characterized by the three primary germ layers (i.e., endoderm, ectoderm, and mesoderm), the blastocoel, and the archenteron.

The early gastrula is two-layered (i.e., ectoderm and endoderm), and shortly afterward, a third layer (i.e., mesoderm) develops.

Gastrulation is followed by organogenesis, when individual organs develop within newly formed germ layers.

C: *morula* is the solid ball of cells from the early cleavage stages in the zygote.

D: *zygote* is the (2N) cell formed by the fusion of two (1N) gametes (e.g., ovum and sperm).

57. B is correct.

Proteases and *acrosin* enzymes degrade the protective barriers around the egg and allow the sperm to penetrate.

Acrosomal reaction by the sperm is hydrolytic enzymes degrading the plasma membrane.

Acrosome is at the tip of the sperm and contains specialized secretory molecules.

Acrosomal reaction is due to a signaling cascade involving the glycoproteins on the egg's surface.

Acrosin digests the zona pellucida and membrane of the oocyte, and the sperm releases its degradation enzymes to penetrate the egg's tough coating and allow the sperm to bind and fuse with the egg.

58. C is correct.

Regarding fertilization, the vagina's acidic environment destroys millions of sperm cells.

59. B is correct.

Zygote is a fertilized egg undergoing rapid cell divisions without significant cellular growth to produce a cluster of cells of the same total size as the original zygote.

The cells derived from cleavage are *blastomeres* and form a solid mass known as a *morula.*

Cleavage ends with the formation of the *blastula.*

Among species, depending mainly on the amount of yolk in the egg, cleavage is:

holoblastic (i.e., total or entire cleavage) or

meroblastic (i.e., partial cleavage).

Egg pole with the highest yolk concentration is the *vegetal pole*, while the opposite is the *animal pole.*

Humans undergo *holoblastic cleavage* (i.e., total cleavage).

Notes for active learning

Frank J. Addivinola, Ph.D.

Dr. Frank Addivinola is the lead author and chief editor for Sterling Education. With his outstanding education, laboratory research, and university science and humanities teaching, he guided the development of this book.

Frank Addivinola earned his BA in biology from Williams College, Master of Liberal Arts with a concentration in biology at Harvard University, Masters in Biotechnology at Johns Hopkins University, Masters of Science in Technology Management, and Masters of Business Administration at the University of Maryland, Juris Doctorate and Masters of Laws at Suffolk University, and Doctorate in Law and Public Policy at Northeastern University.

Dr. Addivinola conducted original research in developmental biology as a doctoral candidate in Molecular and Cell Biology and a pre-IRTA fellow at the National Institutes of Health (NIH), Bethesda, MD. Nobel laureate Marshall W. Nirenberg, Chief of the Biochemical Genetics Laboratory at the National Heart, Lung, and Blood Institute (NHLBI), was his dissertation advisor. His doctoral research focused on sequencing *OG*-2 genomic DNA, cloning and sequencing full-length *OG*-2 cDNAs with splice sites, and determining the cognate nucleotide binding site of the expressed homeodomain and flanking amino acids.

Before his research training at the NIH, Dr. Addivinola researched prostate cancer in the Cell Growth and Regulation Laboratory of Dr. Arthur Pardee at the Dana Farber Cancer Institute of Harvard Medical School. He participated in identifying early genetic markers for prostate cancer.

Frank Addivinola was a teaching fellow in Organic Chemistry at Harvard University and has taught at Johns Hopkins, the University of Maryland, and other universities. Professor Addivinola taught undergraduate and graduate-level courses, including biology, biochemistry, organic chemistry, inorganic chemistry, anatomy and physiology, medical terminology, nutrition, medical ethics, math, and law and public policy. He received several awards for teaching, research, and presentations.

www.ingramcontent.com/pod-product-compliance
Lightning Source LLC
LaVergne TN
LVHW080006210926
848481LV00011B/2954
9798885572392